T0138613

A Computational
Approach to Statistical
Learning

CHAPMAN & HALL/CRC
Texts in Statistical Science Series

Joseph K. Blitzstein, *Harvard University, USA*
Julian J. Faraway, *University of Bath, UK*
Martin Tanner, *Northwestern University, USA*
Jim Zidek, *University of British Columbia, Canada*

Recently Published Titles

Extending the Linear Model with R
Generalized Linear, Mixed Effects and Nonparametric Regression Models, Second Edition
J.J. Faraway

Modeling and Analysis of Stochastic Systems, Third Edition
V.G. Kulkarni

Pragmatics of Uncertainty
J.B. Kadane

Stochastic Processes
From Applications to Theory
P.D Moral and S. Penev

Modern Data Science with R
B.S. Baumer, D.T Kaplan, and N.J. Horton

Generalized Additive Models
An Introduction with R, Second Edition
S. Wood

Design of Experiments
An Introduction Based on Linear Models
Max Morris

Introduction to Statistical Methods for Financial Models
T. A. Severini

Statistical Regression and Classification
From Linear Models to Machine Learning
Norman Matloff

Introduction to Functional Data Analysis
Piotr Kokoszka and Matthew Reimherr

Stochastic Processes
An Introduction, Third Edition
P.W. Jones and P. Smith

Theory of Stochastic Objects
Probability, Stochastic Processes and Inference
Athanasios Christou Micheas

Linear Models and the Relevant Distributions and Matrix Algebra
David A. Harville

An Introduction to Generalized Linear Models, Fourth Edition
Annette J. Dobson and Adrian G. Barnett

Graphics for Statistics and Data Analysis with R
Kevin J. Keen

Statistics in Engineering, Second Edition
With Examples in MATLAB and R
Andrew Metcalfe, David A. Green, Tony Greenfield, Mahayaudin Mansor, Andrew Smith, and Jonathan Tuke

Introduction to Probability, Second Edition
Joseph K. Blitzstein and Jessica Hwang

A Computational Approach to Statistical Learning
Taylor Arnold, Michael Kane, and Bryan Lewis

For more information about this series, please visit: https://www.crcpress.com/go/textsseries

A Computational Approach to Statistical Learning

Taylor Arnold
Michael Kane
Bryan W. Lewis

CRC Press
Taylor & Francis Group
Boca Raton London New York

CRC Press is an imprint of the
Taylor & Francis Group, an **informa** business

CRC Press
Taylor & Francis Group
6000 Broken Sound Parkway NW, Suite 300
Boca Raton, FL 33487-2742

© 2019 by Taylor & Francis Group, LLC
CRC Press is an imprint of Taylor & Francis Group, an Informa business

No claim to original U.S. Government works

Printed on acid-free paper
Version Date: 20181218

International Standard Book Number-13: 978-1-138-04637-5 (Hardback)

Visit the Taylor & Francis Web site at
http://www.taylorandfrancis.com

and the CRC Press Web site at
http://www.crcpress.com

Contents

Preface   xi

1   **Introduction**   **1**
   1.1   Computational approach   1
   1.2   Statistical learning   2
   1.3   Example   3
   1.4   Prerequisites   5
   1.5   How to read this book   6
   1.6   Supplementary materials   7
   1.7   Formalisms and terminology   7
   1.8   Exercises   9

2   **Linear Models**   **11**
   2.1   Introduction   11
   2.2   Ordinary least squares   13
   2.3   The normal equations   15
   2.4   Solving least squares with the singular value decomposition .   17
   2.5   Directly solving the linear system   19
   2.6   ($\star$) Solving linear models using the QR decomposition   22
   2.7   ($\star$) Sensitivity analysis   24
   2.8   ($\star$) Relationship between numerical and statistical error . . .   28
   2.9   Implementation and notes   31
   2.10   Application: Cancer incidence rates   32
   2.11   Exercises   40

3   **Ridge Regression and Principal Component Analysis**   **43**
   3.1   Variance in OLS   43
   3.2   Ridge regression   46
   3.3   ($\star$) A Bayesian perspective   53
   3.4   Principal component analysis   56
   3.5   Implementation and notes   63
   3.6   Application: NYC taxicab data   65
   3.7   Exercises   72

4 Linear Smoothers   **75**
   4.1   Non-Linearity 75
   4.2   Basis expansion 76
   4.3   Kernel regression 81
   4.4   Local regression 85
   4.5   Regression splines 89
   4.6   ($\star$) Smoothing splines 95
   4.7   ($\star$) B-splines 100
   4.8   Implementation and notes 104
   4.9   Application: U.S. census tract data 105
   4.10 Exercises 120

5 Generalized Linear Models   **123**
   5.1   Classification with linear models 123
   5.2   Exponential families 128
   5.3   Iteratively reweighted GLMs 131
   5.4   ($\star$) Numerical issues 135
   5.5   ($\star$) Multi-Class regression 138
   5.6   Implementation and notes 139
   5.7   Application: Chicago crime prediction 140
   5.8   Exercises 148

6 Additive Models   **151**
   6.1   Multivariate linear smoothers 151
   6.2   Curse of dimensionality 155
   6.3   Additive models 158
   6.4   ($\star$) Additive models as linear models 163
   6.5   ($\star$) Standard errors in additive models 166
   6.6   Implementation and notes 170
   6.7   Application: NYC flights data 172
   6.8   Exercises 178

7 Penalized Regression Models   **179**
   7.1   Variable selection 179
   7.2   Penalized regression with the ℓ_0- and ℓ_1-norms 180
   7.3   Orthogonal data matrix 182
   7.4   Convex optimization and the elastic net 186
   7.5   Coordinate descent 188
   7.6   ($\star$) Active set screening using the KKT conditions 193
   7.7   ($\star$) The generalized elastic net model 198
   7.8   Implementation and notes 200
   7.9   Application: Amazon product reviews 201
   7.10 Exercises 206

8 Neural Networks **207**
8.1 Dense neural network architecture 207
8.2 Stochastic gradient descent 211
8.3 Backward propagation of errors 213
8.4 Implementing backpropagation 216
8.5 Recognizing handwritten digits 224
8.6 (⋆) Improving SGD and regularization 226
8.7 (⋆) Classification with neural networks 232
8.8 (⋆) Convolutional neural networks 239
8.9 Implementation and notes 249
8.10 Application: Image classification with EMNIST 249
8.11 Exercises 259

9 Dimensionality Reduction **261**
9.1 Unsupervised learning 261
9.2 Kernel functions 262
9.3 Kernel principal component analysis 266
9.4 Spectral clustering 272
9.5 t-Distributed stochastic neighbor embedding (t-SNE) 277
9.6 Autoencoders 282
9.7 Implementation and notes 283
9.8 Application: Classifying and visualizing fashion MNIST . . . 284
9.9 Exercises 295

10 Computation in Practice **297**
10.1 Reference implementations 297
10.2 Sparse matrices 298
10.3 Sparse generalized linear models 304
10.4 Computation on row chunks 307
10.5 Feature hashing 311
10.6 Data quality issues 318
10.7 Implementation and notes 320
10.8 Application 321
10.9 Exercises 329

A Linear algebra and matrices **331**
A.1 Vector spaces 331
A.2 Matrices 333

B Floating Point Arithmetic and Numerical Computation **337**
B.1 Floating point arithmetic 337
B.2 Computational effort 340

Bibliography **343**

Index **359**

Preface

This book was written to supplement the existing literature in statistical learning and predictive modeling. It provides a novel treatment of the computational details underlying the application of predictive models to modern datasets. It grew out of lecture notes from several courses we have taught at the undergraduate and graduate level on linear models, convex optimization, statistical computing, and supervised learning.

The major distinguishing feature of our text is the inclusion of code snippets that give working implementations of common algorithms for estimating predictive models. These implementations are written in the R programming language using basic vector and matrix algebra routines. The goal is to demystify links between the formal specification of an estimator and its application to a specific set of data. Seeing the exact algorithm used makes it possible to play around with methods in an understandable way and experiment with how algorithms perform on simulated and real-world datasets. This *try and see* approach fits a common paradigm for learning programming concepts. The reference implementations also illustrate the run-time, degree of manual tuning, and memory requirements of each method. These factors are paramount in selecting the best methods in most data analysis applications.

In order to focus on computational aspects of statistical learning, we highlight models that can be understood as extensions of multivariate linear regression. Within this framework, we show how penalized regression, additive models, spectral clustering, and neural networks fit into a cohesive set of methods for the construction of predictive models built on core concepts from linear algebra. The general structure of our text follows that of the two popular texts *An Introduction to Statistical Learning* (ISL) [87] and *The Elements of Statistical Learning* (ESL) [60]. This makes our book a reference for traditional courses that use either of these as a main text. In contrast to both ISL and ESL, our text focuses on giving an in-depth analysis to a significantly smaller set of methods, making it more conducive to self-study as well as appropriate for second courses in linear models or statistical learning.

Each chapter, other than the first, includes a fully worked out application to a real-world dataset. In order to not distract from the main exposition, these are included as a final section to each chapter. There are also many end of chapter exercises, primarily of a computational nature, asking readers to extend the code snippets used within the chapter. Common tasks involve benchmarking performance, adding additional parameters to reference implementations, writing unit tests, and applying techniques to new datasets.

Audience

This book has been written for several audiences: advanced undergraduate and first-year graduate students studying statistical or machine learning from a wide-range of academic backgrounds (i.e., mathematics, statistics, computer science, engineering); students studying statistical computing with a focus on predictive modeling; and researchers looking to understand the algorithms behind common models in statistical learning. We are able to simultaneously write for several backgrounds by focusing primarily on how techniques can be understood within the language of vector calculus and linear algebra, with a minimal discussion of distributional and probabilistic arguments. Calculus and linear algebra are well-studied across the mathematical sciences and benefit from direct links to both the motivation and implementations of many common estimators in statistical learning. While a solid background in calculus-based statistics and probability is certainly helpful, it is not strictly required for following the main aspects of the text.

The text may also be used as a self-study reference for computer scientists and software engineers attempting to pivot towards data science and predictive modeling. The computational angle, in particular our presenting of many techniques as optimization problems, makes it fairly accessible to readers who have taken courses on algorithms or convex programming. Techniques from numerical analysis, algorithms, data structures, and optimization theory required for understanding the methods are covered within the text. This approach allows the text to serve as the primary reference for a course focused on numerical methods in statistics. The computational angle also makes it a good choice for statistical learning courses taught or cross-listed with engineering or computer science schools and departments.

Online references

All of the code and associated datasets included in this text are available for download on our website `https://comp-approach.com`.

Notes to instructors

This text assumes that readers have a strong background in matrix algebra and are familiar with basic concepts from statistics. At a minimum students should be familiar with the concepts of expectation, bias, and variance.

Readers should ideally also have some prior exposure to programming in R. Experience with Python or another scripting language can also suffice for understanding the implementations as pseudocode, but it will be difficult to complete many of the exercises.

It is assumed throughout later chapters that readers are familiar with the introductory material in Chapter 1 and the first four sections of Chapter 2; the amount of time spent covering this material is highly dependent on the prior exposure students have had to predictive modeling. Several chapters should also be read as pairs. That is, we assume that readers are familiar with the first prior to engaging with the second. These are:

- Chapter 2 (Linear Models) and Chapter 3 (Ridge Regression and PCA)

- Chapter 4 (Linear Smoothers) and Chapter 6 (Additive Models)

- Chapter 5 (Generalized Linear Models) and Chapter 7 (Penalized Regression Models)

Within each individual chapter, the material should be covered in the order in which it is presented, though most sections can be introduced quite briefly in the interest of time.

Other than these dependencies, the chapters can be re-arranged to fit the needs of a course. In a one-semester undergraduate course on statistical learning, for example, we cover Chapters 1, 2, 5, 7, and 9 in order and in full. Time permitting, we try to include topics in neural networks (Chapter 8) as final projects. When teaching a semester long course on linear models to a classroom with undergraduates and graduate students in statistics, we move straight through Chapters 1 to 6. For a statistical learning class aimed at graduate students in statistics, we have presented Chapter 1 in the form of review before jumping into Chapter 4 and proceeded to cover all of Chapters 6 through 10.

Completing some of the end-of-chapter exercises is an important part of understanding the material. Many of these are fairly involved, however, and we recommend letting students perfect their answers to a curated set of these rather than having them complete a minimally sufficient answer to a more exhaustive collection.

1

Introduction

1.1 Computational approach

In this text, we describe popular statistical models in terms of their mathematical characteristics and the algorithms used to implement them. We provide a reference implementation, in the R programming environment, for each of the models considered. Code in this book is meant to be read inline, just as you would read the text. Unlike pseudocode, our reference implementations can be run, tested, and directly modified. However, the code is not optimized and does not include the checks and error handling one might expect from production-ready code. It is meant to be simple and readable.

Our computational approach leads to a different presentation compared to traditional approaches in statistical pedagogy where theory is separated from practice (applying functions to datasets). While the two-pronged approach clarifies the capabilities and underpinnings of the field, it is in the space between these prongs that researchers in the field usually find inspiration. New models and approaches are often developed iteratively and empirically while working with real datasets. Each step is justified mathematically only after it is found to be effective in practice. Our approach bridges statistical theory and model building by showing how they are related through their implementation.

Understanding the computational details behind statistical modeling algorithms is an increasingly important skill for anyone who wants to apply modern statistical learning methods. Many popular techniques do not allow for simple algorithms that can be easily applied to any problem. For example, as we show in Chapter 8, training neural networks with stochastic gradient descent is closer to an art form than a push button algorithm that can be obfuscated from the user. Nearly every chapter in this text shows how understanding the algorithm used to estimate a model often provides essential insight into the model's use cases and motivation. Additionally, increasingly large data sources have made it difficult or impossible, from a purely computational standpoint, to apply every model to any dataset. Knowledge of the computational details allows one to know exactly what methods are appropriate for a particular scale of data.

1.2   Statistical learning

Statistical learning is the process of teaching computers to "learn" by automatically extracting knowledge from available data. It is closely associated with, if not outright synonymous to, the fields of pattern recognition and machine learning. Learning occupies a prominent place within artificial intelligence, which broadly encompasses all forms of computer intelligence, both hand coded and automatically adapted through observed data.

We focus in this text on the subfield of *supervised learning*. The goal is to find patterns in available inputs in order to make accurate predictions on new, unseen data. For this reason models used in supervised learning are often called *predictive models*. Take the task of building an automatic spam filter. As a starting point, we could label a small dataset of messages by hand. Then, a statistical learning model is built that discovers what features in the messages are indicative of the message being labeled as spam. The model can be used to automatically classify new messages without manual intervention by the user. Many applications can be written as supervised learning tasks:

- Estimate how much a house will sell for based on properties such as the number of bedrooms, location, and its size.

- Determine whether a mass detected in an MRI scan is malignant or benign.

- Predict whether a flight will be delayed given the carrier, scheduled departure time, the departure airport, and the arrival airport.

- Estimate the number of page views a website will need to handle next month.

- Given a picture of a person, predict their age and mood.

- Determine the correct department to forward an online form request sent to a company's help desk.

- Predict the increased sales resulting from a new advertising campaign.

Domain-specific expertise is essential to the successful construction and deployment of statistical learning algorithms. However, most methods used for learning a predictive model from preprocessed data consist of applying general purpose training algorithms. The algorithms we cover in this text can be applied to a wide range of tasks, including all of the above examples, given the availability of sufficiently representative data for training.

Example 3

1.3 Example

It is useful to look at a concrete example of a statistical learning task. Consider recording the number of capital letters used in the content of 18 text messages and labeling whether the message is spam or "ham" (non-spam). For example, assume we observed the following dataset listing the number of capital letters followed by whether this is a spam message or not:

(0, ham)	(0, ham)	(0, ham)	(1, ham)	(1, ham)	(1, spam)
(2, ham)	(2, spam)	(2, ham)	(2, spam)	(2, ham)	(4, ham)
(5, spam)	(5, spam)	(6, spam)	(6, ham)	(8, spam)	(8, spam)

As one might expect, messages with a large number of capital letters are more likely to be spam. Now, we will use this data to predict whether a new text message is spam based on the number of capital letters that are used. A straightforward model for this prediction task would be to select a cutoff value N, and classify new messages with N or more capital letters as spam. We can apply supervised statistical learning to select a value of N by making use of the observed data.

What is a good value for the parameter N based on the observed text data? If we set N equal to 2, the first row of data will be categorized as ham and the second two will be labeled as spam. This leads to one mistake in the first row and five in the second two rows, for a total of six mistakes. Setting it to 5, where only the last row is labeled as spam, yields a better error rate with only four mistakes. Looking exhaustively through all possible splits illustrates that the choice of 5 leads to the fewest errors, and therefore would be a good choice for N. To test how well this model works, finally, we could acquire an additional dataset of text messages and compute the new error rate for our chosen split value. We will discuss all of these details, with fully worked out examples, throughout the text.

Continuing with our example we illustrate the concept of a reference implementation by writing a best split algorithm for classification. Our function takes two vectors; one gives a numeric set of numbers on which to classify and the second defines the two classes we want to distinguish. It works by exhaustively testing each possible value in the first vector as a possible split point, and returns the best possible split value. If there is a tie, all possible best splits are returned as values.

```
# Compute most predictive split of the training data.
#
# Args:
#     x: Numeric vector with which to classify the data.
#     y: Vector of responses coded as 0s and 1s.
#
# Returns:
```

```
#      The best split value(s).
casl_utils_best_split <-
function(x, y)
{
  unique_values <- unique(x)
  class_rate <- rep(0, length(unique_values))
  for (i in seq_along(unique_values))
  {
    class_rate[i] <- sum(y == (x >= unique_values[i]))
  }
  unique_values[class_rate == max(class_rate)]
}
```

The reference implementation uses common functions available in the base version of R. It should be readable, though perhaps not reproducible from scratch, by anyone familiar with scientific programming. Representing the data displayed in Section 1.3 as objects in R, we can test the implementation to make sure that it returns the same best fit value we arrived at by manually checking each split value.

```
x <- c(0, 0, 0, 1, 1, 1, 2, 2, 2, 2, 2, 4, 5, 5, 6, 6, 8, 8)
y <- c(0, 0, 0, 0, 0, 1, 0, 1, 0, 1, 0, 0, 1, 1, 1, 0, 1, 1)

casl_utils_best_split(x, y)
```

```
[1] 5
```

We see that it does return the correct value of 5. Exercises at the end of this chapter provide an opportunity to further test and tweak the casl_utils_best_split function.

The reference implementations provided in this text, for the most part, accurately approximate the algorithms used to estimate parameters in the respective learning algorithms. While many techniques do have other novel implementations that may be preferred in some applications, you will find that the basic algorithms presented here are at the core of at least one popular R package or Python module. However, looking at the source code of these packages will highlight major differences between the code in these public repositories and the sample code given here. Compare, for example, our function casl_ols_orth_proj on page 24 and the code for R's function lm. Code published for general purpose consumption often has:

- Detailed error checking code to ensure that the user passed valid inputs.

- Checks to ensure that the model converges and the output produces a reasonable result, such as not containing missing or non-finite parameter values.

- Verbose logging code to indicate a model's progress, particularly if the algorithm is known to take a non-trivial amount of time to complete.

- A myriad of additional options, modes, and tuning parameters.

- Code to deal with edge cases resulting from ill-conditioned data or non-standard data types.

- Additional helper functions for predicting new outputs, visualizing the model, and printing model metrics.

- Computationally intensive code chunks re-implemented in a low-level language such as C, C++, or Fortran.

- The ability to pass alternative input formats, such as the formula interface in R.

We mention some of these tweaks in the implementation sections of each chapter, with a focus on the additional options provided by published code. However, our code *intentionally* omits most of these details. The goal is to distill the most important algorithmic core of each technique without being distracted by other details.

1.4 Prerequisites

Readers of this text are expected to be familiar with core concepts of undergraduate-level multivariate calculus (derivatives and gradients), linear algebra (vectors, matrices, and eigenvalues) and introductory statistics (expected value, bias, variance, and confidence intervals). A short review of the required linear algebra material and our notation is given in Appendix A. We also assume some prior experience with scientific computing.

To get the most out of the text, we recommend that readers have also studied calculus-based statistics and are comfortable with applying statistical models and matrix-based operations using the R programming language. Knowledge of calculus-based statistics is most helpful in the discussion of linear regression models within Chapters 2 and 3. Avoiding the starred ($\star$) sections will allow readers to skip the majority of the probabilistically motivated calculations. As mentioned, the reference implementations should be fairly readable without any specific exposure to the R programming language. However, the end-of-chapter exercises require being able to modify and write new R scripts. Many books and tutorials exist for learning the R programming language. We particularly recommend Norm Matloff's book, *The Art of R Programming* [116] and the freely available, no-frills document "An Introduction to R" produced by William Venables and David Smith [165].

Finally, many of the algorithms presented here are motivated by concepts in numerical analysis and the difficulties of working with floating point arithmetic. The details of these challenges are not typically taught in statistics and data science programs. Our core text does not in any way require prior experience in numerical analysis. For readers interested in further motivating many of the algorithms presented here, Appendix B provides a concise, self-contained introduction to the numeric analysis with citations back into the main text.

1.5   How to read this book

Each of the remaining chapters in this text starts by providing an introduction to a new topic in statistical learning. Reference implementations and small simulation studies or data-driven examples are embedded throughout the discussion in line with our computational approach to the subject matter. After the core ideas are presented, notes on published implementations and extensions are briefly discussed. The second part of each chapter presents a longer application of the techniques in question to a full-scale dataset. Here, our reference implementations are typically replaced by their more mature, published variants. At the end-of-the chapter are a set of exercises ranging from basic checks on understanding all the way through to questions that require the construction of fairly involved reference functionalities and extensions to the chapter's application.

The best way to work through this text depends on a reader's prior experience with statistical learning. With no prior experience in the field, it is generally best to work through the book's core sections in order. The implementation and application sections could be skipped or glossed over on a first reading. Sections marked with a star ($\star$) are optional and can also be skipped. When used as part of a reading or lecture course, where an instructor can fill in any missing gaps, there is much more room to reorder the presentation of topics. See the Note to Instructors on page xii for more details. For readers with a reasonably complete background in statistical learning theory, or its application, it should be possible to work through the text in any order based on interest. Such readers may find it instructive to make an initial pass through the following five sections, which lay out the primary computational tools and techniques:

- Section 2.4, *Solving least squares with the pseudoinverse*

- Section 3.2, *Ridge regression*

- Section 5.1, *Classification with linear models*

- Section 7.5, *Coordinate descent*

- Section 8.2, *Stochastic gradient descent*

While not strictly required, we strongly recommend that all readers pay particular attention to the connections between the methods and their reference implementations. Working through several of the end of chapter exercises is a good indicator of how well the core material is understood.

1.6 Supplementary materials

All of the scripts and datasets included in this book are available for download from the book's website at:

> https://comp-approach.com

The website additionally includes the full code to replicate all of the graphics in the text, which are mostly left out from the book itself to reduce clutter. The reference implementations and data are also available within the R package **casl**, which can be installed automatically from CRAN. A list of selected solutions to the end of the chapter exercises are also maintained on the book's website.

1.7 Formalisms and terminology

We conclude this introduction by mathematically formalizing the central elements of supervised learning. The resulting language and terminology will be useful as a reference throughout the text.

Assume that there exists some unknown function f that maps elements from a set Ω into a set $\mathcal{Y}$,

$$f : \Omega \to \mathcal{Y}, \tag{1.1}$$

and consider observing tuples, known as *training data,*

$$\{\omega_i, y_i = f(\omega_i)\}_i, \quad \omega_i \in \Omega, \, y_i \in \mathcal{Y}, \quad i = 1, \ldots n. \tag{1.2}$$

We will avoid writing out a formal definition of the (possibly) random nature of f and the (possibly) random process that generates each ω_i. Doing so here would overly divert from the main discussion; we will define the random nature of the data generation process whenever necessary.

A supervised learning algorithm constructs an estimate $\widehat{f}$ of the function f using the training data. The goal is to minimize errors in estimating f

on new data for some loss function $\mathcal{L}$. When predicting a continuous response variable, such as an expected price, common loss functions include the squared or absolute difference between the observed and predicted values,

$$\mathcal{L}(\widehat{f}(\omega_{new}), f(\omega_{new})) = \left| \widehat{f}(\omega_{new}) - f(\omega_{new}) \right| \tag{1.3}$$

When the set of responses $\mathcal{Y}$ is finite, as it would be when building a spam prediction algorithm, a common choice of $\mathcal{L}$ is to measure the proportion of incorrectly labeled new observations,

$$\mathcal{L}(\widehat{f}(\omega_{new}), f(\omega_{new})) = \begin{cases} 0, & \widehat{f}(\omega_{new}) = f(\omega_{new}) \\ 1, & \text{otherwise} \end{cases} \tag{1.4}$$

Depending on the application, more complex metrics can be used. When evaluating a medical diagnostic algorithm, for instance, it may make more sense to weight incorrectly missing a serious condition more heavily than incorrectly diagnosing a serious condition.

In nearly all supervised learning tasks, the training data will either be given as, or coercible to, a vector of real values. In other words, we can write the set Ω in Equation 1.1 as $\mathbb{R}^p$ for some number p. Similarly, the prediction task can usually be re-written such that $\mathcal{Y}$ is equal to $\mathbb{R}$—in the case of a discrete set, this can be done by associating each category with an integer-based index. Then, by stacking the n training inputs together, we have

$$X = \begin{pmatrix} \omega_1 \\ \vdots \\ \omega_n \end{pmatrix} \in \mathbb{R}^{n \times p}, \quad y = \begin{pmatrix} y_1 \\ \vdots \\ y_n \end{pmatrix} \in \mathbb{R}^n, \quad y = f(X). \tag{1.5}$$

The matrix X is known as the *feature matrix*. This simplification allows for us to draw on techniques from numerical analysis, functional analysis, and statistics in the pursuit of predictive models. A central theme of this text will be building and evaluating supervised learning algorithms motivated by the study of properties of the matrix X and assumptions made regarding the function f.

If the success of a supervised learning algorithm is defined on data that is, by definition, unavailable, how will it be possible to determine how well a predictive model is able to estimate the function f? One approach is to take the observed tuples of data available for building predictive models and partition them into two subsets. Only one of these partitions is used as the *training set* to produce the estimate $\widehat{f}$. The remaining partition, the *testing set*, is used to evaluate how well the estimate can make predictions on new data. More complex schemes operate similarly by splitting the data multiple times (e.g., *cross-validation*) or include a third partition to allow for tuning hyperparameters in the model estimation algorithm.

When considering the construction of a predictive model, a key concern is the desired capacity, or complexity, of a model building algorithm. A formal definition of a training algorithm's complexity is given by its Vapnik–Chervonenkis (VC) dimension [163], though an informal understanding of complexity will suffice here. A model that is overly complex will *overfit* to the training data; that is, $\widehat{f}$ will fit the training data very closely but not be able to generalize well to the testing data. Conversely, a model with low complexity may be too constrained to provide a good approximation to f. In a probabilistic framework, this can be seen as equivalent to a trade-off between *bias* and *variance*. A model building algorithm with high complexity will have a large variance (it may overfit, and is therefore highly sensitive to the training data); if it has a complexity that is too low to approximate f, it will provide systematically biased results for some inputs. The study and control of model complexity, in both its numerical and probabilistic forms, is a guiding theme throughout this text.

1.8 Exercises

1. Change the optimal cutoff value N by flipping two of the ham/spam labels in the example in Section 1.3. Run the function `casl_utils_best_split` to illustrate that the split has changed.

2. Modify the function `casl_utils_best_split` so that false positives (accidentally labeling a true 0 as a 1) are considered twice as bad as false negatives. Does this change the optimal split in our example data?

3. Describe what the spaces Ω and $\mathcal{Y}$ would be for a predictive modeling task that estimates whether an email message is either spam or not spam.

4. Describe what might be some possible columns of the dataset X in a task trying to predict the sale price for a used car.

5. Assume that the feature matrix X has only one column, with values evenly distributed between 0 and 1, and the true function f is deterministic and defined such that to $y_i = x_i^2$. If our modeling algorithm is only allowed to produce a $\widehat{f}$ equal to a linear function of x_i, illustrate using a hand drawn plot how (in this case) the low model complexity directly leads to model bias.

6. Using the same setup as in the previous question, but allowing $\widehat{f}$ to take on any form, illustrate an example of a curve that has been overfit to the data.

7. Look at the source code for the R function `lm`. (Hint: you can see a function's source code by printing the function name without any parentheses

or arguments.) The model fitting routine is done in just one function call; the rest of the function is used to pre- and post-process the results. Identify the model fitting line. What function is being called?

8. The model fitting function called within the `lm` function also contains a large amount of set-up code, with the heart of the algorithm all happening with the code given by

```
z <- .Call(C_Cdqrls, x, y, tol, FALSE)
```

This calls a function written in Fortran. Locate this function in the R source code.

9. Modify the function `cnlp_utils_best_split` so that it also checks whether the best split would classify every value as a 1 if it is less than some cutoff value. Make sure that the output makes it clear whether the cutoff is a less than or greater than value.

2

Linear Models

2.1 Introduction

Linear models are amongst the most well known and often-used methods for modeling data. They are employed to study the outcomes of patients in clinical trials, the price of financial instruments, the lifetimes of fruit flies, and many other responses from a wide range of fields. Their popularity is not unwarranted. In fact, the discussion of linear models and their variants take up a considerable portion of this text.

Consider observing n pairs of data (x_i, y_i) for $i = 1, \ldots n$. A simple linear model would assume that the data are generated according to the equation

$$y_i = \beta_0 + \beta_1 x_i + \epsilon_i, \tag{2.1}$$

where ϵ_i is some unobserved error term and the β_j's are unknown constants. The goal of statistical modeling is to use the observed data to, in some fashion, estimate the parameters β_0 and β_1.

Why are linear models so popular? One important attribute is that linear models provide a concrete interpretation for all of their parameters. Take the two variable model for predicting housing sale prices as a function of total area (in square feet or square meters) and the number of bedrooms,

$$\text{price}_i = \beta_0 + \beta_1 \cdot \text{area}_i + \beta_2 \cdot \text{bedrooms}_i + \epsilon_i. \tag{2.2}$$

The parameters in this model tell us how much the response, price, changes when one of the predictor variables changes with the other variable held fixed. Mathematically, we can describe this precisely using partial derivatives

$$\beta_1 = \frac{\partial \, \text{price}}{\partial \, \text{area}}, \tag{2.3}$$

$$\beta_2 = \frac{\partial \, \text{price}}{\partial \, \text{bedrooms}}. \tag{2.4}$$

The model separates the effect of the total size of a house and the total number of bedrooms. This information is useful to real estate agents, homeowners, construction companies, and economists. Linear models also allow for the interpretation of categorical predictors through the use of indicator variables.

If our housing price data also includes information about whether a given observation is from one of three neighborhoods, say 'uptown,' 'downtown,' and 'suburbia,' we can define variables that are one when observation i is in the given neighborhood and zero otherwise. A linear model with these variables may be written as

$$\text{price}_i = \beta_0 + \beta_1 \cdot \text{area}_i + \beta_2 \cdot \text{bedrooms}_i + \tag{2.5}$$
$$\beta_3 \cdot \text{downtown}_i + \beta_4 \cdot \text{uptown}_i + \epsilon_i.$$

The parameter β_3 can still be viewed as a partial derivative, here representing the difference in the expected price between a house in suburbia and a house in the downtown neighborhood, if both are the same size and have the same number of bedrooms.

The relatively simple form of linear models allows for a great deal of variation in the model assumptions. The x_i's can be treated as fixed values, a *fixed design*, or they may be considered to be random variables themselves, as in a *random design* model. In biological applications the analysis usually depends on strict independence between the errors. In time series data, as commonly seen in finance or macroeconomics, the ϵ_i are often serially correlated with one another. Linear models such as the autoregressive integrated moving average (ARIMA) model and the autoregressive conditional heteroskedasticity (ARCH) model are used to model time series data with serial correlation structures. Longitudinal medical studies, where data is collected on multiple instances from the same cohort of patients over a period of time, may assume that the errors for observations from the same subject correlate differently than errors between different patients. Fixed, random, and mixed effects models—core statistical methods within certain sub-disciplines in the sciences and social sciences—are forms of linear models adapted to handle applications such as resampled data.

Linear models also benefit from a strong theoretical background. The standard estimators, which we will explore in the following two sections, can be described in terms of weighted sums of the original data. Under weak assumptions, we can then draw on the central limit theorem and large sample theory to construct asymptotically valid confidence intervals and hypothesis testing frameworks. Importantly, most of this theory can be extended to the various extensions and complex assumptions often used in practice. Also, these theoretical tools are useful even when the primary task is one of prediction. Hypothesis tests aid in the process of deciding whether to add or delete a certain variable from a model. Confidence intervals, when combined with an estimate of the noise variance, are extensible to prediction intervals. These provide a range of likely values for newly observed data points, in addition to a singular 'best' value. We will see several ways in which these estimates are useful in practice when building predictive models.

The standard estimators for parameters in linear models can be calculated using relatively straightforward computational approaches. For this reason, linear models are often used in applications even when many of the afore-mentioned benefits do not directly apply. Notice that a linear model must be linear only relative to the β terms. If we have pairs of data (x_i, y_i) but believe that there is a non-linear relationship between x and y, we could build the model

$$y_i = \beta_0 + \beta_1 \cdot x_i + \beta_2 \cdot x_i^2 + \cdots + \beta_p \cdot x_i^p + \epsilon_i. \tag{2.6}$$

Here it is difficult to discern a conceptual interpretation of each of the β_j terms. As a result, it is also hard to make use of confidence intervals and hypothesis tests concerning them. However, the linear model framework is incredibly useful as it provides a computationally tractable way of estimating an arbitrarily complex relationship, by setting p as large as possible, between our two variables. Of course, the size of the dataset will limit the ultimate complexity of the model, but this is true regardless of the particular approach taken. We will expand at length on this variable expansion method in Chapters 4 and 6.

2.2   Ordinary least squares

Many of the advantages of linear models concern the beneficial properties of the standard estimators used to compute the unknown parameters β_j from observed data. As a next step we would like to explore the definition of these estimators. To this aim, it will be useful to provide a compact matrix-based description of a linear model. Throughout this text, unless otherwise noted, we use a notation where n is the sample size, p is the number of variables, i is an index over the samples, and j is the index over the variables. With this notation a complete general description of a linear model can be given by

$$y_i = \beta_1 \cdot x_{i,1} + \cdots + \beta_p \cdot x_{i,p} + \epsilon_i, \quad \forall\, i = 1, \ldots, n. \tag{2.7}$$

Or simply

$$y_i = \sum_j \beta_j \cdot x_{i,j} + \epsilon_i, \quad \forall\, i = 1, \ldots, n. \tag{2.8}$$

Notice that we do not need to include an explicit intercept term β_0. If one is required this can be included by setting $x_{i,1}$ equal to one for every single observation i. Using matrix notation, we can write the linear model equation simultaneously for all observations as

$$
\begin{pmatrix} y_1 \\ y_2 \\ \vdots \\ y_n \end{pmatrix}
=
\begin{pmatrix}
x_{1,1} & x_{2,1} & \cdots & x_{p,1} \\
x_{1,2} & \ddots & & x_{p,2} \\
\vdots & & \ddots & \vdots \\
x_{1,n} & x_{2,n} & \cdots & x_{p,n}
\end{pmatrix}
\begin{pmatrix} \beta_1 \\ \beta_2 \\ \vdots \\ \beta_p \end{pmatrix}
+
\begin{pmatrix} \epsilon_1 \\ \epsilon_2 \\ \vdots \\ \epsilon_n \end{pmatrix}
\tag{2.9}
$$

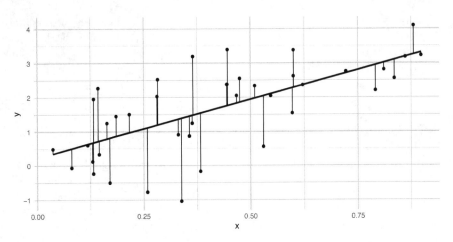

FIGURE 2.1: Visualization of residuals from the linear model $y = \beta_0 + \beta_1 x$.

which can be compactly written in terms of a vector y of the responses, a matrix X of the predictor variables, a vector β of the unknown parameters, and a vector ϵ of the errors

$$y = X\beta + \epsilon. \tag{2.10}$$

Beyond compactness, this notation is also useful as many of the computational properties of linear models can be reduced to linear algebraic properties of the matrix X.

It is desirable for an estimate $\widehat{\beta}$ of the unknown vector β to be able to explain as much variation as possible of the responses y. One way of viewing a linear model is as a decomposition of y into a fixed, deterministic signal $X\beta$ and a stochastic random noise term ϵ. Prediction and inference both benefit from making the signal term as dominant as possible. A good method for measuring this is to construct the vector of residuals

$$r = y - X\widehat{\beta}. \tag{2.11}$$

We can compare estimators by comparing the size of their residual vectors. A graphical representation of residuals from a linear model is given in Equation 2.1.

There are many choices for measuring the size of a regression vector, several of which lead to important, and distinct, estimators. The hinged loss, which penalizes positive residuals more than negative ones (or vice versa), leads to quantile regression. Metrics that penalize all residuals past some large threshold equally lead to robust regression techniques. Metrics that give each sample a weight w_i depending on the specific values of the data x_i can result in kernel regression (Section 4.3), local regression (Section 4.4) and are an important

---

Ordinary Least Squares (OLS)

Given a data matrix X and response vector y, the ordinary least squares estimator is given by

$$\widehat{\beta}_{ols} \in \arg\min_{b} \left\{ ||y - Xb||_2^2 \right\}.$$

When a unique solution exists, we refer to *the* ordinary least squares estimator.

---

intermediate step in solving the generalized linear model problems that arise in Chapters 5, 6, and 7.

In this chapter, we will focus on the most popular choice of metric to measure the size of the regression vector: the sum of squared residuals. Minimizing the sum of squared residuals leads to the *ordinary least squares* (OLS) estimator. Why is this such a popular choice? For one thing, it allows us to write the metric in terms of an inner product or vector norm

$$\sum_i r_i^2 = r^t r = ||r||_2^2, \tag{2.12}$$

a form that is easy to work with both computationally and theoretically. The choice of the sum of squared residuals is also motivated by the maximum likelihood estimator when the ϵ_i's are independent and identically distributed random variables with a normal distribution having zero mean.

2.3   The normal equations

We now have a formal specification of the ordinary least squares estimator. Computing the estimator given a set of observed data requires solving an optimization problem. This particular optimization problem is unconstrained and has a continuous gradient, so an obvious first step would be to find the gradient of the least squares objective function with respect to the vector b

$$\nabla_b \left[||y - Xb||_2^2 \right] = \nabla_b \left[y^t y + b^t X^t X b - 2y^t X b \right] \tag{2.13}$$

$$= 2X^t X b - 2X^t y. \tag{2.14}$$

A necessary condition for minimizing the objective function is to have the gradient equal to the zero vector, $\vec{0}$. If the Hessian matrix is positive definite at this solution, only then are we guaranteed to have a local minimum. The Hessian matrix here is constant everywhere, in that it does not depend on the value of b. Specifically, it is given by

$$H \left(||y - Xb||_2^2 \right) = X^t X. \tag{2.15}$$

For a matrix M to be positive definite, we need to have $z^t M z$ be strictly positive for any vector z not equal to the zero vector. Notice that in our case this matrix product can be written as a vector norm

$$z^t H \left(||y - Xb||_2^2 \right) z = z^t X^t X z \tag{2.16}$$

$$= ||Xz||_2^2. \tag{2.17}$$

The squared ℓ_2-norm is never negative and is only zero at the zero vector.

We see then that the Hessian is positive definite everywhere if and only if a non-zero vector z does not exist such that Xz is the zero vector. This, in turn is true if and only if X is not full rank. In this case, there are many possible values of b that all attain the minimum least squares solution. Such a result should not surprise us. If we have such a z, then there are many parameter vectors b that result in the exact same estimates for y as there are for the true β

$$X(\beta + a \cdot z) = X\beta + aXz \tag{2.18}$$

$$= X\beta + \vec{0} \tag{2.19}$$

$$= X\beta. \tag{2.20}$$

In such cases it is possible to place constraints on the problem to formulate a related problem with a unique solution. For instance, Section 2.4 illustrates how to find the unique OLS solution of minimal norm. Although minimum-norm least squares solutions are widely used in many science and engineering applications, it is more common in statistics to constrain solutions to rank-deficient problems in other ways. In particular, R's lm and glm solvers reformulate rank-deficient problems into full-rank ones by selecting a subset of columns using a heuristic procedure based on the model matrix column order. Other common subset selection approaches include the lasso (see Chapter 7), which solves a penalized version of OLS.

Satisfied that we attain a local minimum wherever the gradient is zero, we return to Equation 2.14. Setting this equal to zero we get what are known as the *normal equations*, a linear system of equations of p variables over p unknowns expressed in matrix form as

$$X^t X b = X^t y. \tag{2.21}$$

Solving systems like the normal equations for b in a numerically stable and efficient manner is an important problem encountered repeatedly in this text. Linear systems of equations are generically solved by Gaussian elimination, but we will see that other more efficient and/or numerically stable methods based on the Cholesky, QR, or SVD decompositions can be used depending on context.

2.4   Solving least squares with the singular value decomposition

Many numerical methods for solving the normal equations rely on decomposing the data matrix X or the Gram matrix $X^t X$ into factors using a variety of standard matrix decomposition algorithms. Here, we specifically make use of the singular value decomposition (SVD). Although more computationally intensive than some other techniques, the SVD gives us detailed insight into properties of the matrix useful for the development of numerically stable solution methods.

Let $X \in \mathbb{R}^{n \times p}$ and let $k = \min\{n, p\}$. Then there exist matrices $U \in \mathbb{R}^{n \times k}$ and $V \in \mathbb{R}^{p \times k}$ with orthonormal columns $U^t U = V^t V = I$ such that

$$U^t X V = \Sigma, \tag{2.22}$$

where Σ is a $k \times k$ diagonal matrix with non-negative entries $\sigma_1 \geq \sigma_2 \geq \cdots \geq \sigma_k \geq 0$ along its main diagonal. This is sometimes called the *thin* SVD. In the case where $n > p$ it is possible to extend the matrix U into a square orthonormal $n \times n$ matrix $\bar{U}$ by adding $n - p$ additional orthonormal columns. Similarly, when $n < p$ we can extend V to a square orthonormal $p \times p$ matrix $\bar{V}$. The extended version is sometimes called the *full* SVD or just the SVD in many references and $\bar{U}^T X \bar{V} = \bar{\Sigma}$ results in an $n \times p$ rectangular diagonal matrix with the same main diagonal entries $\sigma_1 \geq \sigma_2 \geq \cdots \geq \sigma_k \geq 0$ as the thin version. The full SVD is especially useful for analysis, while the thin SVD is more commonly used in computation.

The columns of $\bar{U}$ are called the *left singular vectors* of X and the columns of $\bar{V}$ are called the *right singular vectors*. The σ_i are called *singular values* of X. The SVD breaks matrix vector multiplication into three steps: rotation, scaling, then another rotation. Consider an $n \times p$ matrix X and its product y with a vector $b \in \mathbb{R}^p$. Using the full SVD $y = Xb = \bar{U} \bar{\Sigma} \bar{V}^T b$:

1. Let $\hat{b} = \bar{V}^T b \in \mathbb{R}^p$. Since $\bar{V}$ is an orthonormal matrix, $\hat{b}$ is simply a rotation of the vector b.

2. Now let $s = \bar{\Sigma} \hat{b} \in \mathbb{R}^n$, scaling each entry of $\hat{b}$ by the corresponding σ_i.

3. Finally let $y = \bar{U} s$, simply another rotation by the orthonormal matrix $\bar{U}$.

The SVD reveals a lot of information about the structure of the matrix X. Step 2 tells us how much a vector can be scaled by X, and together with the rotations in steps 1 and 3 about its range and null space. The number of nonzero singular values of X is equal to the *rank* of X—the dimension of the range of X (range means the set of all linear combinations of the columns of X, that is, the span of the columns of X). Section 2.7 llustrates the sensitivity to noise of the solution of least squares problems involving X in terms of the singular values of X.

The SVD can be used to solve general ordinary least squares problems. The following result is adapted from Golub and Van Loan [69, Theorem 5.5.1], a recipe for computing the *unique* ordinary least squares solution of minimal Euclidean norm. Let X be a real $n \times p$ matrix, with full SVD $\bar{U}^T X \bar{V} = \bar{\Sigma}$ using extended matrices $\bar{U} = [u_1, u_2, \ldots, u_n] \in \mathbb{R}^{n \times n}$, $\bar{\Sigma} \in \mathbb{R}^{n \times p}$, and $\bar{V} = [v_1, v_2, \ldots, v_p] \in \mathbb{R}^{p \times p}$, and let $r \leq \min\{n, p\}$ be the rank of X. Then

$$b_{LS} = \sum_{i=1}^{r} \frac{u_i^T y}{\sigma_i} v_i \qquad (2.23)$$

minimizes $\|Xb - y\|^2$ and has the smallest Euclidean norm of all such minimizers.

The proof of the above statement relies on properties of the orthogonal matrices produced by the SVD. For any vector $b \in \mathbb{R}^p$,

$$\begin{aligned}
\|Xb - y\|^2 &= \|\bar{U}\bar{\Sigma}\bar{V}^T b - y\|^2 & \text{(replacing X with its full SVD)} \\
&= \|\bar{U}^T(\bar{U}\bar{\Sigma}\bar{V}^T b - y)\|^2 & \text{(by A.1)} \\
&= \|\bar{\Sigma}\bar{V}^T b - \bar{U}^T y\|^2 \\
&= \sum_{i=1}^{p}(\sigma_i v_i^T b - u_i^T y)^2 + \sum_{i=p+1}^{n}(u_i^T y)^2. & (2.24)
\end{aligned}$$

The columns of $\bar{V}$ for an orthonormal basis of $\mathbb{R}^p$. Express the solution b as a linear combination of the column vectors v_i, $b = \sum_{i=1}^{p} \gamma_i v_i$ (that is, $\gamma_i = v_i^T b$). Since $\text{rank}(X) = r$ then $\sigma_{r+1} = \sigma_{r+2} = \cdots = \sigma_p = 0$ and the corresponding coefficients γ_i may take on any value without affecting the residual norm. The specific choice $\gamma_{r+1} = \gamma_{r+2} = \cdots \gamma_p = 0$ minimizes the norm of any possible solution b. Then the residual norm in Equation 2.24 is minimized by setting the remaining coefficients $v_i^T b = \gamma_i = (u_i^T y)v_i/\sigma_i$ for $i = 1, 2, \ldots, r$.

We can implement an algorithm to solve ordinary least squares using the SVD by calling R's function `svd`.

```
# Compute OLS estimate using SVD decomposition.
#
# Args:
#     X: A numeric data matrix.
#     y: Response vector.
#
# Returns:
#     Regression vector beta of length ncol(X).
casl_ols_svd <-
function(X, y)
{
  svd_output <- svd(X)
  r <- sum(svd_output$d > .Machine$double.eps)
```

```
  U <- svd_output$u[, 1:r]
  V <- svd_output$v[, 1:r]
  beta <- V %*% (t(U) %*% y / svd_output$d[1:r])
  beta
}
```

To test this function, we will first create some random data and set a regression vector β.

```
n <- 1e4; p <- 4
X <- matrix(rnorm(n*p), ncol = p)
beta <- c(1,2,3,4)
epsilon <- rnorm(n)
y <- X %*% beta + epsilon
```

From here, we compute the estimated $\widehat{\beta}$ from casl_ols_svd.

```
beta_h_svd <- casl_ols_svd(X, y)
beta_h_svd
```

```
          [,1]
[1,] 0.9816599
[2,] 1.9938207
[3,] 2.9941449
[4,] 4.0062232
```

The result closely reconstructs the true β, which was set to the vector $(1, 2, 3, 4)$. We should not expect to get the exact solution due to the presence of the noise vector epsilon. We can verify that this is the same solution given by R using the lm function and extracting the coefficients with coef.

```
coef(lm(y ~ X - 1))
```

```
       X1        X2        X3        X4
0.9816599 1.9938207 2.9941449 4.0062232
```

This result matches, at least to the 7th decimal place, with the result from our function.

2.5   Directly solving the linear system

An alternative approach to solving the least squares problem is to first compute the matrix $X^t X$ and vector $X^t y$. This converts the normal equations into a generic linear system of equations

$$Ab = z. \tag{2.25}$$

From here there are many possible approaches to solving the system of equations. As we already know that $A = X^t X$ is positive definite, a reasonable choice is to compute its Cholesky decomposition LL^t, for a lower diagonal matrix L. Then, in order to solve the system, we make use of the back and forward solve algorithms.

The back solve algorithm finds a solution v to the linear system $Av = z$ precisely when A is a triangular matrix. The best way to understand how it works is to write down exactly what it does in a small example. Take the 3×3 linear system given by

$$\begin{pmatrix} R_{1,1} & R_{1,2} & R_{1,3} \\ 0 & R_{2,2} & R_{2,3} \\ 0 & 0 & R_{3,3} \end{pmatrix} \begin{pmatrix} b_1 \\ b_2 \\ b_3 \end{pmatrix} = \begin{pmatrix} v1 \\ v2 \\ v3 \end{pmatrix}. \tag{2.26}$$

Notice that the last equation involves only the factor b_3, and so we can quickly calculate the estimate for this value as

$$b_3 = \frac{v3}{R_{3,3}} \tag{2.27}$$

The second equation involves only b_2 and b_3; as we already know the value of b_3 it is again just arithmetic to solve for the value of b_2

$$b_2 = \frac{v2}{R_{2,2}} + \frac{R_{2,3} \cdot b_3}{R_{2,2}} \tag{2.28}$$

The same logic now applies to the first equation to get a formula for the value of b_1

$$b_1 = \frac{v1}{R_{1,1}} + \frac{R_{1,2} \cdot b_2}{R_{1,1}} + \frac{R_{1,3} \cdot b_3}{R_{1,1}} \tag{2.29}$$

This exact same method applies to right triangular linear systems of arbitrary sizes. The name *back* in *back solve* describes the fact that we solve for b by starting from the last element and moving backwards through to the first element. The analogous algorithm *forward solve* uses the same technique to solve a left triangular system of equations.

Using these two algorithms a solution can then be stably computed by back solving with L and forward solving with L^t. One advantage of this approach over the SVD is that it can be quickly computed even in situations where the matrix X is too large to fit into memory. We will return to this idea in Chapter 10. A potential disadvantage is that directly computing the matrix $X^t X$ effectively squares the values on the left-hand side of the equation and makes the numerical precision of the solution worse.

We can use this second approach to compute the ordinary least squares solution, which we now implement as a new function.

```
# Compute OLS estimate using the Cholesky decomposition.
#
# Args:
#     X: A numeric data matrix.
#     y: Response vector.
#
# Returns:
#     Regression vector beta of length ncol(X).
casl_ols_chol <-
function(X, y)
{
  XtX <- crossprod(X)
  Xty <- crossprod(X, y)
  L <- chol(XtX)

  betahat <- forwardsolve(t(L), backsolve(L, Xty))
  betahat
}
```

In the implementation, we utilize the `crossprod` function to avoid having to
compute the transpose directly. We now call the function `chol` to compute
the Cholesky decomposition. With this lower diagonal matrix, we back solve
and forward solve to get the ordinary least squares estimate. Testing this on
the same data as we used with SVD gives the following solution to the normal
equations:

```
casl_ols_chol(X, y)
```

```
           [,1]
[1,] 0.9813992
[2,] 1.9939830
[3,] 2.9945602
[4,] 4.0064178
```

Again, the result approximately reconstructs the vector β. Notice that this
result matches the SVD's method for computing $\widehat{\beta}$ only up to the third significant digit. The difference is a result of the numerical error from directly
computing $X^t X$.

2.6 ($\star$) Solving linear models using the QR decomposition

The method used by default in statistical software programs, such as R's lm function, is neither the SVD-based approach of Section 2.4 nor the direct approach of Section 2.5. Instead, an orthogonal QR decomposition-based technique is used to solve the normal equations without explicitly forming them. The approach is numerically stable and less computationally expensive than the SVD-based one. While the SVD approach will best motivate and generalize other methods presented throughout this text, we felt it would be lacking to leave out the most common approach to the most common statistical estimator.

We start by taking the QR decomposition of the entire matrix X. As X usually has more rows than columns, the full decomposition takes the form of

$$X = Q \begin{bmatrix} R_1 \\ 0 \end{bmatrix} \tag{2.30}$$

$$= [Q_1, Q_2] \begin{bmatrix} R_1 \\ 0 \end{bmatrix} \tag{2.31}$$

where Q_1 is an $n \times p$ matrix, Q_2 is an $n \times (n - p)$ matrix, R_1 is an upper triangular $p \times p$ matrix, and the columns of both Q_1 and Q_2 (taken together) are all orthogonal. Using this decomposition, consider multiplying the residual vector by the full matrix Q. This simplifies to the following

$$Q^t(y - Xb) = \begin{bmatrix} Q_1^t \\ Q_2^t \end{bmatrix} y - \begin{bmatrix} Q_1^t \\ Q_2^t \end{bmatrix} \cdot [\, Q_1 \quad Q_2 \,] \cdot \begin{bmatrix} R_1 \\ 0 \end{bmatrix} b \tag{2.32}$$

$$= \begin{bmatrix} Q_1^t \\ Q_2^t \end{bmatrix} y - \begin{bmatrix} R_1 \\ 0 \end{bmatrix} b \tag{2.33}$$

The final $n - p$ rows of these rotated residuals do not involve b at all as they are canceled out by the 0 in the decomposition. Therefore, minimizing $r^t r$ is simply a matter of minimizing the first p rows of these rotated coordinates. That is, we have decomposed the sum of squares into two parts, where one does not depend on b

$$||y - Xb||_2^2 = ||Q^t(y - Xb)||_2^2 \tag{2.34}$$

$$= ||Q_1^t y - R_1 n||_2^2 + ||Q_2^t y||_2^2 \tag{2.35}$$

Conveniently, it is possible to pick a vector b that sets the sum of squared residuals on these p rows exactly equal to 0. The task of finding the least squares estimator then reduces to finding the solution to the following system of equations.

$$Q_1^t y = R_1 b \tag{2.36}$$

The final step involves applying the back solve algorithm to the triangular matrix R_1.

The entire procedure outlined here for solving the least squares problem is numerically stable and computationally efficient. The speed comes from by using an implementation of the QR decomposition, such as those using Givens rotations or Householder reflections, that is able to compute the *reduced QR factorization* to get Q_1 without needing to calculate Q_2. The matrix Q_2 is usually significantly larger than any of the other matrices needed in the regression model and not required to solve Equation 2.36.

We can apply this orthogonalization technique directly in R for a given numeric data matrix X and response vector y. To demonstrate, we will first create some random data and set a regression vector β.

```
n <- 1e4; p <- 4
X <- matrix(rnorm(n*p), ncol = p)
beta <- rep(1, p)
epsilon <- rnorm(n)
y <- X %*% beta + epsilon
```

From here, we compute the QR decomposition of X. This is done by first constructing a QR object with qr and then extracting the Q and R components and storing them individually. Note that `qr.Q` and `qr.R` both have a parameter `complete`, set to false by default, that allows for returning the full decomposition in Equation 2.30 rather than the reduced factorization needed here.

```
qr_obj <- qr(X)
Q <- qr.Q(qr_obj)
R <- qr.R(qr_obj)
```

With the decompositions in hand, we proceed to multiply Q^t and y and to back solve the triangular system defined by the matrix R. For efficiency, we will use the `crossprod` function to avoid having to compute and store the transpose of Q.

```
Qty <- crossprod(Q, y)
beta_hat_ols <- backsolve(R, Qty)
beta_hat_ols
```

```
          [,1]
[1,] 0.9870134
[2,] 1.9876739
[3,] 3.0045489
[4,] 4.0102080
```

The result closely reconstructs the true β, which was set to the vector

$(1, 2, 3, 4)$. We should not expect to get the exact solution due to the presence of the noise vector epsilon. We can verify that this is the same solution given by R using the `lm` function and extracting the coefficients with `coef`.

```
coef(lm(y ~ X - 1))
```

```
        X1        X2        X3        X4
0.9870134 1.9876739 3.0045489 4.0102080
```

This result matches to the 7th decimal place because, as mentioned, the reduced QR decomposition is the method used internally by R. As we will do with most methods in this text, we will wrap this functionality up in an minimally documented R function.

```
# Compute OLS estimate using orthogonal projection.
#
# Args:
#     X: A numeric data matrix.
#     y: Response vector.
#
# Returns:
#     Regression vector beta of length ncol(X).
casl_ols_orth_proj <-
function(X, y)
{
  qr_obj <- qr(X)
  Q <- qr.Q(qr_obj)
  R <- qr.R(qr_obj)
  Qty <- crossprod(Q, y)

  betahat <- backsolve(R, Qty)
  betahat
}
```

2.7 ($\star$) Sensitivity analysis

Assume that we have a linear system with a matrix A, and vectors v and z such that

$$Av = z. \tag{2.37}$$

If we are given the matrix A and vector z, what makes it numerically difficult to solve the linear system for the unknown values in v? Skipping the efficient matrix decompositions for a moment, it seems possible to iteratively guess (perhaps intelligently) values $\tilde{v}$, stopping when we have

$$||A\tilde{v} - Av|| = ||A\tilde{v} - z|| \le \epsilon \qquad (2.38)$$

For some vector norm $|| \cdot ||$ and small tolerance factor ϵ. This intuition is reasonable and, with a good choice of an iterative technique such as gradient descent, an acceptable method for solving systems of linear equations. However, numerical issues remain a problem when there exist vectors $\tilde{v}$ that solve Equation 2.38 but are still relatively far away from v. The iterative approach will be unable to choose between $\tilde{v}$ and v. We can of course decrease the value of ϵ, but only to the point where ϵ is equal to the precision of our machine. Past this point no algorithm will be able to estimate v with a high precision.

How can we find an upper bound on the absolute error between the vector v and some solution $\tilde{v}$ to Equation 2.38? Start by defining the quantity c_{min} such that

$$c_{min} = \min_{\delta \ne 0} \left\{ \frac{||A\delta||}{||\delta||} \right\} = \min_{\delta : ||\delta|| = 1} \left\{ ||A\delta|| \right\}. \qquad (2.39)$$

An upper bound on the absolute error can then found by setting $\delta = (v - \tilde{v})$, which gives

$$||v - \tilde{v}|| \le c_{min} \cdot ||A(v - \tilde{v})|| \qquad (2.40)$$
$$\le c_{min} \cdot ||Av - A\tilde{v}|| \qquad (2.41)$$
$$= c_{min} \cdot ||z - A\tilde{v}|| \qquad (2.42)$$
$$\le c_{min} \cdot \epsilon. \qquad (2.43)$$

We see that the quantity c_{min}, which we will investigate further momentarily, is a conversion factor for going from error in Av to error in v.

In a similar derivation, we can provide a bound on the absolute error in v, $||v - \tilde{v}||/||v||$. Define a quantity c_{max} such that

$$c_{max} = \max_{\delta \ne 0} \left\{ \frac{||A\delta||}{||\delta||} \right\} = \max_{\delta : ||\delta|| = 1} \left\{ ||A\delta|| \right\}. \qquad (2.44)$$

Then we have, by setting δ equal to $v \ne 0$, that

$$\frac{||Av||}{||v||} \le c_{max} \qquad (2.45)$$

$$\frac{1}{c_{max}} \cdot ||Av|| \le ||v||. \qquad (2.46)$$

Putting this together with Equation 2.43 gives

$$\frac{||v - \tilde{v}||}{||v||} \le \frac{c_{min}}{c_{max}} \cdot \left(\frac{\epsilon}{||Av||} \right) \qquad (2.47)$$

$$\le \frac{c_{min}}{c_{max}} \cdot \left(\frac{\epsilon}{||z||} \right). \qquad (2.48)$$

As with the absolute error case, we now have a conversion factor that serves as an upper bound on the relative error in v in terms of the relative error in Av. This leads to the quantity we will denote as

$$\kappa(A) = \frac{c_{max}}{c_{min}}. \tag{2.49}$$

Notice that the convention is to define this value so that large numbers correspond with ill-conditioned systems, where good estimation of Av does not guarantee accuracy in measuring v.

If we define the κ using the ℓ_2-norm it is possible to derive a precise description for any matrix A. If A has more columns than rows, or more generally has less than full column rank, notice that c_{min} is equal to zero and κ is infinitely large. Otherwise, consider the coordinate system $\{v_1, \ldots, v_p\}$ described by a set of right singular vectors from the matrix A. We can write the minimization problem over δ in Equation 2.39 in this coordinate system. That is, write

$$\delta = \sum_j \delta_j \cdot V_j \tag{2.50}$$

where $U\Sigma V^t$ is the SVD of A, and V_j is the jth column of V. Squaring the definition of c_{min} gives

$$c_{min}^2 = \min_{\delta:||\delta||=1} \left\{ ||A\delta||_2^2 \right\} \tag{2.51}$$

$$= \min_{\delta:||\delta||=1} \left\{ \sum_j \delta_j \sigma_j^2 \right\} \tag{2.52}$$

$$= \sigma_{min}^2. \tag{2.53}$$

We then see that c_{min} is given by the smallest singular value of A whenever $n \geq p$. By a similar derivation, maximizing δ in Equation 2.44 shows that c_{max} is the largest singular value of A. Therefore, our quantity κ under the ℓ_2-norm, which we will denote as κ_2, is the ratio between the largest and smallest singular values whenever $n \geq p$

$$\kappa_2(A) = \frac{\sigma_{max}}{\sigma_{min}}. \tag{2.54}$$

The quantity 2.54 is called the 2-norm *condition number* of a matrix. This also gives a geometric understanding of what conditions lead to the worst-case sharp bound in Equation 2.48. The worst relative error occurs when the true signal v is proportional to the maximal right singular vector V_1 and the error $v - \tilde{v}$ is proportional to the minimal right singular vector V_{max}.

We can see a nice application of the quantity κ_2 through a toy linear model example with $n = p = 2$. We set X and β as

$$X = \begin{pmatrix} 10^9 & -1 \\ -1 & 10^{-5} \end{pmatrix} \tag{2.55}$$

$$\beta = \begin{pmatrix} 1 \\ 1 \end{pmatrix} \tag{2.56}$$

And if we define $y = X\beta$, this gives

$$y = \begin{pmatrix} 10^9 & -1 \\ -1 & 10^{-5} \end{pmatrix} * \begin{pmatrix} 1 \\ 1 \end{pmatrix} \tag{2.57}$$

$$= \begin{pmatrix} 10^9 - 1 \\ -0.99999 \end{pmatrix} \tag{2.58}$$

As X is a square matrix, we can in this case solve $y = X\beta$ directly. We will do this in R using a direct call to the `solve` function.

```
X   <- matrix(c(10^9, -1, -1, 10^(-5)), 2, 2)
beta <- c(1,1)
y <- X %*% beta
solve(X, y)
```

```
      [,1]
[1,]    1
[2,]    1
```

To the precision of the print out, this perfectly reconstructs the original β vector. What is the value $\kappa_2(X)$? We can compute that be extracting its singular value with the `svd` function.

```
svals <- svd(X)$d
svals
```

```
[1] 1.000e+09 9.999e-06
```

```
max(svals) / min(svals)
```

```
[1] 1.0001e+14
```

The value of κ_2 is fairly large, but its inverse is still two orders of magnitude above the tolerance of double precision floating point arithmetic (which is around $2^{-53} \approx 10^{-16}$). Now, what if we try to compute the normal equations and solve those directly? Doing so throws an error in R.

```
solve( crossprod(X), crossprod(X, y) )
```

```
Error in solve.default(crossprod(X), crossprod(X, y)):
  system is computationally singular: reciprocal
  condition number = 9.998e-29
```

The warning explicitly mentions the reciprocal condition number being too small to compute with. We can verify that our definition of the κ_2, which is equivalent to the condition number here, matches that of R by computing the singular values of $X^t X$ directly.

```
svals <- svd(crossprod(X))$d
svals
```

```
[1] 1.000000e+18 9.998002e-11
```

```
max(svals) / min(svals)
```

```
[1] 1.0002e+28
```

And we see here that this the reciprocal of this value is very close to the one reported by R. We can turn off this error by setting the tolerance parameter `tol` to zero in the call to `solve`.

```
solve( crossprod(X), crossprod(X, y) , tol = 0)
```

```
      [,1]
[1,]    1
[2,]    0
```

However, the result now does not match the original β. This example, then, illustrates how κ_2 serves as a measurement of how numerically stable a linear system is. It also gives an example of where the approach of Section 2.5 leads to worse numerical performance than working directly with the matrix X as in Section 2.4.

2.8 $(\star)$ Relationship between numerical and statistical error

Many texts and research papers on computational statistics, including this one, emphasize the importance of using numerically stable algorithms for solving the normal equations. For one thing, there is no benefit or excuse for using the unstable matrix inverse. From a pragmatic standpoint, however, it is very rare to find an example where ordinary least squares offers a good solution to a system of equations ill-conditioned enough where the choice of method has practical implications. The primary reason to understand the sources and remedies for numerical instability in linear models is that there are close parallels between the numerical noise due to the natural of floating point arithmetic and statistical error due to the noise term in observed data. Measures such as the condition number therefore have a direct role to play in regression analysis, and as we will see in Chapter 3, classical remedies for solving numerically unstable systems lead to important classes of statistical estimators. In this section, we show an explicit example of the link between numerical stability and statistical estimation.

In Equation 2.48, we see that the quantity κ_2 dictates the penalty between an error in estimating Av to an error in estimating v. In the noise-free case this is typically only a minor issue in statistical applications. The quantity Av can be computed to double-precision accuracy and, even for particularly large condition numbers, the quantity v is still typically attainable to single precision accuracy. For predictive modeling, this is usually more than sufficient. The difficulty arises when considering the estimation of a noisy equation. Specifically, assume that we have the linear system

$$Av + e = z \tag{2.59}$$

And our goal is to estimate v knowing only the matrix A and z, with no knowledge of the error vector e. If we find a candidate $\tilde{v}$ such that

$$||Av - A\tilde{v}|| \leq \epsilon \tag{2.60}$$

The estimation error in Av now can be decomposed into two parts:

$$||Av - A\tilde{v}|| = ||(Av - z) + (z - A\tilde{v})|| \tag{2.61}$$
$$\leq ||Av - z|| + ||z - A\tilde{v}|| \tag{2.62}$$
$$= ||e|| + ||A\tilde{v} - z|| \tag{2.63}$$
$$\leq ||e|| + \epsilon. \tag{2.64}$$

We can make the term ϵ small, to within the precision of our machine, by iteratively solving Formula 2.60. The $||e||$ term, in contrast, is a feature of the data and not something that we can control by increasing the numeric precision of our algorithms. Further, it is often quite large in statistical applications.

Even with the addition of a noise vector e, our measurement κ_2 still provides a conversion factor between the relative error in the estimation of Av and the relative error of estimating v

$$\frac{||v - \tilde{v}||}{||v||} \leq \frac{1}{\kappa_2(A) \cdot ||Av||} \cdot (||e|| + \epsilon). \tag{2.65}$$

Notice that now even a relatively small value of $\kappa_2(A)$ can cause estimation problems because it will amplify the variation in the error term e. We see that unlike the numerical problem, where condition numbers are not a problem until they reach their reciprocal approach the precision of the machine, even reasonably small condition numbers make it difficult to estimate v well with a high probability.

To illustrate the relationship between the condition number of the statistical error in estimating a regression vector, we can run a simple simulation. First, construct a regression vector β whose first coordinate is equal to 1, the other 24 coordinates are zeros, and a data matrix X filled with randomly sampled normal variables.

```
n <- 1000; p <- 25
beta <- c(1, rep(0, p - 1))
X <- matrix(rnorm(n * p), ncol = p)
```

The condition number of the matrix X can be computed from its singular values, and is equal to slightly more than 1.36.

```
svals <- svd(X)$d
max(svals) / min(svals)
```

```
[1] 1.369801
```

Now, we will generate data y and see how close, in terms of the ℓ_2-norm, the ordinary least squares estimate is to the true β. We do this ten thousand times and report the mean error rate, a good prediction of the mean squared error.

```
N <- 1e4; 12_errors <- rep(0, N)
for (k in 1:N) {
  y <- X %*% beta + rnorm(n)
  betahat <- casl_ols_svd(X, y)
  12_errors[k] <- sqrt(sum((betahat - beta)^2))
}
mean(12_errors)
```

```
[1] 0.1582981
```

The typical error in this example is quite small, at under 0.16. Now, let us replace the first column of X with a linear combination of the original first column and the second column. The result now has two columns of X being highly correlated and we see this in a significantly increased condition number for the matrix $X^t X$.

```
alpha <- 0.001
X[,1] <- X[,1] * alpha + X[,2] * (1 - alpha)
svals <- svd(X)$d
max(svals) / min(svals)
```

```
[1] 1999.189
```

The condition number is now about 2000, a large increase from the original number of 1.36. Running the simulation again, we notice that the error rate has increased by a factor of over 200.

```
N <- 1e4; l2_errors <- rep(0, N)
for (k in 1:N) {
  y <- X %*% beta + rnorm(n)
  betahat <- solve(crossprod(X), crossprod(X, y))
  l2_errors[k] <- sqrt(sum((betahat - beta)^2))
}
mean(l2_errors)
```

```
[1] 35.95446
```

So we see here that the condition number of X does serve as an indicator of the statistical error in estimating β. The bounds in that equation are only the worst-case scenario, and in practice the increase in the error will be less extreme than the increase in the condition number.

2.9   Implementation and notes

As would be expected given their ubiquity within statistics, there are many readily available functions for fitting linear models with pre-existing libraries in R. The function `stats::lm` accepts a model via the formula interface. Options exist to allow for a set of sample weights and sample offsets. There are several S3 methods for working with the output. The summary method returns a regression table, complete with standard errors and goodness of fit tests. Plotting the object displays several classic model diagnostics and calling the function `coef` returns the predicted regression vector. Perhaps most importantly from a learning perspective, the `predict` method accepts a new data frame containing the independent variables in the original formula and returns the predicted values from the model. The lower level function `stats::lm.fit`, which is actually called by the `stats::lm`, can also be useful. The primary differences are that it accepts a data matrix in place of a formula object and that it returns a raw unclassed list. This can be useful when working with large datasets, particularly if we already have converted data frames into data matrices.

There are two alternative implementations of linear model algorithms in R worth noting for their potential computational benefits. The package **Matrix-Models** offers `lm.fit.sparse`, which behaves similarly to the **stats** package version but accepts sparse matrix types from the **Matrix** package [18]. It also provides a formula based `glm4` capable of mimicking `lm`, but using a sparse matrix internally. Both of these can provide considerable improvements in the memory required as well as decrease the computational cost of fitting regression models. Similarly, the package **biglm** provides the function **biglm** for fitting linear models on data too large to fit into the available memory of a single machine [108]. We explore the algorithms behind this approach further in Chapter 10. There are of course many packages offering extensions to the classic regression model. We explore many of these extensions through the following chapters and will highlight some notable implementation examples as their applications arise.

We mentioned that one major strength of linear models is that they have many well-understood and optimal theoretical properties. With our focus on the task of predictive modeling and a desire to avoid too many prerequisites, we have not developed this underlying theory in this text. There are several excellent treatments of this theory for interested readers. Fumio Hayashi's *Econometrics* assumes a thorough background in statistics and probability theory but closely follows our notation and approach [72]. It provides finite sample results in the first chapter and the asymptotic theory in the second. Rao and Toutenburg's *Linear Models* is another recommended classic reference that mixes a matrix-based approach with statistical theory [135]. Some self-contained references that require no prior exposure to probability theory include [141] and [59]. These, however, focus on the simple linear model with only an intercept and single independent variable.

2.10   Application: Cancer incidence rates

2.10.1   Data and setup

We now show an example of how linear models can be used for predictive modeling. The task is to predict the incidence rate of cancer diagnoses as a function of spatial and socioeconomic variables across United States counties. The data we use takes cancer incidence rates from data published by the Centers for Disease Control and Prevention (CDC) and the Census Bureau for the year 2015. We have cleaned the data to join the incidence rates with economic variables and have discarded counties with missing or incomplete data.

We read the dataset into R using the `read.csv` function.

```
cancer <- read.csv("data/cancer_inc_data.csv")
names(cancer)
```

```
> names(cancer)
```

```
[1]  "name"       "state"      "breast"
[4]  "colorectal" "prostate"   "lung"
[7]  "melanoma"   "poverty"    "income"
[10] "region"     "lat"        "lon"
```

There is a variable giving the spatial information in the form of the county name, the state containing the county, the region of the country the state is in, and the latitude and longitude of the county centroid. Economic variables describe the median household income in US dollars and the percentage of households living below the poverty line. Cancer incidence rates are given for breast, colorectal, prostate, lung, and skin (melanoma) cancer types. These are presented in the expected number of diagnoses per one hundred thousand adults, or in the case of prostate cancer, diagnoses per one hundred thousand adult males.

We will delay a discussion of melanoma at the moment, focusing on building predictive models for the remaining four cancer types. Figure 2.2 shows a scatter plot of these incidence rates with respect to the income and poverty variables. Linear regression lines with only a slope and intercept are included in the plot. While our focus is on prediction, it is clear from this plot that linear regression is a useful graphical tool as well. The lines help illustrate the general trend of the plots. For all cancer types, the sign of the slope from the income predictor variable is the opposite of the slopes for the poverty variable. This should not be surprising because a high median income should generally correspond to a low percentage of households living below the poverty line. We see that breast cancer seems to be positively correlated with income whereas colorectal and lung cancer are negatively correlated with income. Prostate cancer rates seem to be only weakly related to income.

In order to evaluate a predictive model, we need to split our dataset into training and testing sets. In this case, we will do so by splitting the data randomly into two (approximately) equally sized subsets. We set the random seed in R so that the training and testing sets are created consistently between sessions.

```
set.seed(1)
cancer$train <- runif(nrow(cancer)) > 0.5
```

Using this new variable, we will set models on the training set and evaluate their performance on the testing set.

2.10.2   Socioeconomic predictors

We can use the lm function to fit the linear model predicting breast cancer rates as a function of the percentage of households living below the poverty line. This corresponds to the upper-right plot in Figure 2.2. The option subset to the function lm makes it easy to only use the training data in estimating the slope and intercept of the model. We will then call the predict function

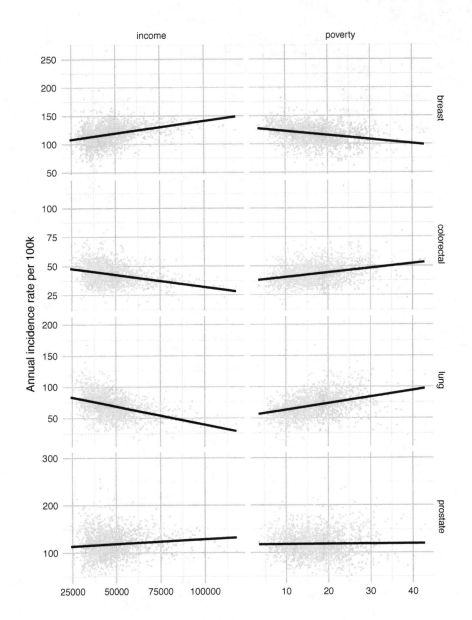

FIGURE 2.2: Cancer incidence rates from US counties as a function of demographic variables with regression lines fit using ordinary least squares from 2015 data. Incidence rates are given as annual counts per one hundred thousand adults (adult males for prostate cancer). Income is median household income in dollars and poverty is percentage of households falling below the federal poverty line.

on the entire dataset and return the root mean squared error on the training set. Depending on the application, it may be better to weight the counties by population, but we will weight them equally in this analysis for simplicity.

```
model <- lm(breast ~ poverty,
            data = cancer, subset = train)
prd <- predict(model, cancer)
sqrt(mean((prd - cancer$breast)[cancer$train]^2))
```

```
[1] 17.16241
```

It is hard to make much of just this single number without some context. The standard deviation of breast cancer incidence rates on the training data is 17.49, and offers a good point of comparison. We see that this one variable seems to offer only marginal predictive power. It is also more reasonable to look at the performance of the model on the test set, which has not been used in the estimation process.

```
sqrt(mean((prd - cancer$breast)[!cancer$train]^2))
```

```
[1] 16.36905
```

Perhaps surprisingly, the error rate on the test set is smaller. Have we perhaps made a mistake somewhere? In fact, this behavior is quite common in predictive modeling. The reason for the lower rate is that the variation in the response is somewhat smaller in the test set to begin with. The standard deviation in the test set is 17.03. Compared to its standard deviation, we again see affirmation that the model is giving a noticeable but small amount of predictive power.

In Figure 2.2, it visually appears that the income variable more strongly correlates with the incidence rate of breast cancer. We can re-do our analysis using this variable as a predictor. Going forward, we will restrict our evaluation to performance on the test set.

```
model <- lm(breast ~ income,
            data = cancer, subset = train)
prd <- predict(model, cancer)
sqrt(mean((prd - cancer$breast)[!cancer$train]^2))
```

```
[1] 15.99735
```

As expected from the plots, we see that the income level provides a stronger predictive model. Relative to the standard deviation, we see that it approximately doubles the reduction in the root mean squared error. Now, what happens if we add both variables into the model? Both variables measure similar quantities, but given their different predictive powers clearly do not do so in exactly the same fashion.

```
model <- lm(breast ~ poverty + income,
            data = cancer, subset = train)
prd <- predict(model, cancer)
sqrt(mean((prd - cancer$breast)[!cancer$train]^2))
```

```
[1] 16.00985
```

Here, putting both variables into the model yields a nearly similar rate to the income-only model. At least in the case of breast cancer, the poverty rate seems to give no additional information about incidence levels. We will, however, keep it in the following models in order to make comparison to the other cancer types more straightforward.

2.10.3   Geographic region predictors

Now, let us try to include some geographic information into the model. One variable that has been given to us is the region in which the county's state resides; here, there are nine spatial regions specified in the data. We can easily include this in the model using the formula interface within R. Before using it for prediction, we will first look at the estimated regression vector.

```
model <- lm(breast ~ poverty + income + region,
            data = cancer, subset = train)
coef(model)
```

```
> coef(model)
              (Intercept)                      poverty
           94.8164843905                 0.2048696354
                   income  regioneast_south_central
            0.0004117657                 0.9544261074
    regionmiddle_atlantic              regionmountain
            5.9327755582                -6.0609225249
        regionnew_england                regionpacific
           12.5145949123                 3.0480059485
      regionsouth_atlantic  regionwest_north_central
            3.6872254140                 2.8989287260
  regionwest_south_central
           -8.0784780668
```

One of the regions is left out of the design matrix to avoid perfect multi-collinearity. In this case, the region is 'east_north_central'. The other region coefficients describe how different the breast cancer incidence rates are in each region relative to the east north central region. It seems that New England has the highest rates and the south central the lowest. With the model constructed, we can again compute the error rate on the test set.

```
v <- prd - cancer$breast
sqrt(mean((v)[!cancer$train]^2))
```

```
[1] 15.68953
```

Even though income and poverty rates are strongly correlated with geographic regions, the root mean squared error decreases considerably given the additional geographic data.

2.10.4 State indicator variables

The region variable offers additional predictive power so it is reasonable to assume that including an intercept for a county's state into the model would be useful. We will remove the region variable as region is fully determined by state and will therefore create a perfectly collinear data matrix. It is easy to modify the formula to include state in the model, but when predicting values over the the testing set, R throws an error.

```
model <- lm(breast ~ poverty + income + state,
            data = cancer, subset = train)
pred <- predict(model, cancer)
```

```
Error in model.frame.default(Terms, newdata,
  na.action = na.action, xlev = object\$xlevels) :
  factor state has new levels hi
```

The problem is that Hawaii ('hi') has only four counties and, by chance, all four of these have been put into the test set. Therefore, the way in which R constructed the model matrix in the call to lm will not work correctly on the testing set because there is no Hawaii indicator variable. Note that the District of Columbia, regarded as a state in this dataset, has only a single county, but this is included in the training set and therefore does not cause any errors. One way to deal with this, common in social science applications, is to stratify the training and testing sets. That is, we explicitly make the states as balanced as possible across the two sets. Problems with new levels in a categorical variable often occur when working with datasets that are implemented in a streaming environment or stored across several machines. In these cases it may not be possible to stratify the data and other techniques such as feature hashing must be used. We will explore these function in later chapters. For simplicity, in this situation we will simply remove those four observations from counties in the state of Hawaii.

```
cancer <- cancer[cancer$state != "hi", ]
model <- lm(breast ~ income,
            data = cancer, subset = train)
```

	Poverty	Income	Both	Region	State
Breast	16.39	16.02	16.03	15.69	15.68
Colorectal	7.78	7.69	7.69	7.12	6.67
Prostate	25.71	25.65	25.40	25.13	22.46
Lung	15.32	14.86	14.87	12.43	11.30

TABLE 2.1: These are the root mean squared errors on the testing set for various linear models in the prediction of county-level cancer incidence rates (excluding those counties in Hawaii). The poverty and income columns are univariate models with only the respective variable as a predictor. The *both* column includes the poverty and income variables. Region and state include factors for the respective geographic level as well as both poverty and income variables.

```
prd <- predict(model, cancer)
v <- prd - cancer$breast
sqrt(mean((v)[!cancer$train]^2))
```

```
[1] 15.67717
```

In this case we see that the state variable offers only a negligible improvement over the region variable when predicting breast cancer incidence rates.

Root mean squared errors using these five approaches to predict the other cancer types are shown in Table 2.1. For consistency, Hawaii was removed from all of the testing sets. For all cancer types, the median income offered a more predictive model than the poverty percentage. Only for prostate cancer was there any benefit of combining these both into the same model and even in this case the benefit is minimal. The region variable improves all of the models, with a large improvement in all but the prostate cancer incidence. Replacing the region variable with the state variable offers noticeable improvements in the other three cancer types, unlike with the breast cancer incidence rates. Prostate cancer in particular is seen to have a greatly improved model when adding state-level effects.

2.10.5   Melanoma and latitude

As a final model, we will look at melanoma (skin cancer) incidence rates. One may initially suspect that melanoma is negatively correlated with a county's latitude. Locations farther north receive less harmful UVA and UVB light and it seems reasonable to suspect lower levels of skin cancer in more northern locations. Fitting a model to the data gives a different story.

```
model <- lm(melanoma ~ lat, data = cancer)
coef(model)
```

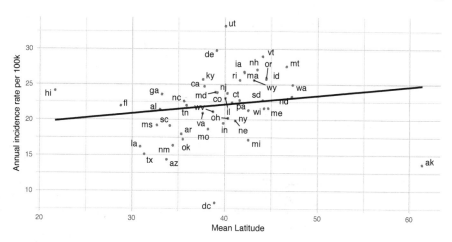

FIGURE 2.3: A scatter plot of the mean state level melanoma incidence rates against the average latitude of the state's counties. States are labeled with their standard two-letter abbreviations and a regression line fit using ordinary least squares is included.

```
(Intercept)          lat
  8.375773      0.350673
```

We see that latitude is actually positively correlated with skin cancer rates. To understand why, we plot the average melanoma rate at the state level with the average latitude of all its counties in Figure 2.3. The District of Columbia is an interesting outlier in this plot and points to the reason behind the counter-intuitive result. Skin cancer rates are highly influenced by skin color, and in the United States northern states have much higher concentrations of people who identify as white, the most susceptible demographic for melanoma. The District of Columbia has a particularly high proportion of African American residents, and this most likely causes it to be such a noticeable outlier in this plot. Utah is likely a particularly positive outlier given its relatively high number of white residents and its high average altitude. Notice that the extreme points of Alaska and Hawaii seem to more closely follow our expected negative relationship between melanoma and latitude.

From the analysis of the melanoma model, along with a simple graphic, we can see the need to include other covariates such as ethnicity and altitude into the model. Such conclusions were made possible largely because it was particularly easy to understand the implied relationship between the predictor variable and the response. Most of the more complex models we will explore throughout this text will not offer such a straightforward understanding. As mentioned in Section 2.1 this is one of the main reasons for using linear models within a predictive context. A common procedure is to use linear models as a first step in understanding the relationship between the variables before moving on to other techniques.

2.11 Exercises

1. Generate a data matrix X with 100 rows, 10 columns, and values sampled from the continuous uniform distribution. Construct β as a vector with all 0's. Using a simulation with 1000 replications, produce various response vectors y by adding Gaussian noise with variance 0.5 to the projected data and record the ℓ_2-norm error in prediction of β. How does the average error change if the noise variance is doubled? How about if it is multiplied by 4 or 8? What seems to be the general relationship between σ^2 and the error?

2. Write a function `ols_grad_decent` with the same inputs and output as `casl_ols_orth_proj`, but replace the orthogonal projects with gradient descent. As a naïve stopping criterion, simply let the algorithm run for 100 iterations, starting at $\hat{\beta}$ equal to all zeros. Test the function on some input data and compare to the orthogonal projection code.

3. Modify the code in `ols_grad_decent` to accept arguments `eps` and `max_iter`. Terminate the iteration whenever the maximum number of iterations has been reached or the ℓ_∞-norm change between iterations is less than `eps`. As a test, print the number of iterations to the console before returning the result with the function `print`. Compare the performance to the gradient descent function once again.

4. Generate a 5-by-1000 dimensional data matrix X with independent standard random normal variables and set $\beta = (1, 1, 1, 0, 0)$. For 1000 replications each, construct a response vector y with each of the following random variables:

   - `runif(1000, min = -sqrt(3), max = sqrt(3))`
   - `rnorm(1000, mean = 0, sd = 1)`
   - `rcauchy(1000, scale = 1 / 1.4826)`

   The parameters have been chosen so that each has a mean (or median, in the case of the Cauchy) of zero and similar scales. Using these replications, estimate the mean squared error in estimated β with ordinary least squares. The classic theory gives no convergence results regarding Cauchy errors. How well does it perform in practice?

5. Consider the simple regression model with only a scalar x and intercept:

$$y = \beta_0 + \beta_1 \cdot x \qquad (2.66)$$

Using the explicit formula for the inverse of a 2-by-2 matrix, write down the least squares estimators for $\widehat{\beta_0}$ and $\widehat{\beta_1}$.

3

Ridge Regression and Principal Component Analysis

3.1 Variance in OLS

In Chapter 2 we saw how the simplicity and extensibility of linear models make them powerful tools for predictive modeling. However, once we have constructed the data matrix X, there are no tuning parameters left to control the output of the model. Put another way, we have to decide whether a particular predictor variable should be wholly included in the model or entirely excluded from it. This is a particular problem if we have a large collection of weakly predictive variables. There may not be enough data to accurately measure a large β vector, but artificially throwing out useful data will degrade the predictive power of the model. In this chapter we will see several extensions that offer solutions to this problem.

We have already stated that the ordinary least squares estimator is unbiased under the assumption that the random vector ϵ has a zero mean; this is a weak assumption because we will generally include an intercept column in X and would expect this intercept to capture the expected value of y. What we would now like to do is to trade some bias to reduce the variance of the ordinary least squares solution. This will allow us to fit larger models without requiring more data. A naïve approach to trade bias for variance would be to scale the β vector by a linear factor a. Specifically, we would have

$$\widehat{\beta}^a_{scaled} = a \cdot \widehat{\beta}_{ols}, \quad 0 < a < 1. \tag{3.1}$$

If the model has an intercept, we could specifically choose not to scale that term. In fact, we could pick different scaling terms for each component of β. How might we pick the values of a? Generally, this is a hard problem with as many free parameters as the original estimation task. If we stick with a single a, estimating the best value for it by some form of validation, this will shrink components of β for which we have a strong signal equal to those for which we have a very weak signal.

It might make more sense to shrink $\widehat{\beta}_{ols}$ in such a way as to reduce the variance as fast as possible given a fixed 'budget' by which we are comfortable reducing the bias. To do this we need to better understand where the variance of $\widehat{\beta}_{ols}$ is coming from. Specifically, let us look at the *total variation*, or squared

ℓ_2-norm loss, of the ordinary least squares estimator. This can be written compactly as the trace of the variance-covariance matrix of $\widehat{\beta}_{ols}$

$$\mathbb{E}\left[||\beta - \widehat{\beta}_{ols}||_2^2\right] = \sum_j Var(\widehat{\beta}_j) \tag{3.2}$$

$$= tr\left(Var(\widehat{\beta}_{ols})\right). \tag{3.3}$$

Assuming that the matrix X is non-random, this quantity can be re-written in terms of the variance-covariance matrix of y, or equivalently the error vector ϵ

$$Var(\widehat{\beta}_{ols}) = \mathbb{E}\left[\widehat{\beta}\widehat{\beta}^t\right] \tag{3.4}$$

$$= \mathbb{E}\left[(X^t X)^{-1} X^t y y^t X (X^t X)^{-1}\right] \tag{3.5}$$

$$= (X^t X)^{-1} X^t \cdot \mathbb{E}\left[y y^t\right] \cdot X (X^t X)^{-1} \tag{3.6}$$

$$= (X^t X)^{-1} X^t \cdot Var(y) \cdot X (X^t X)^{-1} \tag{3.7}$$

$$= (X^t X)^{-1} X^t \cdot Var(\epsilon) \cdot X (X^t X)^{-1} \tag{3.8}$$

In order to simplify this further we need to make a simplifying assumption about the nature of the error terms ϵ. The spherical error assumption sets the variance of ϵ to a diagonal matrix with the same value, denoted σ^2, along the entire diagonal. Plugging in this assumption we can re-write the variance of $\widehat{\beta}_{ols}$ as

$$Var(\widehat{\beta}_{ols}) = \sigma^2 (X^t X)^{-1} X^t I_p X (X^t X)^{-1} \tag{3.9}$$

$$= \sigma^2 (X^t X)^{-1}. \tag{3.10}$$

This result is very useful when using ordinary least squares to make inferential claims about the β vector. Most statistical software programs use Equation 3.10 in constructing the standard errors of the components of the ordinary least squares estimator.

The total variation is given by the trace of the variance-covariance matrix. The easiest way to describe the trace in this situation is to decompose the matrix X into its singular value decomposition $U\Sigma V^t$:

$$tr\left(Var(\widehat{\beta})\right) = \sigma^2 \cdot tr\left((X^t X)^{-1}\right) \tag{3.11}$$

$$= \sigma^2 \cdot tr\left((V\Sigma U^t U\Sigma V^t)^{-1}\right) \tag{3.12}$$

$$= \sigma^2 \cdot tr\left((V\Sigma^2 V^t)^{-1}\right) \tag{3.13}$$

$$= \sigma^2 \cdot tr\left(V\Sigma^{-2} V^t\right) \tag{3.14}$$

Inside the trace operator we can cyclically rotate a matrix product so that, for example, $tr(ABC)$ is equal to $tr(BCA)$. Using this trick here the V matrices cancel and we find an explicit formula for the total variation

$$tr\left(Var(\widehat{\beta})\right) = \sigma^2 \cdot tr\left(\Sigma^{-2}\right) \tag{3.15}$$

$$= \sigma^2 \cdot \sum_j \frac{1}{\sigma_j^2} \tag{3.16}$$

where σ_j are the singular values of X. From this formula we see that the largest contributor of variance is coming from the smallest singular value. If one singular value is significantly smaller than the others, this can easily represent the vast majority of the variation in $\widehat{\beta}_{ols}$.

It is insightful to simulate and numerically validate the variance formulas we have derived. The process of doing so will also mirror that of testing the performance of the estimators we will study that attempt to control for this variance. We start by setting the simulation parameters; we will have 200 observations, 20 variables, 100 different condition numbers, and 20 replications of each scenario. The regression vector β is set to all 1's.

```
n <- 200; p <- 20; N <- 100; M <- 20
mu <- rep(0, p)
beta <- rep(1, p)
```

The data matrix X that we generate will come from a multivariate normal distribution. The means of each column will be 0 and their variance will be 1. The correlation between any two columns will be set to some value $0 < \rho < 1$, which we will change in order to control how ill-conditioned the matrix $X^t X$ is expected to be. When ρ is close to 1, Equation 3.16 tells us that an ordinary least squares estimator will have a very high variance. The most interesting results occur when we let the values of ρ be picked uniformly on a *log scale*, which is accomplished by the following block of code.

```
mu <- rep(0, p)
rhoVals <- log(seq(exp(.9), exp(1), length.out = N + 1))[-(N+1)]
```

Finally, we now create empty matrices to store the output of the simulation. We want to compute the expected variance of $\widehat{\beta}$ and the observed variance. Within each of the 20 sub-trials, in order to compute the observed variance, we need to store all the values of $\widehat{\beta}$ and also construct a matrix to hold these.

```
var_exp <- rep(NA_real_, N)
var_obs <- rep(NA_real_, N)
betahat <- matrix(NA_real_, nrow = M, ncol = p)
```

Creating matrices of the correct size and type ahead of time can significantly improve the run-time of a simulation as it eliminates the copying of data that occurs if R functions are used such as c, cbind, and rbind.

We are now ready to run the actual simulation. This involves a double loop, with the outer loop cycling over values of ρ and the inner loop replicating the results 20 times. The expected variance is calculated with Equation 3.16, and the estimated least squares solution uses the function `solve` to directly solve the least squares problem.

```
for (j in 1:N) {
  Sigma <- matrix(rhoVals[j], nrow = p, ncol = p)
  diag(Sigma) <- 1
  X <- MASS::mvrnorm(n, mu, Sigma)
  svals <- svd(X)$d
  var_exp[j] <- sum(1 / svals^2)

  for (m in 1:M) {
    eps <- rnorm(n, sd = 1)
    y <- X %*% beta + eps
    betahat[m,] <- solve(crossprod(X), crossprod(X, y))
  }
  var_obs[j] <- sum(apply(betahat, 2, var))
}
```

We see that the worst relative error in the simulation is just 25%.

```
max(abs(var_obs - var_exp) / var_exp)
```

```
[1] 0.2510161
```

A complete scatter plot of the results is shown in Figure 3.1, where the increase in both the expected and observed variances with increased values of ρ is clearly seen.

In Section 2.7 we saw that the smallest singular value gives a measurement of how hard it is to solve a linear system. Here, in another connection between numerical error and statistical error, we see that this smallest singular value contributes a disproportionate amount of variance to the estimator $\hat{\beta}_{ols}$. In the remainder of this chapter we show how to intelligently shrink our parameter estimates in such a way as to reduce the impact of the smallest singular values without unnecessarily scaling along the direction of the largest singular values. These approaches fulfill the promise of allowing us to include a larger set of highly correlated predictors into a single linear model.

3.2   Ridge regression

Consider once again a simple system of linear equations defined by

$$Ab = z. \qquad (3.17)$$

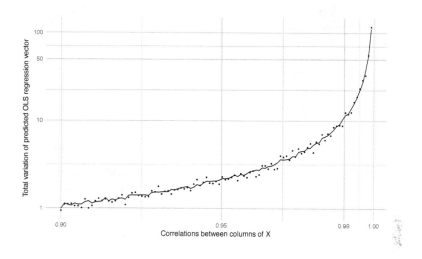

FIGURE 3.1: A plot of the relationship between the expected (line) and observed (points) total variation of the OLS regression vector with increasing correlation between the columns of the design matrix. Both axes are shown on the log scale.

For a square matrix A. We have seen that the difficulty of solving this system can be described by the ratio between the largest and smallest singular values of A. A reasonable approach to fixing this problem is to replace A with a new matrix A' having the same singular vectors, but modifying the specific singular values. Specifically, we can increase the singular values all by a fixed factor of λ:

$$\sigma_j \to \sigma_j + \lambda, \quad \lambda > 0. \tag{3.18}$$

If A is written as its singular value decomposition $U\Sigma V^t$, this modified system of equations becomes

$$A'b = z \tag{3.19}$$

$$U(\Sigma + 1_p \cdot \lambda)V^t b = z. \tag{3.20}$$

This can be solved by making use of the pseudoinverse, as was done in Section 2.4. The condition number of A' is given by

$$cond(A') = \frac{\sigma_{max} + \lambda}{\sigma_{min} + \lambda}. \tag{3.21}$$

Even with a relatively small choice for λ, this new condition number can usually be made to be smaller than the inverse of the machine's precision. This technique is known as Tikhonov regularization, after the Russian mathematician Andrey Tikhonov, and is a well-known and popular technique for solving ill-conditioned linear systems. Notice that we can rewrite Equation 3.20 directly in terms of the original matrix A

$$U(\Sigma + 1_p \cdot \lambda)V^t b = z \tag{3.22}$$

$$(A + U \cdot V^t \cdot \lambda)b = z. \tag{3.23}$$

Adding a small factor to each of the singular values is equivalent to adding a multiple of the rotation product UV^t.

We will use this regularization technique on the matrix $X^t X$ within the normal equations. Notice that if $A = X^t X$ then the U and V in Equation 3.23 are both equal to the right singular vectors of X, which we will now set to V. Also, the singular values of A are the squared singular values of X. We will denote the singular values of X as $\Sigma = \text{Diag}(\sigma_1, \ldots, \sigma_k)$. Using this regularization technique to modify the normal equations yields

$$(X^t X + 1_p \cdot \lambda)b = X^t y. \tag{3.24}$$

And the condition number of the matrix on the left-hand side is equal to

$$cond(X^t X + 1_p \cdot \lambda) = \frac{\sigma_{max}^2 + \lambda}{\sigma_{min}^2 + \lambda}. \tag{3.25}$$

Notice that even if σ_{min} is equal to zero, the condition number will be finite if λ is larger than zero. Equation 3.24 can be simplified by expanding the matrix X by its SVD as

$$V(\Sigma + 1_p \cdot \lambda)V^t b = X^t y \tag{3.26}$$

$$b = V(\Sigma + 1_p \cdot \lambda)^{-1}V^t V \Sigma U^t y \tag{3.27}$$

$$b = V(\Sigma + 1_p \cdot \lambda)^{-1}\Sigma U^t y \tag{3.28}$$

$$b = V \cdot \text{Diag}\left(\frac{\sigma_1}{\sigma_1^2 + \lambda}, \ldots, \frac{\sigma_p}{\sigma_p^2 + \lambda}\right) \cdot U^t y. \tag{3.29}$$

Solutions to this modified equation will have a lower variance, but increased bias, compared to the ordinary least squares estimator. As we have seen, the condition number becomes a problem for statistical noise at much lower levels than it does for numerical error. Therefore we often need to set λ to a relatively large value in order to find the most predictive model.

The ordinary least squares estimator was defined by the solution to an optimization problem. The normal equations were simply a consequence of this definition. We would again like to define our modified estimator in terms of an optimization problem. To do so, we need to describe Equation 3.24 as the solution to some minimization task. The correct choice is to add the squared ℓ_2-norm of the regression vector to the sum of squared residuals

$$f(b) = ||y - Xb||_2^2 + \lambda||b||_2^2. \tag{3.30}$$

Ridge Regression

Given a data matrix X and response vector y, ridge regression is defined as the solution to the following optimization task

$$\widehat{\beta}^{ridge}(\lambda) \in \arg\min_{b} \left\{ ||y - Xb||_2^2 + \lambda \cdot ||b||_2^2 \right\}$$

with a unique, well-defined solution for every value of $\lambda > 0$.

To verify this we take the gradient of f

$$\nabla_b f(b) = \nabla_b \left[(y^t - b^t X^t)(y - Xb) + \lambda b^t b \right] \tag{3.31}$$

$$= \nabla_b \left[y^t y + b^t (X^t X) b - 2 y^t X b + \lambda b^t b \right] \tag{3.32}$$

$$= 2(X^t X) b - 2 X^t y + 2\lambda b. \tag{3.33}$$

And set it equal to 0

$$(X^t X) b + \lambda b = X^t y \tag{3.34}$$

$$(X^t X + \lambda 1_p) b = X^t y. \tag{3.35}$$

This yields the same formula given in Equation 3.24. This estimation technique is called *ridge regression*. As with ordinary least squares, the Hessian matrix is constant everywhere

$$H(f) = 2 X^t X + 2\lambda 1_p. \tag{3.36}$$

In ordinary least squares we showed that the Hessian matrix is positive definite whenever X has full column rank. Here, as long as $\lambda > 0$, we see that the Hessian is positive definite regardless of the rank of X. To see this notice that we can write the quadratic form $v^t H v$, for any $v \neq \vec{0}$ as

$$2 \cdot v^t H(f) v = v^t X^t X v + 2\lambda v^t 1_p v \tag{3.37}$$

$$= ||Xv||_2^2 + \lambda \cdot ||v||_2^2. \tag{3.38}$$

We know that the ℓ_2-norm of Xv is non-negative and the norm of v is positive unless v is the zero vector and therefore the positive definiteness of $H(f)$ follows. Ridge regression can therefore by applied to any matrix X, including cases where there may even be more variables than there are observations.

Our initial task was to find an intelligent way of reducing the size of the prediction vector $\widehat{\beta}$ while maintaining as much predictive power as possible. In retrospect, the optimization form of ridge regression simply makes this explicit. The ridge regression vector minimizes a combination of the sum of squares and the size of the regression vector, with the parameter λ giving the relative balance between these two tasks. Of course, it is possible to choose alternative measurements for the size of complexity of β. Other choices lead to the class of penalized estimators, some of which we will explore in Chapter 7. The choice of the ℓ_2-norm, though, has two key advantages. Because it pairs an ℓ_2-norm with the squared norm of the residuals, it admits the direct solution given in Equation 3.24, a benefit not shared by other penalized least squares estimators. Also, ridge regression maintains the rotational invariance of ordinary least squares. Consider rotating X by an orthogonal matrix Q. The ridge regression vector $\widehat{\beta}^\lambda_{new}$ on this new term is simply a rotated version of the ridge regression vector for the original data matrix X. We see this quickly from the penalized sums of squares

$$||y - (XQ)\gamma||_2^2 + \lambda \cdot ||\gamma||_2^2 = ||y - X(Q\gamma)||_2^2 + \lambda \cdot \gamma^t \gamma \tag{3.39}$$
$$= ||y - X(Q\gamma)||_2^2 + \lambda \cdot b^t Q^t Q b \tag{3.40}$$
$$= ||y - X(Q\gamma)||_2^2 + \lambda \cdot ||(Q\gamma)||_2^2 \tag{3.41}$$
$$= ||y - Xb||_2^2 + \lambda \cdot ||b||_2^2. \tag{3.42}$$

Unlike ordinary least squares, however, ridge regression is not scale invariant. Consider the case where X is a two column matrix with two highly correlated columns both centered around zero. If the first column has a standard deviation ten times that of the second column, the vector $\widehat{\beta}^\lambda_{ridge}$ will be concentrated on X_1 because it can get a larger signal $X_1\beta_1$ at a lower cost β_1^2. Due to this issue with scale, the columns of the data matrix are typically centered to zero and scaled to have a sample variance of 1 before applying ridge regression. The result is then scaled back to give $\widehat{\beta}^\lambda_{ridge}$ in the original coordinate system if desired.

Another benefit of the direct solution provided implied by Equation 3.24 is that ridge regression can be solved relatively efficiently for a sequence of tuning values λ. If we calculate and store the SVD decomposition of X, then the ridge regression solution given by Equation 3.29 is a straightforward matrix multiplication. The complexity of each new λ value then is just $O(p^2)$. Computing the values of the ridge regression quickly for many values of λ is important when performing cross-validation.

We will now use this approach to compute the ridge regression vector for a large set of tuning parameters λ. We will work with a small dataset with 4 modestly correlated variables.

```
n <- 200; p <- 4; N <- 500; M <- 20
beta <- c(1, -1, 0.5, 0)
mu <- rep(0, p)
Sigma <- matrix(0.9, nrow = p, ncol = p)
diag(Sigma) <- 1
X <- MASS::mvrnorm(n, mu, Sigma)
y <- X %*% beta + rnorm(n, sd = 5)
```

We will also generate a separate testing set,

```
X_test <- MASS::mvrnorm(n, mu, Sigma)
```

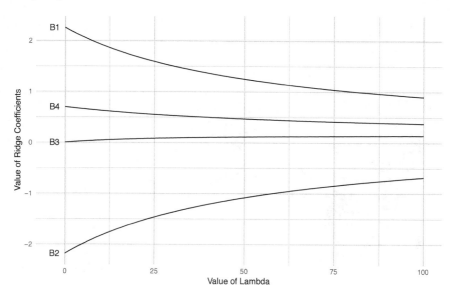

FIGURE 3.2: Visualization of the coordinates of the ridge regression vector for varying values of the tuning parameter λ. At the value of 0, the coefficients match the ordinary least squares solution.

```
y_test <- X_test %*% beta + rnorm(n, sd = 5)
y_test <- as.numeric(y_test)
```

The test set will be used to tune the value of λ in the ridge regression algorithm.

We can create an efficient implementation of ridge regression for a sequence of values λ by storing the singular value decomposition of X. From here, we cycle over a range of tuning parameters and fill the estimates $\widehat{\beta}^{\lambda}$ into a single matrix.

```
# Compute ridge regression vector.
#
# Args:
#     X: A numeric data matrix.
#     y: Response vector.
#     lambda_vals: A sequence of penalty terms.
#
# Returns:
#     A matrix of regression vectors with ncol(X) columns
#     and length(lambda_vals) rows.
casl_lm_ridge <-
function(X, y, lambda_vals)
```

```
{
  svd_obj <- svd(X)
  U <- svd_obj$u
  V <- svd_obj$v
  svals <- svd_obj$d
  k <- length(lambda_vals)

  ridge_beta <- matrix(NA_real_, nrow = k, ncol = ncol(X))
  for (j in seq_len(k))
  {
    D <- diag(svals / (svals^2 + lambda_vals[j]))
    ridge_beta[j,] <- V %*% D %*% t(U) %*% y
  }

  ridge_beta
}
```

We can now apply this to our training data to get the regression vector for a sequence of tuning parameters.

```
lambda_vals <- seq(0, n*2, length.out = N)
beta_mat <- casl_lm_ridge(X, y, lambda_vals)
```

A visualization of these paths is given in Figure 3.2. Notice that the coefficients all limit towards 0 as the penalty increases towards infinity. While difficult to see given the scale of the plot, the sign of the fourth parameter actually changes signs, from negative to positive, around λ equal to 200.

How do we pick a value of λ? It is usually set using some form of validation or cross-validation. That is, we do prediction on a validation set and find the value that has the lowest prediction error. We now produce estimated predictions $\hat{y}$ on the testing set and calculate the mean squared error prediction loss for each value of λ.

```
y_hat <- tcrossprod(X_test, beta_mat)
mse <- apply((y_hat - y_test)^2, 2, mean)
lambda_vals[which.min(mse)]
```

```
[1] 6.464646
```

The best value of λ is at about 6.5. It helps to visualize the mean squared error for ridge regression, as the shape is fairly typical of many datasets. Figure 3.3 shows this curve zoomed into the smaller values of λ. As the penalty increases, the mean squared error decreases sharply until it reaches a minimum value and then slowly increases again. If we continued the plot for larger penalties, the mean squared error would limit towards the variance of y as the predicted regression vector β approaches the zero vector.

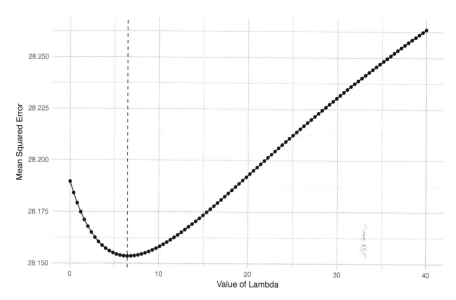

FIGURE 3.3: A plot of the validation mean squared error for ridge regression using varying values for the tuning parameter λ. The optimal value of 6.46 is shown with the dashed line.

3.3 (⋆) A Bayesian perspective

Looking at Figure 3.2, we see that applying the ridge regression for increasing values of λ tends to shrink the values of $\hat{\beta}$ towards zero. While a particular coefficient may locally grow in size, it is an easy derivation to show that the ℓ_2-norm of the predicted regression vector is in fact monotonically decreasing in λ. This shrinking behavior, and the corresponding decrease in variance that we were initially after, is commonly associated with Bayesian estimation. In fact, ridge regression can be formulated as Bayesian estimator. Doing so provides both an additional helpful way of understanding the underlying behavior of ridge regression as well as indicating avenues for practically useful extensions of the model.

In Bayesian estimation we start with some presumed knowledge about the parameter β we are attempting to estimate. This knowledge is given in the form of a probability distribution $f(\beta)$, called the *prior*, over the unknown quantity. Estimation proceeds by observing some data y and updating our understanding of the unknown parameter by computing the conditional density $f(\beta|y)$. This coincides well with the intuitive notion of conditional distributions: if $f(\beta)$ tells us our initial understanding of β then $f(\beta|y)$ represents our knowledge once we account for the data y. In order to calculate this conditional density we typically also have knowledge of the distribution of the data given a specific value of β; in other words, we know the likelihood function $f(y|\beta)$. Using the Bayes' theorem, hence the name, we then see that it is possible to update our understanding by the formula

$$f(\beta|y) \propto f(y|\beta) \cdot f(\beta). \tag{3.43}$$

In complex Bayesian models this equation must be approximated by numerical techniques. Conveniently, in the case of the Bayesian interpretation of ridge regression this can be solved analytically.

Ridge regression can be understood as a Bayesian estimator to the prior distribution and likelihood functions implied by assigning independent, identically distributed, normally random variables to the components of β and setting the likelihood function to that of the standard regression model.

$$\beta_j \sim_{i.i.d.} N(0, \sigma^2 \cdot \lambda^{-1}) \tag{3.44}$$

$$y_i|\beta \sim_{i.i.d.} N(x_i^t \beta, \sigma^2) \tag{3.45}$$

For simplicity we have assumed that the noise variance σ^2 is some known constant. From these distributions we can use Bayes' theorem to compute the density function $f(\beta|y)$; because y and X are assumed to be known in the conditional probability, we can simplify the calculation by dropping any multiplicative terms that do not depend on β. From the independence of the y_i's we then have

$$f(\beta|y) \propto \prod_{j=1}^{p} \exp\left\{-\frac{\lambda}{2\sigma^2}\beta_j^2\right\} \times \prod_{i=1}^{n} \exp\left\{-\frac{1}{2\sigma^2}(y_i - x_i^t\beta)^2\right\} \tag{3.46}$$

$$= \exp\left\{-\frac{\lambda}{2}||\beta||_2^2\right\} \times \exp\left\{-\frac{1}{2\sigma^2}||y - X\beta||_2^2\right\} \tag{3.47}$$

$$= \exp\left\{-\frac{1}{2\sigma^2} \cdot \left(||y - X\beta||_2^2 + \lambda||\beta||_2^2\right)\right\} \tag{3.48}$$

The posterior in this form closely resembles a multivariate normal distribution. We need to simplify the term inside the exponential to see the exact form of this multivariate normal. We then have

$$||y - X\beta||_2^2 + \lambda||\beta||_2^2 = y^ty - \beta^t X^t X y - y^t X\beta + \beta^t X^t X\beta + \lambda\beta^t\beta \tag{3.49}$$

$$= y^ty - \beta^t(X^t X\lambda I_p)(X^t X\lambda I_p)^{-1}X^ty$$
$$\quad - y^t X(X^t X + \lambda I_p)^{-1}(X^t X + \lambda I_p)\beta$$
$$\quad + \beta^t(X^t X + \lambda I_p)\beta \tag{3.50}$$

$$= y^ty - \beta^t(X^t X\lambda I_p)\widehat{\beta}_\lambda - \widehat{\beta}_\lambda^t(X^t X + \lambda I_p)\beta$$
$$\quad + \beta^t(X^t X + \lambda I_p)\beta \tag{3.51}$$

$$= y^ty - y^t X(X^t X + \lambda I_p)^{-1}X^ty$$
$$\quad + \left[\beta - \widehat{\beta}_\lambda\right]^t (X^t X + \lambda I_p)\left[\beta - \widehat{\beta}_\lambda\right], \tag{3.52}$$

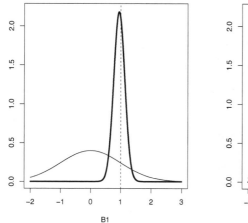
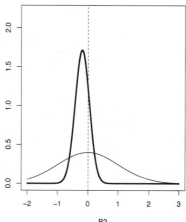

FIGURE 3.4: A plot of the prior density (thin line) and posterior density (dense line) of the two coordinates of β in an application of Bayesian ridge regression.

where $\widehat{\beta}_\lambda$ is the ridge regression with penalty λ. Plugging this into the posterior density, and throwing away the terms that depend only on y and X, we finally have

$$f(\beta|y) \propto \exp\left\{ -\frac{1}{2\sigma^2} \cdot \left[\beta - \widehat{\beta}_\lambda\right]^t (X^t X + \lambda I_p) \left[\beta - \widehat{\beta}_\lambda\right] \right\} \qquad (3.53)$$

Therefore, this is in fact a multivariate normal distribution with the mean and variance given by

$$\beta|y \sim MVN(\widehat{\beta}_\lambda, \sigma^2 \cdot (X^t X + \lambda I_p)^{-1}). \qquad (3.54)$$

We see from this that the mean of the posterior distribution, known as the Bayes estimator, is just the ridge regression. We also see that the updated uncertainty in the location of β is given by the same term that controls the variance of the ridge regression estimate: $(X^t X + \lambda I_p)^{-1}$.

One benefit of the Bayesian perspective is that it gives us an entire distribution for the parameters β rather than a single value. We can simulate this with a small example having only two predictor variables. First, construct some data X, y and β.

```
n <- 20; beta <- c(1, 0)
X <- MASS::mvrnorm(n = n, mu = c(0, 0), Sigma = diag(1,2))
y <- X %*% beta + rnorm(n, sd = 2)
```

Setting λ equal to 0.5, we have the prior mean for each value of β_j equal to zero. The mean and variance from Equation 3.54 can then be calculated from the following.

```
lambda <- 0.5
bhat <- solve(crossprod(X) + diag(lambda, 2), crossprod(X, y))
bhat_sd <- sqrt(diag(solve(crossprod(X) + diag(lambda, 2))))
```

Using these, we can plot the posterior distribution of β, coordinate-wise, as is done in Figure 3.4. Notice that the posterior distribution has a significantly lower variance. The mean for β_1 has also moved closer to the true value of 1, whereas the mean for β_2 has moved only slightly off of the true value of 0.

A benefit of the Bayesian model is that it is easy to add further tweaks by inserting new parameters into the hierarchical model. For example, one popular tweak is to modify the setup to have another prior distribution for the noise variance

$$\sigma \sim IG(a, b) \tag{3.55}$$

$$\beta_j | \sigma \sim_{i.i.d.} N(0, \sigma^2 \cdot \lambda^{-1}) \tag{3.56}$$

$$y_i | \beta \sim_{i.i.d.} N(x_i^t \beta, \sigma^2) \tag{3.57}$$

where IG is the inverse gamma distribution. With a bit more arithmetic rearranging, it can be shown that $\beta | \sigma^2, y$ has the exact same form as Equation 3.54 in this new setup. Similarly, it also possible to set up a separate prior distribution for λ or to add more complex covariance structures between the values of y_i or β_j. We will explore these Bayesian connections again in Chapter 7 in relationship to a more general framework for penalized linear regression.

3.4   Principal component analysis

We have seen that ridge regression acts by scaling the singular values of $X^t X$ in such a way that the smaller singular values are increased in order to stabilize the solution to the normal equations. Another reasonable approach to deal with these small singular values is to drop them entirely, solving the problem projected into the space spanned by the largest singular vectors. In this section we explore this approach, known as principal component analysis (PCA), and compare its results to that of ridge regression.

It is possible to motivate principal components outside of the formal context of singular values and condition numbers. Consider a standard regression model with a data matrix X having just two columns.

$$y = X\beta + \epsilon, \quad \beta = \begin{pmatrix} 1 \\ 0 \end{pmatrix}. \tag{3.58}$$

We will assume that the first column of X, X_1, is very similar to the second column X_2. In this case, notice that the expected value of Y can be estimated very well by several very different values of β

$$\mathbb{E}y = X\beta = X_1 \cdot 1 + X_2 \cdot 0 = X_1 \tag{3.59}$$
$$\approx X_1 \cdot 0 + X_2 \cdot 1 \tag{3.60}$$
$$\approx X_1 \cdot 100 - X_2 \cdot 99 \tag{3.61}$$
$$\approx X_1 \cdot 1000 - X_2 \cdot 999. \tag{3.62}$$

Specifically, as long as β_1 and β_2 sum up to one, we would expect to get a reasonable prediction for y. This is just a particular special case of an ill-conditioned matrix and the problems that arise. We know that ridge regression should solve the problem. What happens in the case of the ridge estimator?

We can simulate this scenario to find out what ridge regression would estimate for the parameters of β by first generating some test data.

```
n <- 100; beta <- c(1, 0)
Sigma <- matrix(0.999, nrow = 2, ncol = 2)
diag(Sigma) <- 1
X <- MASS::mvrnorm(n = n, mu = c(0, 0), Sigma = Sigma)
y <- X %*% beta + rnorm(n)
```

We then solve for the ridge regression estimator.

```
solve(crossprod(X) + diag(5, 2), crossprod(X, y))
```

```
          [,1]
[1,] 0.5005489
[2,] 0.5230596
```

The ridge regression vector, for this value of λ, correctly predicts that the sum of β_1 and β_2 should be close to 1. The specific configuration that it produces has weights spread evenly between the two components. If you want two numbers β_1 and β_2 to sum to one, the choice that minimizes the ℓ_2-norm of the vector β is that which puts equal weight on both components. Therefore, this simulated outcome is quite reasonable.

It is relatively easy to estimate the sum of the components of β and relatively difficult to estimate their difference. One way to formalize this is to transform the data matrix X into a new matrix Z according to the following

$$Z_1 = X_1 + X_2 \tag{3.63}$$
$$Z_2 = X_1 - X_2. \tag{3.64}$$

Or, in matrix form, the following

$$\begin{pmatrix} Z_1 & Z_2 \end{pmatrix} = \begin{pmatrix} X_1 & X_2 \end{pmatrix} \cdot \begin{pmatrix} 1 & 1 \\ 1 & -1 \end{pmatrix}. \tag{3.65}$$
$$Z = X \cdot A, \tag{3.66}$$

denoting the transformation matrix as A. In this parametrization the variable Z_1 will have a relatively high variance and Z_2 will have very low variance. If we wanted to solve the transformed linear system,

$$y = X\beta + \epsilon = ZA^{-1}\beta + \epsilon \equiv Z\gamma + \epsilon, \tag{3.67}$$

It would be reasonable to set $\widehat{\gamma}_2$ equal to zero and attempt only to solve for $\widehat{\gamma}_1$. Doing so, we would expect to get $\widehat{\gamma}$ approximately equal to $\binom{0.5}{0}$. In this particular example, translating back into the original quantity would yield

$$\widehat{\beta} = A\widehat{\gamma} = \begin{pmatrix} 1 & 1 \\ 1 & -1 \end{pmatrix} \cdot \begin{pmatrix} 0.5 \\ 0 \end{pmatrix} = \begin{pmatrix} 0.5 \\ 0.5 \end{pmatrix}, \tag{3.68}$$

approximately equal to what we saw as the solution to the simulated ridge regression estimator.

The ideas behind this small toy example exemplify those of the principal component analysis, with the matrix Z serving as the principal components of the matrix X. To extend these concepts to a formal definition, we will consider now an arbitrary data matrix X and derive a formula for its first principal component. We remarked previously in the example that Z_1 contains most of the variance in the data. Here we use this as a definition, by first defining the quantity W_1 by

$$W_1 = \underset{v}{\arg\max} \left\{ \frac{v^t X^t X v}{v^t v} \right\} \tag{3.69}$$

$$= \underset{v:||v||_2=1}{\arg\max} \left\{ v^t X^t X v \right\}. \tag{3.70}$$

The first principal component is then the projection of X by the value W_1, called the (first) *loading vector*, which is written as

$$Z_1 = X \cdot W_1. \tag{3.71}$$

Recall that we have already seen the quantity in Equation 3.70 when deriving the condition number of a matrix. In fact, from Equation 2.44, W_1 is equal to the maximal right singular vector of X, V_1. Then, the first principal component Z_1 has the form

$$Z_1 = U^t \Sigma V^t \cdot V_1 \tag{3.72}$$

$$= U^t \Sigma \cdot \begin{pmatrix} 1 \\ 0 \\ \vdots \\ 0 \end{pmatrix} = U^t \cdot \begin{pmatrix} \sigma_1 \\ 0 \\ \vdots \\ 0 \end{pmatrix} = U_1 \cdot \sigma_1 \tag{3.73}$$

where U_1 is the first left singular vector.

The second principal component should be defined similarly: specifying a loading vector W_2 and then computing the component by projecting X

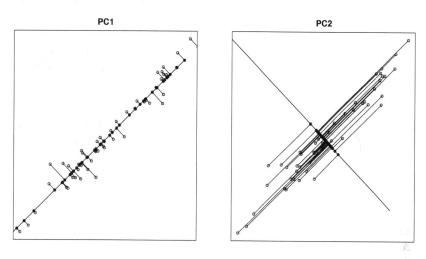

FIGURE 3.5: The first (left) and second (right) principal components of a two-dimensional dataset. Notice that the first principal component has the largest variance among all available projections and the second component is perpendicular to the first.

through this vector. As with the toy example, we do not want to duplicate any information found in the first component in the second. Mathematically, this can be described by forcing W_1 and W_2 to be orthogonal to one another. Subject to this restriction, we again want the component to capture the maximal amount of variation possible. From this, we come to a general formula for the kth loading vector

$$W_k = \underset{v:||v||_2=1}{\arg\max} \left\{ v^t X^t X v \; : \; v^t W_m = 0, m < k \right\}. \tag{3.74}$$

Using the same logic as we did for the maximal right singular vector, we can show that W_k is the right singular vector corresponding to the kth largest singular value. Once we have constructed all p loading vectors, the entire matrix of principal components is given by

$$Z = X \cdot W. \tag{3.75}$$

Equivalently, Z is equal to $U\Sigma$ in terms of the SVD of X. By construction, and the fact that we know that W will be equal to the right singular vectors of X, we know that the matrix W is orthogonal and that Z is nothing more than a special rotation of the matrix X. Figure 3.5 visualizes the principal components of a small design matrix as the projection into the principal component basis.

It may seem strange at first glance that we are explicitly interested in

---

Principal Components

Let X be any $n \times p$ with an SVD written as $U\Sigma V^t$. The matrix V is called the *loading matrix* for X and the principal component matrix Z is given by

$$Z = X \cdot V.$$

The first column of Z is called the *first principal component*, the second column the *second principal component*, and similarly, through all $\min(n, p)$ components.

---

working with those components of X which have a particularly high variance. In statistical learning, variance is often described as something to be avoided. Section 3.1, where we tracked down sources of error in OLS, in fact directly supports this idea. The fallacy here is that while variance in the response y is not good, variation in the data matrix X is generally very useful. For example, if we observe data points all concentrated in a very small part of the input space in $\mathbb{R}^p$ it will be very difficult to estimate new values anywhere outside of this region. We see this, in fact, in the term $(X^tX)^{-1}$ that comes in the variance of the OLS estimators from Equation 3.10. If the columns of X have a high variance the diagonal elements of X^tX will be relatively high, leading to small coefficients in its inverse. This is the exact opposite of the effect resulting from an increase in the variation of y, as given by σ^2.

Now that we have constructed the principal components of X, how will these be useful in statistical learning? A popular application is to use the first 2 or 3 components to visualize a high-dimensional matrix X. We will use this technique in order to better understand the data in many applications throughout the remainder of this text. Another usage of principal components is to keep only the first k columns of Z, working only with these in fitting a predictive model. We used this approach in our toy example, with k equal to 1. As in the example, it is always possible to convert a model in the Z coordinate system back into parameters in terms of the X coordinate system by taking the inverse of the loading matrix (an easy and stable inverse to compute as it was constructed to be orthonormal). While this technique may be used within any class predictive model, it has a particularly nice interpretation in the context of regression analysis, where it is referred to as *principal component regression*. As with the λ parameter in ridge regression, the tuning parameter k in principal component regression can be set via heuristics or a more formal validation framework.

With some manipulation, we can directly relate the form of the principal component regression vector to that of the ridge and OLS regression vectors.

Principal Component Regression

Let X be a data matrix with principal components given by the matrix Z. Defining $Z_{1:k}$ as a matrix consisting of the first k columns of Z, the *principal component regression vector* of order k, denoted by $\widehat{\beta}_k$, is defined by the solution to the following equation

$$\widehat{\beta}_k = V_{1:k} \cdot (Z_{1:k}^t Z_{1:k})^{-1} Z_{1:k}^t y$$

where V is the loadings matrix, the right singular vectors of X.

First of all, we see that $Z^t Z$ is just a diagonal matrix of the singular values

$$Z^t Z = V^t X^t X V \tag{3.76}$$

$$= V^t (V\Sigma^2 V^t) V \tag{3.77}$$

$$= \Sigma^2. \tag{3.78}$$

Using the notation $M_{1:k}$ to indicate taking columns 1 through k of the matrix M, then $Z_{1:k}$ is just $XV_{1:k}$ and $Z_{1:k}^t Z_{1:k}$ just a diagonal matrix with the first k singular values. Therefore, the principal component regression vector is:

$$\widehat{\beta}_k = V_{1:k} \cdot (Z_{1:k}^t Z_{1:k})^{-1} Z_{1:k}^t y \tag{3.79}$$

$$= (V_{1:k}\Sigma_{1:k}^{-2}V_{1:k}^t) X^t y. \tag{3.80}$$

Finally, notice that the terms grouped together in the parentheses can be rewritten if we replace $\Sigma_{1:k}^{-1}$ with a $p \times p$ matrix having zeros along the diagonal for the last $p - k$ spots

$$\widehat{\beta}_k = V \cdot \text{Diag}\left(\frac{1}{\sigma_1^2}, \ldots, \frac{1}{\sigma_k^2}, 0, \ldots, 0\right) \cdot V^t X^t y.d \tag{3.81}$$

This is nearly equivalent to the formula for ridge regression given in Equation 3.29; the only difference is the values along the diagonal. The ridge estimator replaces σ_j^2 with $\sigma_j^2 + \lambda$, and the principal component regression estimator replaces the last $p - k$ values of σ_j^2 with zero. Similar to the ridge regression estimation, this formula shows how we can quickly solve the principal component regression problem for many values of k using a minimal amount of additional work at each step.

As we did with ridge regression, we will simulate the process of using a testing set to find the prediction optimal value for the principal component regression vector. We will construct a dataset with 30 columns, 500 observations, and with each column of X having a correlation of 0.9. The vector β will be set to have all components equal to 1.

```
n <- 500; p <- 30
```

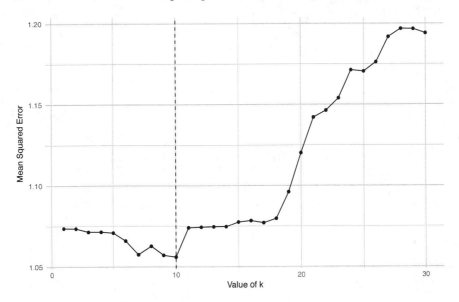

FIGURE 3.6: A plot of the validation mean squared error for principal component regression with varying numbers of components. The optimal value of 10 components is shown with the dashed line.

```
mu <- rep(0, p)
Sigma <- matrix(0.9, nrow = p, ncol = p)
diag(Sigma) <- 1
X <- MASS::mvrnorm(n = n, mu = rep(0, p), Sigma = Sigma)
y <- X %*% rep(1, p) + rnorm(n, sd = 1)
```

To validate the model, we construct a separate testing set using the same underlying generating model.

```
Xtest <- MASS::mvrnorm(n = n, mu = rep(0, p),
                Sigma = Sigma)
ytest <- Xtest %*% rep(1, p) + rnorm(n, sd = 1)
```

Now, we can implement the principal component regression function for an input X.

```
# Compute PCA regression vector.
#
# Args:
#     X: A numeric data matrix.
#     y: Response vector.
#     k: Number of components to use in the fit.
#
```

```
# Returns:
#      Regression vector beta of length ncol(X).
casl_lm_pca <-
function(X, y, k)
{
  svd_obj <- svd(X)
  U <- svd_obj$u
  V <- svd_obj$v
  dvals <- rep(0, ncol(X))
  dvals[seq_len(k)] <- 1 / svd_obj$d[seq_len(k)]

  D <- diag(dvals)
  pca_beta <- V %*% D %*% t(U) %*% y

  pca_beta
}
```

As with the ridge regression vector, it is possible to modify this code to efficiently calculate the regression vector for a range of values k. In the interest of simplicity we will avoid adding this enhancement.

With these computed, we cycle over the integers 1 through p, computing first $\widehat{\gamma}^k$, translating back to $\widehat{\beta}^k$, and then computing the mean squared error.

```
mse <- rep(NA_real_, p)
for (k in 1:p) {
  beta_hat <- casl_lm_pca(X, y, k)
  mse[k] <- mean((Xtest %*% beta_hat - ytest)^2)
}
```

The results of this simulation are shown in Figure 3.6. The optimal value occurs at k equal to 10, with reasonably good performance for any k between 7 and 10. The curve is not as smooth as ridge regression, in part because we can only compute 30 specific values.

3.5   Implementation and notes

The matrix decomposition functions available in base R, such as svd, and solve, call the low level BLAS and LAPACK libraries [9]. There are several implementations of these, including Armadillo [142], the AMD Core Math Library, ScaLAPACK [24], and ATLAS [175]. Commercial versions such as the Math Kernel Library hand-tuned for Intel processors [169] and cuBLAS [15] for GPU-programming also exist. It is possible to compile R to interface with alternative BLAS and LAPACK implementations. The function `stats::prcomp` provides a wrapper to the `svd` function for computing principal components using, internally, the formula provided in Section 3.4. It accepts either a data matrix or a formula object and provides a `predict` method for converting new datasets using the principal component loadings learned via the training data.

The **Matrix** package provides comparable functions for sparse matrices [19]. Rather than providing a full LAPACK implementation, it instead directly provides functions for working with matrix decompositions in a way that does not break the sparsity of the input matrix. These functions are available by wrapping Tim Davis' SuiteSparse Library [42]. Similarly, the **irlba** package provides functions for fitting truncated forms of the singular value decomposition [13]. These truncated forms include only the top k singular values and their corresponding vectors. We have already seen how these can be useful in the context of principal components. There are many other applications of these truncated decompositions, known as low-rank approximations; for a good survey of applications see [115].

For fitting ridge regression, the **MASS** package offers the function `lm.ridge` with a similar formula interface [166]. Users are able to provide a sequence of tuning parameters. These are fit using the efficient algorithm described in Section 3.2. The returned object has some, but not all, of the methods available for `lm` objects such as `predict` and `coef`. The **glmnet** package, which we will explore further in Chapters 5 and 7, provides the function `glmnet` as an alternative method for fitting ridge regression by setting the parameter `alpha` to zero [61]. It only supports a matrix-based input but does efficiently handle sparse matrix inputs and is able to pre-select a sequence of interesting values for the tuning parameter λ. There is even a wrapper function `cv.glmnet` that will automatically perform cross-validation. Note that direct comparison of these various implementations with our own requires some care as there are implementation differences in the exact form of the optimization problem. For example, **glmnet** solves

$$\underset{b}{\arg\min} \left\{ \frac{1}{2n} ||y - Xb||_2^2 + \lambda \cdot ||\beta||_2^2 \right\} \tag{3.82}$$

which requires scaling the penalty by a factor of $2n$ when compared to Equation 3.24. There are also different defaults regarding the treatment of intercept terms and scaling of the data matrix columns.

For further study of the details behind the numerical issues of dense matrix decompositions and the solving of linear systems see the canonical text *Matrix Computations* by Golub and Van Loan [69]. The text *Direct Methods for Sparse Linear Systems*, which is written by the author of SuiteSparse, has an excellent description of the algorithmic issues of extending matrix computations to the sparse setting [43]. We will explore indirect methods for solving optimization problems in subsequent chapters, many of which can be used to find approximate solutions to large dense and sparse linear systems.

3.6 Application: NYC taxicab data

3.6.1 Data and setup

In this section we will look at a dataset from the NYC Taxi and Limousine Commission. Each observation corresponds to a ride taken in a yellow taxi cab; while data exists for billions of rides across several years, we will look at just a subset of data from May 2015. We have cleaned the raw data, merged it with spatial data from the NYC Department of City Planning, and removed any unneeded variables. An example of the dataset is given as follows.

```
taxi <- read.csv("data/taxi.csv")
head(taxi)
```

```
    tip_perc pickup_hour pickup_wday pickup_NTACode
1 0.25000000          18            4           MN15
2 0.13600000           7            4           MN24
3 0.29500000          18            6           MN21
4 0.08333333           2            6           MN17
5 0.23920000          17            6           MN17
6 0.21391304           7            4           MN25
  dropoff_NTACode pair_NTACode fare_amount
1            MN19    MN15-MN19         7.0
2            MN17    MN24-MN17        12.5
3            MN24    MN21-MN24        10.0
4            MN33    MN17-MN33        12.0
5            QN98    MN17-QN98        37.5
6            MN27    MN25-MN27        11.5
```

Our response of interest is the tip percentage, the amount tipped to the driver as a ratio of the total fare amount. The data has been filtered to contain only those transactions paid for with a credit card. Predictor variables include the hour and day of the week that the cab was picked up, neighborhood codes for the neighborhood tabulation area where the cab was picked up, and the neighborhood where the passenger was dropped off. We also have the total fare amount and a variable giving the pickup drop-off neighborhood pairs.

In most applications in this text we use consistent training and testing sets in order to directly compare different featurization and modeling approaches. Here we will break this pattern in order to create increasingly large training sets as the number of variables increases. As ridge regression is most interesting when there are slightly too many variables given the number of training observations, this allows for the most exposition across several different feature sets.

To simplify the following code, we introduce a function for computing the root mean squared error given a set of predictions and responses.

```
# Root mean squared error of a prediction.
#
# Args:
#     y: A vector of responses to predict.
#     y_hat: A vector of predicted values.
#     by: Optional categories to stratify results.
#
# Returns:
#     A table of the root mean squared error for each
#     group in "by".
casl_util_rmse <-
function(y, y_hat, by)
{
  if (missing(by))
  {
    ret <- sqrt(mean((y - y_hat)^2))
  } else {
    ret <- sqrt(tapply((y - y_hat)^2, by, mean))
  }

  ret
}
```

Typically, we will use the variable `flag` as an input to the stratification variable in order to see how well our estimators perform on both the training and testing sets. These two numbers will help to identify when we are over-fitting to the training data and how well each approach generalizes to new data.

3.6.2   Fare and temporal data

As we have mentioned, ridge regression is particularly helpful when we do not have enough data to accurately learn the regression vector using ordinary least squares. In order to illustrate this we will create a training set here that has only 100 samples.

```
set.seed(3)
taxi$flag <- 2
taxi$flag[sample(1:nrow(taxi), 100)] <- 1
```

Our first model will use just the fare amount, the hour that the cab was picked up, and the day of the week that the cab was picked up. To use ridge regression, we will manually construct the data matrix and response vector using calls to `model.frame` and `model.matrix`. We will code the hour and weekday as factors, which forces the design matrix to have a large number of zeros in it. Because `glmnet` accepts sparse matrix inputs, and handles them efficiently, we will construct the model matrix with `sparse.model.matrix` from the **Matrix** package.

```
mf <- model.frame(tip_perc ~ fare_amount +
                  factor(pickup_hour) +
                  factor(pickup_wday), data = taxi)
mt <- attr(mf, "terms")
X <- Matrix::sparse.model.matrix(mt, mf)
y <- model.response(mf)
```

With these created for all samples, we can now fit ridge regression to those 100 rows corresponding to the training set.

The **glmnet** package provides a cross-validation function, making it straightforward to apply ridge regression and to determine the best value for the tuning parameter λ. We will set the parameter α to zero in order to get a ridge penalty and manually indicate that we want to cross-validate over a set that includes 0 as a minimal value by setting `lambda.min.ratio`.

```
set.seed(1)
ridge <- glmnet::cv.glmnet(X[taxi$flag == 1,], y[taxi$flag == 1],
                           alpha = 0, nfolds = 3,
                           lambda.min.ratio = 0)
beta_ridge <- coef(ridge, s = ridge$lambda.min)
beta_ols <- coef(ridge, s = min(ridge$lambda))
```

Note that the result in `beta_ols` is only an approximate version of the ordinary least squares fit; it is technically a ridge regression vector with a very small λ as a result of the internal logic within `cv.glmnet`.

As ridge regression penalizes the ℓ_2-norm of the beta vector, we know that the ridge vector should have a smaller norm than the ordinary least squares fit. We can test this on our specific example to confirm.

```
sum((beta_ridge)^2)
```

```
[1] 0.05221988
```

```
sum((beta_ols)^2)
```

```
[1] 0.1712353
```

Indeed, the cross-validated solution has a squared norm that is over three times smaller than the (approximated) ordinary least squares solution.

Given that the value of λ was chosen by cross-validation, and the OLS solution was an option in the validation process, we would expect the ridge regression to have better predictive power over the testing set.

```
y_hat <- predict(ridge, X, s = ridge$lambda.min)
casl_util_rmse(y, y_hat, taxi$flag)
```

```
         1         2
0.1081203 0.2047438
```

```
y_hat <- predict(ridge, X, s = min(ridge$lambda))
casl_util_rmse(y, y_hat, taxi$flag)
```

```
          1          2
0.09156523 0.21519093
```

Just as expected, we see that the ordinary least squares solution has a lower error on the training set but a higher error on the testing set compared to the ridge solution. Notice that the ridge solution with the minimal cross-validation error still fits the training data much better than the testing data; the RMSE is nearly doubled when moving from the training to the testing set. The sample size here is small and the difference may just be due to a lower variance in the training set. Even if this were not the case, observing this behavior is in fact quite normal when using datasets with a large number of variables relative to the number of samples. Typically a moderate degree of over-fitting to the training data offers the most predictive model on an external testing set. See the appendix to the paper of Wasserman and Roeder for an excellent theoretical description of why this occurs [171].

3.6.3 Pickup and drop-off neighborhoods

We will now try to incorporate some of spatial data into our predictive model. In order to explore the performance of ridge regression in this context, we will pick a training set having only 400 observations.

```
set.seed(3)
taxi$flag <- 2
taxi$flag[sample(1:nrow(taxi), 400)] <- 1
```

We again construct a data matrix using the original predictor variables, adding in the neighborhood tabulation codes.

```
mf <- model.frame(tip_perc ~ fare_amount +
                  factor(pickup_hour) +
                  factor(pickup_wday) +
                  factor(pickup_NTACode) +
                  factor(dropoff_NTACode), data = taxi)
mt <- attr(mf, "terms")
X <- Matrix::sparse.model.matrix(mt, mf)
```

```
y <- model.response(mf)
dim(X)
```

```
[1] 411829   377
```

Notice that due to the relatively large number of neighborhoods, the resulting data matrix has a large number of columns (377).

Fitting the data once again with `cv.glmnet`, we will compare the cross-validated solution to the solution with the minimal value of λ.

```
set.seed(1)
ridge <- glmnet::cv.glmnet(X[taxi$flag == 1,],
                           y[taxi$flag == 1],
                           alpha = 0, nfolds = 3,
                           lambda.min.ratio = 0)
beta_ridge <- coef(ridge, s = ridge$lambda.min)
beta_ols <- coef(ridge, s = min(ridge$lambda))
```

As before, the size of the regression vector at the cross-validated tuning parameter is much smaller than that of the approximate ordinary least squares solution.

```
sum((beta_ridge)^2)
```

```
[1] 0.05841448
```

```
sum((beta_ols)^2)
```

```
[1] 0.5789292
```

This time, the ordinary least squares solution is nearly ten times as large as the cross-validated ridge value. Notice that when compared to the size of the vector in the model not containing the spatial data, the cross-validated ridge vector was only slightly smaller than the one here (0.052 vs. 0.058). The ridge penalty has stopped the model from using too much information from the spatial model in an attempt to reduce overfitting.

Looking at the predictive power of these two methods, we again see that the ridge regression outperforms the ordinary least squares solution.

```
y_hat <- predict(ridge, X, s = ridge$lambda.min)
casl_util_rmse(y, y_hat, taxi$flag)
```

```
         1          2
0.08322745 0.20458129
```

```
y_hat <- predict(ridge, X, s = min(ridge$lambda))
casl_util_rmse(y, y_hat, taxi$flag)
```

```
         1          2
0.07418478 0.20978430
```

Both models have improved from the prior section, though this is hard to directly compare because we have different training sets.

Recall that in Section 2.10, we ran into trouble when some factor levels were only included in the testing set. Here, we do have many neighborhoods only included in the testing set, but we seem to have no particular warnings arising from this issue. The reason for this is two-fold. First of all, we made the model matrix in one single step using the entire dataset. This means that every neighborhood is given a valid indicator variable regardless of whether it was in the training set. Secondly, we were able to fit a model on a training set where some columns have all zeros without running into numerical errors because of the strict convexity of the ridge regression problem for $\lambda > 0$. This is why we did not run a *true* ordinary least squares solution, instead approximating it with a ridge regression having a very small penalty.

3.6.4   Neighborhood pairs

As a final application, we will look at pairs of neighborhoods corresponding to where each cab picked up a passenger and then the neighborhood where they dropped the passenger off. As this feature set has many more levels, we will use a training set with ten thousand training samples.

```
set.seed(3)
taxi$flag <- 2
taxi$flag[sample(1:nrow(taxi), 10000)] <- 1
```

Creating the dataset once again, we see that the dataset now has over four thousand columns.

```
mf <- model.frame(tip_perc ~ fare_amount +
                    factor(pickup_hour) +
                    factor(pickup_wday) +
                    factor(pair_NTACode), data = taxi)
mt <- attr(mf, "terms")
X <- Matrix::sparse.model.matrix(mt, mf)
y <- model.response(mf)
dim(X)
```

```
[1] 411829    4494
```

As with the neighborhood codes, we do not expect every single pair to occur in the training set but this is okay given the ability of ridge regression to work with non-full column rank design matrices.

Fitting cross-validated ridge regression, we see that the overall size of the optimal ridge regression vector has increased nearly six-fold from the other two feature sets.

```
set.seed(1)
ridge <- glmnet::cv.glmnet(X[taxi$flag == 1,],
                           y[taxi$flag == 1],
                           alpha = 0, nfolds = 3,
                           lambda.min.ratio = 0)
beta_ridge <- coef(ridge, s = ridge$lambda.min)
beta_ols <- coef(ridge, s = min(ridge$lambda))
sum((beta_ridge)^2)
```

```
[1] 0.2928718
```

The increase in size is partially a result of a large feature set but is also due to a larger sample size providing more information to help overcome the complexity penalty. How does this compare with the (approximate) ordinary least squares solution?

```
sum((beta_ols)^2)
```

```
[1] 378.8414
```

The discrepancy between the two models here is quite extreme. The squared ℓ_2-norm here is over one thousand times larger than the cross-validated solution.

Given the large difference in magnitude between the two solutions, it is no surprise that their performance on the testing dataset is also significantly different.

```
y_hat <- predict(ridge, X, s = ridge$lambda.min)
casl_util_rmse(y, y_hat, taxi$flag)
```

```
        1         2
0.3787119 0.1982126
```

```
y_hat <- predict(ridge, X, s = min(ridge$lambda))
casl_util_rmse(y, y_hat, taxi$flag)
```

```
        1         2
0.2896623 0.3244322
```

The ridge regression performs much better than the ordinary least squares solution. Although it is unfair to directly compare the different training sets, the neighborhood pairs data has improved the ridge model compared to the other feature sets, but has degraded that of the ordinary least squares model.

3.7 Exercises

1. Being careful about the handling of scale, intercepts, and penalties, verify that our function `casl_lm_ridge` produces similar results to `MASS::lm.ridge` and `glmnet::glmnet` for specific values of λ.

2. Add functionality to `casl_lm_ridge` to pick a reasonable sequence of values λ when none is supplied. Include an option `nlam`, set to 100 by default, to set the number of lambda values to be created. Reasonable values can be inferred from the range of the singular values of X as shown in Equation 3.29. Note that it makes sense to select lambda values on the log scale.

3. Now, add additional input parameters to `ridge_reg` for the validation data: `X_valid` and `y_valid`. If supplied, return only the best value of β.

4. Construct a function `cv.ridge_reg` that performs 10-fold cross-validation to select the optimal value of λ for ridge regression.

5. Modify `ridge_reg` to include an option `scale` that, when set to `TRUE`, centers and scales the columns of X before running the regression. Make sure to return the result in the original scale.

6. The original implementation of `ridge_reg` should work when given a sparse matrix X; the function `Matrix::crossprod` is used in place of the default. Generally this is okay because $X^t X$ will usually be much smaller than X itself. However, if we center and scale the columns of X this causes a problem as the matrix will be converted to a dense format. Implement the `scale` option in an intelligent way so that the matrix never needs to be converted to a dense format.

7. Modify the function `casl_lm_pca` so that it efficiently handles returning results for a sequence of values k.

8. Modify the function `casl_lm_pca` so that it accepts a sequence of values k and uses the **irlba** package to avoid computing all of the singular values. Note: Your first step should be checking for the largest value k.

9. Redo the simulation analysis in Section 2.8 using the condition number of X as the upper bound. How closely does the simulation hit the implied bound?

10. There is a well-known theoretical result showing that there must exist a positive λ such that the training mean squared error of ridge regression dominates that of the ordinary least squares fit. Design a simulation to test this claim empirically and describe the results.

4

Linear Smoothers

4.1 Non-Linearity

Linear regression has excellent theoretical properties and, as we have seen, can be readily computed from observed data. Using ridge regression and principal component analysis we can tune these models to optimize for predictive error loss. Indeed, linear models are used throughout numerous fields for predictive and inferential models. One situation in which linear models begin to perform non-optimally is when the relationship between the response y and the data is not linear nor can it be approximated closely by a linear relationship.

As an example of a non-linear model consider observing a variable y_i governed by

$$y_i = \cos(\beta_1 \cdot x_i) + e^{-x_i \cdot \beta_2} + \epsilon_i \qquad (4.1)$$

for some scalar value x_i, unknown constants β_1 and β_2, and the random noise variable ϵ_i. A common approach for estimating the unknown parameters given a set of observations is to again minimize the sum of squared residuals. This sum is a well-defined function over the set of allowed β_j's and often, as in this case, twice differentiable. While there is no analogous closed-form solution to the linear case, the minimizing estimate values can usually be found using a general purpose first, or second-order optimization technique. This approach is known as *non-linear least squares* and has significant theoretical guarantees over a wide class of problem formulations.

What happens when we do not know a specific formula for y_i that can be written down in terms of a small set of unknown constants β_j? Models of the form seen in Equation 4.1 often arise in engineering and science applications where the specific causal mechanism for the response y_i is well understood. In statistical learning this is rarely the case. More often we just know that

$$\mathbb{E}y_i = f(x_i) \qquad (4.2)$$

holds for some unknown function f. We may suspect that f has some general properties; depending on the application it may be reasonable to assume that f is continuous, has a bounded derivative, or is monotonically increasing in x_i. As we do not know a specific formula for f in terms of parameters β_j, the model given in Equation 4.1 is known as non-parametric regression. Common estimators for estimating the non-parametric regression function f are the topic of this chapter.

In parametric regression it is clear that a point estimator will yield a single prediction $\widehat{\beta}$ inside of $\mathbb{R}^p$ for the unknown regression vector. For non-parametric regression it is not even clear what an estimator $\widehat{f}$ would look like. Two methods we evaluate will produce an explicit formula in terms of x_i as a prediction of y_i. The other two instead provide an algorithm for computing $\widehat{f}(x)$ for any input value of x. While it is possible to do this for a large set of x's, these techniques will not yield an estimated parametric model such as that given in Equation 4.1. The added computational time for predictions, which often require non-trivial computations for every value of x should be taken into account when considering the value added to the predictive performance by a non-parametric model. We will discuss techniques for minimizing the burden of the prediction time.

As with most predictive models, non-parametric regression techniques have tuning parameters to control the trade-off between variance and bias. Often this can be visualized in terms of the smoothness of the function $\widehat{f}$. If the regression function is penalized from changing too fast, this reduces the variance in the estimates but introduces additional bias for values at x_i where it *should* vary more. Likewise, if the function is allowed to change rapidly this will generally give nearly unbiased estimates at the price of a high variance in the estimated values. Much of our discussion about non-parametric regression will focus on methods for setting tuning parameters and the performance of these methods on various types of data.

4.2   Basis expansion

We have already discussed one technique for dealing with non-linearity back in Section 2.1. Recall that linear regression requires only that the relationship with respect to the parameter β_j be linear. The terms multiplied by each parameter may be any known quantity derivable from the data matrix X. For example, with a scalar value of x_i both

$$y_i = \sum_{k=0}^{K} x_i^k \cdot \beta_j + \epsilon \tag{4.3}$$

and

$$y_i = \sum_{k=1}^{K} sin(x_i/(2\pi K)) \cdot \beta_j + \epsilon \tag{4.4}$$

are valid linear regression models. The first corresponds to the first K terms of the polynomial basis and the second to the first K odd terms of the Fourier basis. If we construct a new matrix Z consisting of columns that are copies of x taken to various powers or to varying applications of the sine function, an estimate of the relationship between y and x can be determined using the standard techniques for calculating a linear regression model. In general we model the relationship

$$y_i = \sum_{k=1}^{K} B_{k,K}(x_i) \cdot \beta_j + \epsilon \tag{4.5}$$

for some basis function $B_{k,K}$. This method is known as a *basis expansion*.

While in theory any collection of functions can serve for describing a basis expansion, there are several properties that make for particularly useful choices. For one thing, it is nice to have a formula for constructing K basis functions such that in the limit of very large K this basis is able to approximate any continuous function over the domain of x. Such a basis makes it possible to learn increasingly accurate values for f as more data is observed. The second useful property is to form an orthonormal basis. Specifically, this means that when we take the constructed matrix Z, the inner product $Z^t Z$ will be equal to the identity matrix. This has an immediate computational benefit in that the least squares estimator becomes

$$\widehat{\beta} = n^{-1} \cdot (Z^t y). \tag{4.6}$$

This formula simplifies such that we can directly compute

$$\widehat{\beta}_j = n^{-1} \cdot \sum_i y_i \cdot B_{j,K}(x_i) \tag{4.7}$$

The final useful property of a set of basis functions is to have $B_{j,K}$ not depend on K. That is, the basis consists simply of a sequence of functions. This is useful because if we have already calculated the model for K, the model for $K + 1$ will have the same first K terms followed be a new $K + 1$th term. Specifically, we solve Equation 4.7 for $K + 1$ and add it into the predicted model for $\widehat{f}$.

We will implement the basis expansion in Equation 4.7 over an orthonormal polynomial basis. This special case is sometimes described as *polynomial regression*. To construct this polynomial basis we make use of the R function `stats::poly`, which finds an orthonormal polynomial basis over a set of inputs.

```
x <- 1:10
poly_5 <- poly(x, n = 5)
crossprod(poly_5)
```

	1	2	3	4	5
1	1.00e+00	0.00e+00	-2.78e-17	-1.11e-16	1.67e-16

```
2  0.00e+00  1.00e+00 -2.78e-17 -5.55e-17 -1.25e-16
3 -2.78e-17 -2.78e-17  1.00e+00 -8.33e-17  2.22e-16
4 -1.11e-16 -5.55e-17 -8.33e-17  1.00e+00 -5.55e-17
5  1.67e-16 -1.25e-16  2.22e-16 -5.55e-17  1.00e+00
```

The cross product matrix is very close to the identity, up to an error of about only 10^{-16}. This is certainly close enough to ignore the $Z^t Z$ term in the calculation of $\widehat{\beta}$. It is also easy in R to compute the same polynomial basis for a new set of data using the `predict` function.

```
x_new <- runif(100)
poly_5_new <- predict(poly_5, x_new)
crossprod(poly_5_new)
```

```
        1       2       3       4      5
1    30.2   -40.3    52.5   -73.7    119
2   -40.3    54.4   -72.1   103.1   -168
3    52.5   -72.1    97.9  -143.3    239
4   -73.7   103.1  -143.3   214.8   -364
5   118.6  -168.5   238.7  -364.5    627
```

Notice that the basis is only orthonormal for the original dataset X, and not for a new sample of data.

With the orthonormal polynomials, we will write a short function to find the regression vector for polynomial regression.

```
# Apply one-dimensional polynomial regression.
#
# Args:
#     x: Numeric vector of the predictor variables.
#     y: Numeric vector of the responses.
#     n: An integer giving the order to the polynomial.
#
# Returns:
#     A length n numeric vector of coefficients.
casl_nlm1d_poly <-
function(x, y, n=1L)
{
  Z <- cbind(1/length(x), poly(x, n=n))
  beta_hat <- crossprod(Z, y)
  beta_hat
}
```

We also need a special prediction function to take care of constructing the polynomial basis for new data. Notice that this prediction function requires the original data x in order to compute the polynomial basis. A more involved

solution would store the coefficients as an attribute in the output object of
`casl_nlm1d_poly`.

```
# Predict values from one-dimensional polynomial regression.
#
# Args:
#     beta: A length n numeric vector of coefficients.
#     x: Numeric vector of the original predictor variables.
#     x_new: A vector of data values at which to estimate.
#
# Returns:
#     A vector of predictions with the same length as x_new.
casl_nlm1d_poly_predict <-
function(beta, x, x_new)
{
  pobj <- poly(x, n=(length(beta) - 1L))
  Z_new <- cbind(1.0, predict(pobj, x_new))
  y_hat <- Z_new %*% beta
  y_hat
}
```

Putting this prediction function together with the estimation function allows
for easy cross-validation to select the order of the polynomial n.

To test the polynomial regression function, we will create some training
and validation data that has a non-linear relationship between the input and
output.

```
n <- 100
x <- seq(0, 1, length.out = n)
y_bar <- sin(x * 3 * pi) + cos(x * 5 * pi) + x^2
y <- y_bar + rnorm(n, sd = 0.25)
x_test <- sort(runif(n))
y_bar <- sin(x_test * 3 * pi) + cos(x_test * 5 * pi) +
         x_test^2
y_test <- y_bar + rnorm(n, sd = 0.25)
```

We then fit a polynomial regression to this data of order 2.

```
casl_nlm1d_poly(x = x, y = y, n = 2L)
```

```
    [,1]
  0.547
1 2.206
2 4.529
```

And, finally, fit a model with order 4.

```
casl_nlm1d_poly(x = x, y = y, n = 4L)
```

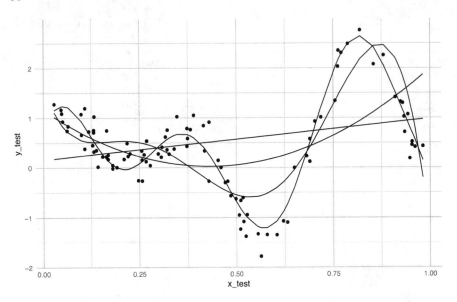

FIGURE 4.1: Predicted values from several polynomial regression curves. Includes all polynomials up to order 6.

```
     [,1]
   0.547
1  2.206
2  4.529
3 -2.158
4 -4.981
```

We see that the intercept and first two coefficients are unchanged, as expected.

The next step is to use the test set to pick a good value for the order of the polynomial. We do this by cycling over several orders and recording the mean squared error on the test set for each value of K.

```
N <- 20
mse <- rep(NA_real_, N)
for (k in 1:N)
{
  beta <- casl_nlm1d_poly(x = x, y = y, n = k)
  y_test_hat <- casl_nlm1d_poly_predict(beta, x, x_test)
  mse[k] <- mean((y_test - y_test_hat)^2)
}
```

A plot showing the curve produced for several values is shown in Figure 4.1.

```
best_k <- which.min(mse)
sprintf("mse: %01.03f  k = %d", mse[best_k], best_k)
```

```
[1] "mse: 0.078  k = 9"
```

We see that the best fit is found at k equal to 9, though higher-order fits tend to be similarly predictive.

4.3   Kernel regression

We are not constrained to only using techniques derived from ordinary least squares for addressing non-parametric regression problems. It is useful to consider more straightforward approaches as alternatives. We observe pairs of data (x_i, y_i) and are told that the relationship between x and y is some unknown function; the task is to predict the value of y_{new} for some newly observed x_{new}. The most straightforward approach would probably be to predict that the new value of y will most likely be equal to the mean of those values previously observed. With small datasets or functions f that are either changing very quickly or very slowly, this naïve approach may perform quite well. Although simple, with just a minor refinement this approach becomes a powerful technique known to be one of the most predictive in several problem domains.

Using the overall mean for prediction assumes that new observations of y tend to be similar to those previously observed. If we believe that f tends to be continuous, a modified approach might instead use a weighted mean. Observations i that have an x_i value near x_{new} could be given larger weights and those farther away can be given a lower weight. If the weights are set to 1 for the closest n points and to 0 for all of the remaining observations, this leads to the k-nearest neighbors estimator. More generally we have

$$\widehat{y}_{new} = \sum_i w_i \cdot y_i \tag{4.8}$$

where the weights generally depend on x_{new} and any of the training data points x_i but should not depend on any of the responses y_i. This will be a weighted mean whenever the sum of the weights are equal to 1.

One way of choosing a good weighting function is by way of a *kernel* function. A kernel function is a non-negative, real valued function that is symmetric around the origin; it is also typically scaled to have an integral of 1, in order to make Equation 4.8 correspond to a weighted mean. One simple example is the uniform kernel

$$K(x) = \frac{1}{2} \cdot 1_{\{|x| \leq 1\}}. \tag{4.9}$$

Nadaraya–Watson kernel regression

Given a training dataset with a scalar independent variable x and continuous
response y, the Nadaraya–Watson kernel regression prediction with bandwidth
$h > 0$ at a new point (x_{new}, y_{new}) is defined as

$$\widehat{y}_{new} = \frac{\sum_i K_h(|x_i - x_{new}|) \cdot y_i}{\sum_i K_h(|x_i - x_{new}|)}$$

where K is a non-negative, real valued function and $K_h(x)$ is equal to $K(\frac{x}{n})$.

With a kernel in hand, a natural weighting scheme is to set the weight equal
to the kernel function evaluated at the distance of the ith observation away
from the new point, or

$$w_i = K\left(|x_{new} - x_i|\right). \tag{4.10}$$

In the case of the uniform kernel, this behaves similarly to the k-nearest neighbors estimator, but instead of cutting off at the top proportion of closest
observations, all values within a distance of 1 are included. Averaging points
within a cutoff distance is reasonable but there is no particular reason to make
this cutoff value equal to 1. To make this value more flexible we introduce a
bandwidth into the kernel, usually denoted with the variable h and placed as
a subscript on K. Generally, the bandwidth scales the input to the kernel
function by an inverse multiplicative factor

$$K_h(x) = K\left(\frac{x}{h}\right). \tag{4.11}$$

So when h is large, weights are spread out more evenly and when h is small
the weights are more concentrated around the point x_{new}. Weighted non-
parametric regression with a kernel function and bandwidth is known as
Nadaraya–Watson kernel regression after Élizbar Nadaraya and Geoffrey Watson, both of whom independently published the technique in 1964.

The uniform kernel is conceptually simple but non-ideal in practice. We
will want to perform validation to learn the best bandwidth parameter h. In
the uniform kernel, the predictions are not continuous with respect to the
bandwidth because all of the weights are either 0.5 or 0. When a data point
falls out of the bandwidth it goes immediately from a large weight to a small
one. A better choice is the Epanechnikov kernel, given by

$$K(x) = \frac{3}{4} \cdot (1 - x^2) \cdot 1_{\{|x| \leq 1\}}, \tag{4.12}$$

which is continuous in h. It also has several optimality results in the case of
Gaussian errors. We can code this in R with an option for a specific bandwidth.

```
# Evaluate the Epanechnikov kernel function.
#
# Args:
#     x: Numeric vector of points to evaluate the function at.
#     h: A numeric value giving the bandwidth of the kernel.
#
# Returns:
#     A vector of values with the same length as x.
casl_util_kernel_epan <-
function(x, h=1)
{
  x <- x / h
  ran <- as.numeric(abs(x) <= 1)
  val <- (3/4) * ( 1 - x^2 ) * ran
  val
}
```

With this kernel, it is also relatively easy to write a function that computes the kernel regression function for a newly observed data point.

```
# Apply one-dimensional (Epanechnikov) kernel regression.
#
# Args:
#     x: Numeric vector of the original predictor variables.
#     y: Numeric vector of the responses.
#     x_new: A vector of data values at which to estimate.
#     h: A numeric value giving the bandwidth of the kernel.
#
# Returns:
#     A vector of predictions for each value in x_new.
casl_nlm1d_kernel <-
function(x, y, x_new, h=1)
{
  sapply(x_new, function(v)
    {
      w <- casl_util_kernel_epan(abs(x - v), h=h)
      yhat <- sum(w * y) / sum(w)
      yhat
    })
}
```

More sophisticated techniques allow for a speed-up in the case of large datasets, where we can safely ignore observations that are far away from x_{new}.

Using the same training and testing dataset that we looked at with the basis expansion data, we will again perform cross-validation to find an optimal bandwidth.

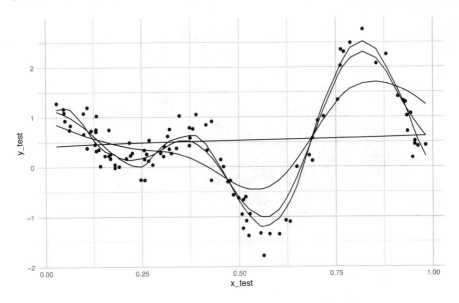

FIGURE 4.2: Predicted values from kernel regression using Epanechnikov kernels and various values of the bandwidth parameter.

```
N <- 100
bwidths <- seq(0.01, 0.1, length.out = N)
mse <- rep(NA_real_, N)
for (k in 1:N) {
  y_test_hat <- casl_nlm1d_kernel(x = x, y = y,
                                  x_new = x_test,
                                  h = bwidths[k])
  mse[k] <- mean((y_test - y_test_hat)^2)
}
```

A figure showing the predicted fits for several values of the bandwidth are given in Figure 4.2.

```
best_h <- which.min(mse)
sprintf("mse: %01.03f  h = %f", mse[best_h],
        bwidths[best_h])
```

```
[1] "mse: 0.071  h = 0.051"
```

We see that the minimal mean squared error occurs at $h = 0.051$ and is slightly lower compared to the optimal value found for basis expansion.

Kernel regression and basis expansion solve the same problem and in this small example perform similarly. Several important differences do exist. A primary benefit of basis expansion is that, as a form of linear regression, it can be used as is in combination with other predictor variables. Basis expansion also does not need to store the entire dataset after the model has been fit (in our simple example we did store this data in order to use the prediction function from `stats::poly`, but this can be avoided). Kernel regression shines when used with a regression function f that is unpredictably varying over the range of x. These functions are hard to fit with polynomials even without the presence of noise.

4.4 Local regression

Starting with a weighted sum of the observed data points y_i, the concept behind kernel regression seems to be a natural method for setting the weights. Taking a close look at Equation 4.8, notice that ordinary least squares is also a weighted sum of the response variables. Specifically, we have

$$w_i = \left[X_{new} (X^t X)^{-1} X^t \right]_i \tag{4.13}$$

for a training set X and a matrix of new values for prediction at X_{new}. So, there are many more reasonable choices for the weights that do not fit within the kernel regression framework. The broad class of estimators that can be written as a weighted sum of the responses are known as *linear smoothers*.

Linear smoothers have several nice theoretical properties that generalize some results from the theory of ordinary least squares. One particular property that is useful for predictive modeling is a generalized concept of the degrees of freedom. Consider trying to predict the values of new observations at the same locations as the training set X. If we have a linear smoother we will be able to write this as

$$\widehat{y} = Hy \tag{4.14}$$

for some matrix H that depends only on the value of X. Again assuming that the random vector ϵ are independent and identically distributed normal random variables, we can compute the variance matrix for $\widehat{y}$

$$Var(\widehat{y}) = Var(Hy) \tag{4.15}$$
$$= H \cdot Var(yy^t) \cdot H^t \tag{4.16}$$
$$= \sigma^2 \cdot HH^t \tag{4.17}$$

and the total variation is just

$$TV(\widehat{y}) = \sigma^2 \cdot \text{tr}(HH^t). \tag{4.18}$$

In the case of linear regression this total variation is equal to the noise variance times p

$$TV(\widehat{y}) = \sigma^2 \cdot \text{tr}(X(X^tX)^{-1}X^tX(X^tX)^{-1}X^t) \qquad (4.19)$$

$$= \sigma^2 \cdot \text{tr}(X(X^tX)^{-1}X^t) \qquad (4.20)$$

$$= \sigma^2 \cdot \text{tr}((X^tX)^{-1}X^tX) \qquad (4.21)$$

$$= \sigma^2 \cdot \text{tr}(I_p) = \sigma^2 \cdot p. \qquad (4.22)$$

The total variation tells us how much the predicted values are able to vary; higher values indicate a model with more implicit parameters that are able to vary. In fact, after dividing by the noise variance σ^2, the total variation gives the *effective degrees of freedom* of a linear smoother. We can use this measurement to compare the mean squared error of different linear smoothers across their respective tuning parameters.

Putting linear regression and kernel regression into one larger class of models suggests that there may be a way of combining these two approaches. In *local regression* this is done by fitting a linear model at each point x_{new} with sample weights given by a kernel function centered on x_{new}. The theoretical justification of this is to consider the Taylor series of f around some point x_0

$$f(x_0 + \delta) = f(x_0) + \delta f'(x_0) + \frac{1}{2}\delta^2 f''(x_0) + O(\delta^3) \qquad (4.23)$$

Kernel regression assumes that f can be reasonably approximated by the constant term. Local regression includes the first-order term as well. Combining local regression with polynomial regression, this approach can include as many higher-order terms as desired. In practice there is usually no reason to go beyond the first order approximation.

Solving for local regression requires calculating the same weights as in the implementation of kernel regression and then using the standard method for solving a weighted least squares problem.

```
# Apply one-dimensional local regression.
#
# Args:
#     x: Numeric vector of the original predictor variables.
#     y: Numeric vector of the responses.
#     x_new: A vector of data values at which to estimate.
#     h: A numeric value giving the bandwidth of the kernel.
#
# Returns:
#     A vector of predictions for each value in x_new.
casl_nlm1d_local <-
function(x, y, x_new, h=1) {
  X <- cbind(1, x)
  sapply(x_new, function(v) {
```

Local regression

Given a training dataset with a scalar independent variable x and continuous response y, kth order local regression with the kernel function K solves the following optimization problem at any given x_{new} for which prediction is required

$$\widehat{\beta}(x_{new}) = \arg\min_{b} \left\{ \sum_{i} K_h(|x_{new} - x_i|) \cdot (y_i - \sum_{j=0}^{k} x_i^k \cdot \beta_k)^2 \right\}$$

The predicted value is then given by

$$\widehat{y}_{new} = \sum_{j=0}^{k} x_{new}^k \cdot \widehat{\beta}_k(x_{new}).$$

The output $\widehat{y}_{new}$ can be computed for any value x_{new}, and thus this procedure defines a prediction $\widehat{f}(x)$ any value of x.

```
    w <- casl_util_kernel_epan(abs(x - v), h=h)
    beta <- solve(t(X) %*% diag(w) %*% X,
                  t(X) %*% diag(w) %*% y)
    yhat <- cbind(1, v) %*% beta
    yhat
  })
}
```

Using the same dataset as in the previous sections, we use cross-validation to estimate the optimal bandwidth.

```
N <- 100
bwidths <- seq(0.015, 0.1, length.out = N)
mse <- rep(NA_real_, N)
for (k in 1:N) {
  y_test_hat <- casl_nlm1d_local(x = x, y = y,
                               x_new = x_test,
                               h = bwidths[k])
  mse[k] <- mean((y_test - y_test_hat)^2)
}
best_h <- which.min(mse)
sprintf("mse: %01.03f  h = %f", mse[best_h],
        bwidths[best_h])
```

```
[1] "mse: 0.075  h = 0.055354"
```

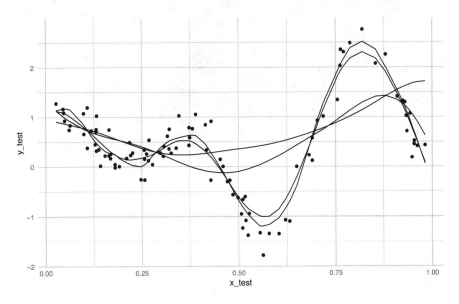

FIGURE 4.3: Predicted values from local regression using Epanechnikov kernels, a linear fit, and various values of the bandwidth.

We see that the best mean squared error is very similar to that of kernel regression. The resulting curves for several bandwidths are shown in Figure 4.3. Notice that these curves look smoother than those from the kernel regression, which is a result of using a linear approximation rather than a constant one at each point x_{new}.

A popular variation of local regression as defined here is the *LOESS* procedure. LOESS also fits a local regression at each point, but replaces the kernel with a k-nearest neighbors selection. Regression is done by fitting a linear model to the closest k points. One advantage of this technique is that reduces variance in areas of the x-space where there is a low density of training data. LOESS also makes it much faster to predict a large number of new data points, because the local regression only needs to be done once for every training data point. Also, as the weights are only ever 0 or 1, by sorting the input points x_i the weighted regression values can be updated rather than having to compute the entire matrix $X^t X$ from scratch. The main downside of LOESS is that there are discontinuities in $\widehat{f}$ whenever we hit a point where the k-nearest neighbors changes.

What are the benefits and drawbacks of local regression compared to kernel regression? Local regression produces smoother prediction functions and has a theoretically lower asymptotic mean squared error. Both of these properties come at the cost of higher computational times. Typically in practice the cross-validated bandwidth is small enough that only a small proportion of larger datasets have non-zero weights in either approaches. This means that the difference in computational time is generally negligible, but so are the benefits of using higher order approximations.

4.5 Regression splines

In Section 4.2 we explored the idea of using an expanded set of basis functions in order to represent non-linear functions in one dimension. Here we discuss an important set of basis functions, known as splines, that are particularly well-suited to this task. While technically just a specific application of basis expansion, the derivation of the splines is subtle enough and their application sufficiently important to warrant a separate treatment of their form and usage.

When using a polynomial or Fourier basis to represent a non-linear function, small changes in a coefficient will lead to changes in the predicted values at every point of the unknown regression function f. The global nature of the estimation problem in these cases leads to poor local performance in the presence of high noise variance or with regression functions f that have many critical points. Local regression, as we have seen, offers a solution to this problem by using a relatively small basis expansion locally at each point. A downside of this approach, however, is that it requires fitting a regression model for every desired point where a prediction is needed. The computational load of this calculation can become considerable and in the case of larger datasets, the requirement that all training data must be accessible to run this regression may also limit its application. Fortunately there is a way of doing basis expansion that mimics the primary benefits of local regression.

We start by picking a point k within the range of the data points x_i. A natural choice would be the median or mean of the data. In lieu of a higher-order polynomial fit, imagine fitting two linear polynomials to the data: one for points less than k and another for points greater than k. Using indicator functions, we can describe this approach with a specific basis expansion, namely

$$B_0(x) = I(x \leq k) \tag{4.24}$$
$$B_1(x) = x \cdot I(x \leq k) \tag{4.25}$$
$$B_2(x) = I(x > k) \tag{4.26}$$
$$B_3(x) = x \cdot I(x > k). \tag{4.27}$$

It will be useful going forward to re-parameterize this in terms of a baseline intercept and slope for $x \leq k$ and changes in these values for points $x > k$

$$B_0(x) = 1 \tag{4.28}$$
$$B_1(x) = x \tag{4.29}$$
$$B_2(x) = I(x > k) \tag{4.30}$$
$$B_3(x) = (x - k) \cdot I(x > k). \tag{4.31}$$

It is left as an exercise to show that these are equivalent bases.

A shortcoming of the space spanned by these splines is that at the point k, known as a *knot*, the predicted values will generally not be continuous. It is possible to modify our original basis to force continuity at the knot k by removing the secondary intercept described by $B_2(x)$ in Equations 4.28–4.31. The basis now becomes

$$B_0(x) = 1 \tag{4.32}$$
$$B_1(x) = x \tag{4.33}$$
$$B_2(x) = (x - k) \cdot I(x > k). \tag{4.34}$$

Notice that forcing one constraint, continuity at k, has reduced the degrees of freedom by one, from 4 down to 3. How might we generalize this to fitting separate quadratic term on the two halves of the data? One approach would be to use the basis functions

$$B_0(x) = 1 \tag{4.35}$$
$$B_1(x) = x \tag{4.36}$$
$$B_2(x) = x^2 \tag{4.37}$$
$$B_3(x) = (x - k) \cdot I(x > k) \tag{4.38}$$
$$B_4(x) = (x - k)^2 \cdot I(x > k). \tag{4.39}$$

The number of parameters here works out correctly; we have two quadratic polynomials (2×3) minus one constraint, for a total of $6 - 1 = 5$ degrees of freedom. What will a function look like at the knot k using the basis from Equations 4.35–4.39? It will be continuous at the knot but is not constrained to have a continuous derivative at the point. This is easy to accomplish, however, by removing the $B_3(x)$ basis. Notice that once again the inclusion of an additional constraint, a continuous first derivative, reduces the degrees of freedom by one.

Defining the positive part function $(\cdot)_+$ as

$$(x)_+ = \begin{cases} x, & x \geq 0 \\ 0, & \text{otherwise} \end{cases} \tag{4.40}$$

we may generalize to an arbitrarily large polynomial of order M by using the basis

$$B_0(x) = 1 \qquad (4.41)$$

$$B_j(x) = x^j, \quad j = 1, \ldots, M \qquad (4.42)$$

$$B_{M+1}(x) = (x - k)_+^M \qquad (4.43)$$

This basis results in a function with continuous derivatives of orders 0 through $M - 1$. We can further generalize this by considering a set of P knots $\{k_p\}_{p=1}^P$, given by

$$B_0(x) = 1 \qquad (4.44)$$

$$B_j(x) = x^j, \quad j = 1, \ldots, M \qquad (4.45)$$

$$B_{M+p}(x) = (x - k_p)_+^M, \quad p = 1, \ldots, P \qquad (4.46)$$

Equations 4.44–4.46 defines the *truncated power basis* of order M. It yields piecewise Mth order polynomials with continuous derivatives of order 0 through $M - 1$. Note that once again the degrees of freedom math works out as expected. There are $P + 1$ polynomials of order M and P sets of M constraints; the truncated power basis has $(P+1)(M+1) - PM$, or $1 + M + P$, free parameters.

Now that we have defined these basis functions, we can fit a regression model to learn the representation of the unknown function $f(x)$ by minimizing the sum of squared residuals over all functions spanned by this basis. This is equivalent to the basis expansion we used in Section 4.2. When used over the spline basis, the resulting estimator is known as a *regression spline*. As with any basis expansion, we can compute the solution by explicitly constructing a design matrix G as

$$G_{i,j} = B_{j-1}(x_i), \quad i = 1, \ldots n, \, j = 1, \ldots, 1 + M + P. \qquad (4.47)$$

Then, to calculate $\widehat{f}(x_0)$, we simply compute the basis expansion at x_{new}

$$g_i = B_{j-1}(x_0), \quad i = 1, \ldots n, \, j = 1, \ldots, 1 + M + P. \qquad (4.48)$$

The regression spline can be written in this basis using a vector $\beta \in \mathbb{R}^{1+M+P}$

$$\widehat{f}(x) = \sum_j^{1+M+P} \widehat{\beta}_j B_{j-1}(x) \qquad (4.49)$$

where $\widehat{\beta}$ is given by

$$\widehat{\beta} = (G^t G)^{-1} G^t y. \qquad (4.50)$$

We can then compute $\widehat{f}$ this for any new point x_0, thus providing an estimate of the entire function f.

By far the most commonly used truncated power basis functions are those with M equal to three. These are justified by the empirical evidence that higher order rarely offer performance gains and that human observers are unable to detect changes in the third derivative of a function (the idea being that you will not be able to point out the knots in a cubic spline). Let us now construct a function that produces the matrix G in Equation 4.47.

```
# One-dimensional regression using a truncated power basis.
#
# Args:
#     x: Numeric vector of values.
#     knots: Numeric vector of knot points.
#     order: Integer order of the polynomial fit.
#
# Returns:
#     A matrix with one row for each element of x and
#     (1 + length(knots) + order) columns.
casl_nlm1d_trunc_power_x <-
function(x, knots, order=1L)
{
  M <- order
  P <- length(knots)
  n <- length(x)
  k <- knots

  X <- matrix(0, ncol=(1L + M + P), nrow = n)
  for (j in seq(0L, M))
  {
    X[, j+1] <- x^j
  }
  for (j in seq(1L, P))
  {
    X[, j + M + 1L] <- (x - k[j])^M * as.numeric(x > k[j])
  }


  X
}
```

With this matrix defined, we can then apply regression splines using the ordinary least squares equations on the design matrix given by `casl_nlm1d_trunc_power_x` to our non-parametric simulated dataset.

```
knots <- seq(0.2, 0.8, 0.2)
G <- casl_nlm1d_trunc_power_x(x, knots, order = 3L)
y_hat <- G %*% solve(crossprod(G)) %*%
          crossprod(G, y)
```

Figure 4.4 shows this solution, along with examples of changing the order of the polynomial.

An observed behavior when using splines as a form of basis expansion for regression is that predicted values often behave erratically for values less than the smallest knot and larger than the largest knot. Natural splines solve this problem by enforcing another constraint on the truncated power basis. Specifically, the function is made to only have order $(M - 1)/2$ outside of the extremal knots. In the case of M equal to three, natural cubic splines are linear outside of these boundary knots. A formula for the natural cubic splines is given by

$$B_0(x) = 1 \tag{4.51}$$
$$B_1(x) = x \tag{4.52}$$
$$B_{1+j}(x) = d_j(x) - d_{P-1}(x), \quad j = 1, \ldots, k - 2 \tag{4.53}$$

where

$$d_j(x) = \frac{(x - k_j)_+^3 - (x - k_P)_+^3}{k_P - k_j}. \tag{4.54}$$

While this formula may seem complicated, it reduces to a scaled version of the truncated power basis for $x \leq k_{P-1}$. Specifically, we have

$$B_{1+j}(x) = \frac{(x - k_j)_+^3}{k_P - k_j}, \quad j = 1, \ldots, k - 2, \quad x \leq k_{P-1}. \tag{4.55}$$

It seems reasonable that the formula becomes more complex at the boundary of the largest knots as this is the location where we want to enforce an additional boundary condition.

We can construct the natural cubic spline basis to verify that the fitted values are in fact linear beyond the extremal knots. Our formula for the basis functions will explicitly use the full notation in Equations 4.51–4.53 rather than the special case for interior points in order to simplify the code.

```
# Natural cubic spline basis.
#
# Args:
#     x: Numeric vector of values.
#     knots: Numeric vector of knot points.
#
# Returns:
#     A matrix with one row for each element of x and
#     length(knots) columns.
casl_nlm1d_nat_spline_x <-
function(x, knots)
{
  P <- length(knots)
```

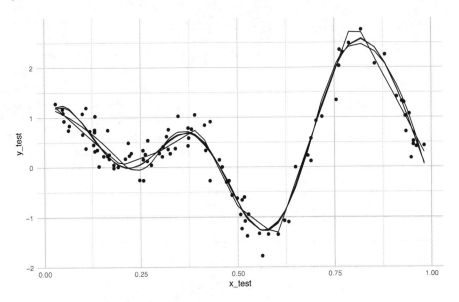

FIGURE 4.4: Predicted values from regression splines using a truncated power basis with knots at 0.2, 0.4, 0.6, and 0.8. We show the fits with linear, quadratic, and cubic basis functions.

```
n <- length(x)
k <- knots

d <- function(z, j)
{
  out <- (x - k[j])^3 * as.numeric(x > k[j])
  out <- out - (x - k[P])^3 * as.numeric(x > k[P])
  out <- out / (k[P] - k[j])
  out
}

X <- matrix(0, ncol=P, nrow=n)
X[, 1L] <- 1
X[, 2L] <- x
for (j in seq(1L, (P-2L)))
{
  X[, j + 2L] <- d(x, j) - d(x, P - 1L)
}

X
}
```

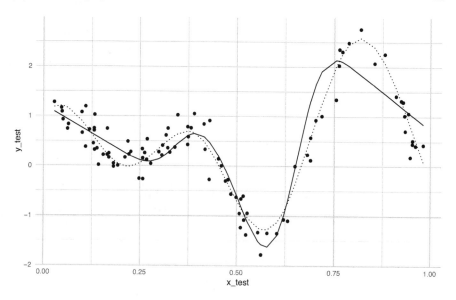

FIGURE 4.5: Using knots at 0.2, 0.4, 0.6, and 0.8, the dashed curve shows regression splines using cubic splines and the solid curve shows regression splines using natural cubic splines. Notice that the solid curve is a line for points less than 0.2 and greater than 0.8.

We can then, as before, use this basis to solve our non-parametric regression problem.

```
knots <- seq(0.2, 0.8, 0.2)
G <- casl_nlm1d_nat_spline_x(x, knots)
y_hat <- G %*% solve(crossprod(G)) %*%
          crossprod(G, y)
```

In Figure 4.5, we see that the solution is very similar for the interior points to the truncated power basis, but differs noticeably on the boundaries of the plot.

4.6 (⋆) Smoothing splines

Regression splines enforce many desirable properties on solutions to non-parametric regression problems. One difficulty with them, however, is the need to choose a set of knot points. Smoothing splines offer an alternative that selects every training point as a knot.

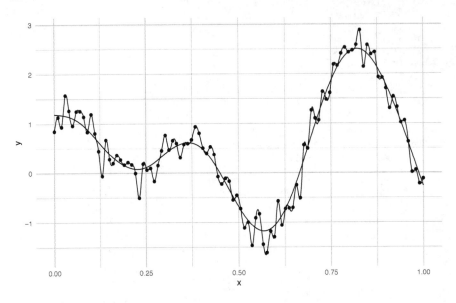

FIGURE 4.6: A natural cubic spline interpolating every training point (wiggly) and a cubic smoothing spline with smoothing parameter $S = 5$ (smooth).

Simply interpolating all points by cubic splines is generally not a good idea. Interpolation by cubic spline functions using every perturbed point can be very wiggly. Figure 4.6 shows an interpolating cubic spline on the training data, along with an example cubic smoothing spline on the same data. Examples like this are effectively over-fit and noise may exhibit excessive influence on the fit. *Smoothing splines* address this problem by relaxing the interpolation condition, trading data fitting for smoothness.

Given data $\{x_i, y_i\}$, $i = 1, 2, \ldots, n$, with distinct, ordered x-values $x_1 < x_2 < \cdots < x_n$, and smoothing parameter $S \geq 0$, a cubic smoothing spline $\widehat{f}$ is defined by

$$\widehat{f} = \arg\min_{f \in C^2} \int_{x_1}^{x_n} f''(x)^2 dx \text{ such that } \sum_{i=1}^{n}(f(x_i) - y_i)^2 \leq S. \qquad (4.56)$$

Setting S to zero in Equation 4.56 reduces to interpolation by cubic spline functions (the wiggly curve in Figure 4.6). Values of $S > 0$ relax the interpolation condition, striking a balance between interpolation and smoothness (the smoother curve above).

The constrained optimization problem in Equation 4.56 was proposed by Reinsch [136] and solved using the Lagrange multiplier method; an outline of Reinsch's approach follows. The Lagrange problem introduces multiplier μ

and auxiliary variable z and seeks a minimum of the functional

$$\int_{x_1}^{x_n} f''(x)^2 dx + \mu \left\{ \sum_{i=1}^{n} (f(x_i) - y_i)^2 + z^2 - S \right\}. \qquad (4.57)$$

In the interesting cases where a straight-line ordinary least squares fit does not satisfy Equation 4.56, the auxiliary variable $z = 0$. Solution of the corresponding Euler-Lagrange equations shows that the optimal function f satisfies:

$$f''''(x) = 0, \ x_i < x < x_{i+1}, \ i = 1, \ldots, n - 1, \qquad (4.58)$$

$$f^{(k)}(x_i)_- - f^{(k)}(x_i)_+ = \begin{cases} 0 & k = 0, 1, \ 2 \le i \le n - 1 \\ 0 & k = 2, \ 1 \le i \le n \\ 2\mu(f(x_i) - y_i) & k = 3, \ 1 \le i \le n, \end{cases} \qquad (4.59)$$

where $f^{(k)}(x_i)_\pm = \lim_{h \to 0} f^{(k)}(x_i \pm h)$ and limits beyond the interval endpoints are defined to be zero. See Reinsch [136] for full details.

Equations 4.58 and 4.59 show that the optimal function f is made up of cubic functions, say with coefficients $\{a_i, b_i, c_i, d_i\}$ for $i = 1, 2, \ldots, n - 1$:

$$f(x) = a_i + b_i(x - x_i) + c_i(x - x_i)^2 + d_i(x - x_i)^3, \ x_i \le x < x_{i+1}, \qquad (4.60)$$

with continuity of f, f', and f'' at their common endpoints–that is, f is a cubic spline.

We can use Equation 4.59 to establish relationships between the coefficients, setting $c_1 = c_n = 0$ and denoting the $n - 2$-element vector $\hat{c}$ to contain the elements $c_2, c_3, \ldots, c_{n-1}$:

$$d_i = (c_{i+1} - c_i)/3h_i, \ i = 1, 2, \ldots, n - 1 \ (\text{from } k = 2) \qquad (4.61)$$

$$b_i = (a_{i+1} - a_i)/h_i - c_i h_i - d_i h_i^2, \ i = 1, 2, \ldots, n - 1 \ (\text{from } k = 0) \qquad (4.62)$$

$$T\hat{c} = Q^t a \ (\text{from } k = 1) \qquad (4.63)$$

$$Q\hat{c} = \mu(y - a) \ (\text{from } k = 3), \qquad (4.64)$$

where $h_i = x_{i+1} - x_i$, T is a symmetric positive definite matrix of order $n - 2$ with diagonal entries $2(h_{i-1} + h_i)/3$ and super-/sub-diagonal entries $h_i/3$ ($i = 2, 3, \ldots, n - 1$) and Q is a tridiagonal matrix with n rows and $n - 2$ columns with main diagonal entries $1/h_i$, first subdiagonal entries $-1/h_i - 1/h_{i+1}$, and second subdiagonal entries $1/h_{i+1}$ for $i = 1, 2, \ldots, n - 2$.

Solving Equations 4.63 and 4.64 for $\hat{c}$ yields the linear systems

$$\hat{c} = \mu(Q^t Q + \mu T)^{-1} Q^t y, \qquad (4.65)$$

$$a = y - \mu^{-1} Q\hat{c}. \qquad (4.66)$$

The remaining spline coefficients b_i and d_i can be determined from Equations 4.61 and 4.62. The linear system to be solved in Equation 4.65 is defined by a banded symmetric positive definite matrix $Q^t Q + \mu T$ with five non-zero diagonals. Its solution requires knowing the Lagrange parameter μ. Given a smoothing parameter S in Equation 4.56, solve the following quadratic equation to determine μ:

$$\|Q(Q^tQ + \mu T)^{-1}Q^ty\|_2 = +\sqrt{S}. \qquad (4.67)$$

The following R code implements the Reinsch algorithm. For simplicity, the code forgoes performance optimizations that take advantage of the banded structure of T and Q. First, we define a function to make it easy to modify the banded diagonals of a matrix.

```
# Pick out and set matrix diagonals.
#
# Args:
#     x: A matrix.
#     k: Integer describing the diagonal number.
#
# Returns:
#     A vector of indices corresponding to diagonal
casl_dgnl <-
function(x, k=0)
{
  N <- prod(dim(x))
  i <- seq(-k + 1, N, by=nrow(x) + 1)
  i[i > 0 & i <= N]
}
```

Next, we use this helper function in our implementation of the smoothing spline.

```
# Natural cubic spline basis.
#
# Args:
#     x: Numeric vector of values.
#     y: Numeric vector of the responses.
#     S: Non-negative smoothing parameter.
#
# Returns:
#     A model object that works with coef() and predict().
casl_smspline <-
function(x, y, S)
{
  h <- diff(x)
  n <- length(x)
  m <- length(h)
  T <- matrix(0, n - 2, n - 2)
  T[casl_dgnl(T)] <- 2 * (h[-m] + h[-1]) / 3
```

```
T[casl_dgnl(T, 1)] <- h[-c(1, m)] / 3
T[casl_dgnl(T, -1)] <- h[-c(1, m)] / 3
Q <- matrix(0, n, n-2)
Q[casl_dgnl(Q)] <- 1/h[-m]
Q[casl_dgnl(Q, -1)] <- -1/h[-m] - 1/h[-1]
Q[casl_dgnl(Q, -2)] <- 1/h[-1]
F <- function(mu)
{
  u <- solve(crossprod(Q) + mu * T, crossprod(Q, y))
  drop(sqrt(crossprod(Q %*% u)))
}
mu <- nlm(function(mu, S) abs(F(mu) - sqrt(S)), 0, S,
          stepmax=1e4*S + (S < 5)*1e9)$estimate
c <- drop(solve(crossprod(Q) + mu * T, mu * crossprod(Q, y)))
a <- y - (1/mu) * (Q %*% c)
c <- c(0, c, 0)
d <- (c[-1] - c[-n]) / (3*h)
c <- c[-n]
b <- (a[-1] - a[-n]) / h - c * h - d * h^2
coef <- data.frame(a=a[-n], b=b, c=c, d=d)
structure(function(newdata) { # predict function
  vapply(newdata, function(z) {
    i <- tail(which(z >= x), 1)
    if(length(i) < 1) i <- 1
    if(i > nrow(coef)) i v nrow(coef)
    dx <- z - x[i]
    coef$a[i] + coef$b[i] * dx +
      coef$c[i] * dx ^ 2 + coef$d[i] * dx ^ 3
  }, pi)}, class="smspline"
)
}
```

Finally, we can implement the prediction and coefficient methods in R to make it easy to use the smoothing spline function on data.

```
predict.smspline <- function(object, newdata) object(newdata)
coef.smspline <- function(object) environment(object)$coef
```

The smoothing parameter S in the `casl_smspline` function above corresponds to the constraint in equation 4.56 and differs from the `spar` smoothing parameter used by the `smooth.spline` function in R's `stats` package. Instead, R's parameter more directly references the Lagrange multiplier version of the problem.

The requirement of unique abscissae in equation 4.56 can be relaxed with the consequence of a more complicated algorithm. The original Reinsch algorithm, and many other implementations including the `smooth.spline` function in R's **stats** package, support individual weights on the y_i values. We omit weighting above for simplicity.

Many derivations of smoothing splines begin with a version of the Lagrange form shown in equation 4.57, but with a Lagrange parameter on the integral smoothing term instead of the least squares term. But the derivation is effectively the same. You can see this, for instance, in the interesting cases when $z = 0$ and $\mu > 0$ in equation 4.57 by multiplying through by $\lambda = 1/\mu$.

B-splines can be used to derive smoothing splines in a similar way and are often used in software implementations. The banded matrix structure noted above is a consequence of the derivative continuity condition on f in equation 4.56, and B-spline implementations similarly yield banded systems. However, they exhibit potentially better numerical stability in edge cases.

4.7   ($\star$) B-splines

Both the truncated power basis and its natural spline variant are designed to allow for local changes to the predicted function without needing to modify its global behavior. Looking at the basis functions, however, we can see that the basis function corresponding to the kth knot is non-zero for all input points greater than the knot. The leading polynomial terms are non-zero for all inputs. From a theoretical view there is nothing inherently wrong with this approach. The basis allows for local behavior because the coefficient at the $(k+1)$th knot can be thought of as canceling out the Mth order term from the prior knot. In other words, the knot-based coordinates in the truncated power basis give the changes in the Mth derivative rather than their local values. This leads to two computational challenges, both of which can be solved by switching to an alternative basis spanning an equivalent space.

Consider generating data points between 0 and 1 and setting 9 evenly distributed knots

```
n <- 1000
M <- 3
P <- 9
x <- runif(n)
y_bar <- sin(x * 3 * pi) + cos(x * 5 * pi) + x^2
y <- y_bar + rnorm(n, sd = 0.25)
k <- seq(from = 0.1, to = 0.9, by = 0.1)
X <- casl_nlm1d_trunc_power_x(x, knots = k, order = 3)
singular_values <- svd(X)$d
max(singular_values) / min(singular_values)
```

```
[1] 78337
```

The condition number of the design matrix is quite large, particularly given that this is a relatively small dataset. Also, notice that the design matrix X has mostly non-zero elements.

```
mean(X != 0)
```

```
[1] 0.6486154
```

We see, therefore, that our choice of basis forces us to construct a large, dense design matrix that is poorly conditioned. As the specific basis we choose should not affect the final result, it would seem like a good idea to try to find an alternative representation.

The *B-spline basis* is a re-parametrization of the truncated power basis that simultaneously solves both of these computational issues. It can be defined by setting $B_{j,0}$ to

$$B_{j,0}(x) = \begin{cases} 1 & \text{if } k_j \leq x \leq k_{j+1} \\ 0 & \text{otherwise} \end{cases} \tag{4.68}$$

and defining $B_{j,m}$ recursively as

$$B_{j,m}(x) = \frac{x - k_j}{k_{j+m} - k_j} \cdot B_{j,m-1}(x) + \frac{k_{j+m+1} - x}{k_{j+m+1} - k_{j+1}} \cdot B_{j+1,m-1}(x). \tag{4.69}$$

Note that the Mth order basis uses only the basis functions $B_{j,M}$; the lower-order terms are not needed or used. The computational benefits of B-splines come from their having *local support*, in other words they are only non-zero for a small subset of the regions defined by the knots. Specifically, the first order B-splines are just indicator functions for the regions between each knot and the second order B-splines are symmetric triangles with a maximum value at an interior knot and equal to zero at the neighboring knots. Higher-order terms follow a similar pattern. Given the local support property, the inner product of the corresponding design matrix $G^T G$ will, assuming the input data are ordered, be a banded matrix. That is, the only non-negative values are along the main diagonal and a small number of diagonals off of it.

We can use this formula to produce the design matrix using the B-spline basis. As with the natural cubic splines, here we will use the explicit definition to simplify the code. Being careful to only compute those non-zero values would yield a significantly faster solution.

```
# B-spline basis.
#
# Args:
#     x: Numeric vector of values.
#     knots: Numeric vector of knot points.
#     order: Integer order of the polynomial fit.
#
```

```
# Returns:
#     A matrix with one row for each element of x and
#     (length(knots) + order) columns.
casl_nlm1d_bspline_x <-
function(x, knots, order=1L)
{
  P <- length(knots)
  k <- knots
  k <- c(rep(min(x), order + 1L), k,
         rep(max(x), order + 1L))

  # Note: this is NOT vectorized over x
  b <- function(x, j, m)
  {
    if (m == 0L) {
      return(as.numeric(k[j] <= x & x <= k[j + 1L]))
    } else {
      r <- (x - k[j]) / (k[j + m] - k[j]) *
             b(x, j, m - 1)
      if (is.nan(r)) r <- 0
      r <- r + (k[j + m + 1] - x) /
             (k[j + m + 1] - k[j + 1]) *
             b(x, j + 1, m - 1)
      if (is.nan(r)) r <- 0
      r
    }
  }

  X <- matrix(0, ncol=(P + order), nrow=length(x))
  for (j in seq(1L, (P + order)))
  {
    for (i in seq(1L, length(x)))
    {
      X[i, j] <- b(x[i], j, order)
    }
  }

  X
}
```

Applying this to a toy example, we see how the banded structure forms in the design matrix.

```
x <- seq(0, 16, by = 1)
knots <- seq(2, 14, by = 2)
G <- casl_nlm1d_bspline_x(x, knots, order = 2)
```

```
round(crossprod(G), 2)
```

```
     [,1] [,2] [,3] [,4] [,5] [,6] [,7] [,8] [,9]
[1,] 1.06 0.16 0.03 0.00 0.00 0.00 0.00 0.00 0.00
[2,] 0.16 1.41 1.17 0.02 0.00 0.00 0.00 0.00 0.00
[3,] 0.03 1.17 2.59 1.19 0.02 0.00 0.00 0.00 0.00
[4,] 0.00 0.02 1.19 2.59 1.19 0.02 0.00 0.00 0.00
[5,] 0.00 0.00 0.02 1.19 2.59 1.19 0.02 0.00 0.00
[6,] 0.00 0.00 0.00 0.02 1.19 2.59 1.19 0.02 0.00
[7,] 0.00 0.00 0.00 0.00 0.02 1.19 2.59 1.19 0.02
[8,] 0.00 0.00 0.00 0.00 0.00 0.02 1.19 2.59 1.14
[9,] 0.00 0.00 0.00 0.00 0.00 0.00 0.02 1.14 1.16
```

There will be one additional band for each order of the B-spline, so cubic splines have four diagonal bands.

Applying the B-spline basis to our original problem, we see that this basis does in fact lead to a significantly better-conditioned matrix.

```
n <- 1000
M <- 3
P <- 9
x <- sort(runif(n))
y_bar <- sin(x * 3 * pi) + cos(x * 5 * pi) + x^2
y <- y_bar + rnorm(n, sd = 0.25)
k <- seq(from = 0.1, to = 0.9, by = 0.1)
G <- casl_nlm1d_bspline_x(x, k, order = 3)
singular_values <- svd(G)$d
max(singular_values) / min(singular_values)
```

```
[1] 7.426937
```

The B-splines have a condition number almost 4 orders of magnitude smaller than the truncated polynomial basis. Furthermore, we can see that it yields a very similar set of predictions.

```
X <- casl_nlm1d_trunc_power_x(x, k, order = 3)
G <- casl_nlm1d_bspline_x(x, k, order = 3)
yt <- X %*% solve(crossprod(X)) %*% crossprod(X, y)
yb <- G %*% solve(crossprod(G)) %*% crossprod(G, y)
max(abs(yt - yb))
```

```
[1] 0.08350365
```

For larger problems, these two approaches will begin to diverge as the truncated power basis runs into numerical issues. While outside the scope of our analysis here, the banded nature of the B-spline basis can be solved particularly quickly using custom solvers for banded systems of linear equations.

The B-spline basis may also be used, and most frequently is, for smoothing splines. We will not derive this here as the calculation of the matrix Ω in the B-spline basis is quite involved. However, notice that as each basis function has local support, the penalty matrix will also be banded. This allows the cubic smoothing spline to be computed efficiently in $O(n)$ time and $O(n)$ space.

4.8   Implementation and notes

There is excellent support for linear smoothers in R. Given the importance of non-parametric regression within statistics and the popularity of splines during the 1990s in the early development of R, much of the core functionality is included in the packages loaded by default or in the recommended suite. The **stats** package provides `smooth.spline` for cubic smoothing splines using an efficient B-spline basis with the computationally intensive parts written in Fortran. The recommended package **splines** gives the function `bs` for generating the design matrix corresponding to the B-spline basis of an arbitrary order and given collection of knots. The function `splines::ns` gives an specific variation for natural cubic splines; we show in Section 4.9 how to use these directly within the formula interface for implementing regression splines. For kernel smoothing, the function `stats::ksmooth` offers reasonable speed and functionality; the recommended package **KernSmooth** provides several functions for more control over the fit [168]. The recommended package **class** gives the function `knn` for implementing k-nearest neighbors but only supports classification. For regression, the **FNN** package gives the function `knn.reg` [22]. Finally, the functions `stats::loess` and `stats::lowess` implement local regression; the first uses a formula interface while the second takes plain numeric vectors.

As with linear regression, one reason for the popularity of linear smoothers is the availability of extensive theoretical results underlying many of the techniques. Cosma Shalizi's text *Advanced Data Analysis From an Elementary Point of View* provides an excellent survey of these results requiring no prerequisites beyond those required by this text and an introductory course on probability and statistics [147]. A full technical treatment requires knowledge of basic functional analysis. An excellent textbook treatment at this level is Alexandre Tsybakov's *Introduction to Nonparametric Estimation* [158]. While establishing convergence properties of linear smoothers, these sources also illustrate the non-optimality of their approach to estimating functions with varying levels of smoothness over the domain. Important non-linear smoothers include wavelets [40], locally adaptive regression [113], and trend filtering [156]. Empirical and computational challenges have, however, so far prevented these modifications from enjoying the same popularity as those presented in this chapter.

4.9 Application: U.S. census tract data

4.9.1 Data and setup

The United States Census Bureau reports data from the annual American Community Survey (ACS) in several different formats. One of the most granular sets is the five-year averages giving estimated population data at the level of a census tract. Tracts traditionally correspond to neighborhoods in urban areas and ideally have around 4000 people in each, though their sizes can range from 1200 to over 8000 individuals. These are used only for statistical purposes, with tract boundaries having no specific political or legal implications. In this section we will look at the five year tract-level ACS data published in 2015. The goal will be to predict a response variable from the survey as a function of univariate demographic variables. In order to illustrate the various methods introduced in this chapter, we will investigate several different combinations of response and predictor variables.

The raw ACS data has thousands of variables; we have reduced the dataset to include only those needed in this section. Reading it in we see several interesting covariates for several economic and spatial quantities of interest.

```
tract <- read.csv("data/acs_tract.csv")
names(tract)
```

```
 [1] "total_households" "median_age"       "income_q1"
 [4] "median_income"    "public_transit"   "housing_costs"
 [7] "center_dist"      "state"            "cbsa_name"
[10] "lat"              "lon"
```

The meaning of many of these variables should be straightforward. `income_q1` gives the first quartile of annual household-level income in dollars, and `public_transit` gives the percentage of workers commuting by public transit. Geospatial data includes the distance to the centroid of the metropolitan area in kilometers, stored as `center_dist`, and the tract's centroid latitude and longitude. Spatial data were computed using the shapefiles also provided as a supplement from the US Census Bureau. We have removed any rows with missing data.

The Census Bureau breaks the United States into regions known as community based statistical areas (CBSA), roughly corresponding to metropolitan areas. These may, and often do, cross county and state borders. As with tracts, these have no formal political meaning, however the CBSA regions do often map to often-cited social constructs of place. For example, the 'New York-Newark-Jersey City, NY-NJ-PA' CBSA roughly corresponds to the New York City Tri-State area. These regions are listed in the ACS data under the variable cbsa_name. As some of our classification tasks will use these regions, we want to be careful this time to stratify our training set to include a balanced number of tracts from each region. This avoids the issue we had with states in Section 2.10. As a first step, we draw a random number between 0 and 1 for each row of the dataset. We want to construct a flag variable equal to 1 for the training set (40%), 2 for the validation set (40%), and 3 for the testing set (20%).

```
set.seed(1)
rnum <- runif(nrow(tract))
tract$flag <- 1
```

Setting all of the data to the training set by default, we now compute the 40th percentile of rnum for each CBSA. Any tract with a value greater than this is moved from the training set into the validation set.

```
coff <- tapply(rnum, tract$cbsa_name, quantile, probs = 0.4)
coff <- coff[match(tract$cbsa_name, names(coff))]
tract$flag[rnum > coff] <- 2
```

And, finally, we compute the 80th percentile rnum for each CBSA and move any points above this into the test set.

```
coff <- tapply(rnum, tract$cbsa_name, quantile, probs = 0.8)
coff <- coff[match(tract$cbsa_name, names(coff))]
tract$flag[rnum > coff] <- 3
```

The result leads to a dataset with the desired breakdown (40/40/20), evenly distributed across the CBSA regions. We can verify this by looking at a table of regions crossed with the flag variable.

```
tab <- table(tract$cbsa_name, tract$flag)
head(tab)
```

```
              1   2   3
Aberdeen, SD  4   4   2
Aberdeen, WA  7   6   3
Abilene, TX  17  17   9
Ada, OK       4   4   2
Adrian, MI    9   9   5
Akron, OH    68  67  34
```

The counts in column one are not exactly equal to column two for every region given the discrete nature of the data, but they are at most 1 away from one another. Splitting our dataset in this way makes tuning, evaluating, and understanding our models significantly easier.

We will again use the `rmse` function from Section 3.6.1 to compare the performance of predicted values on the training, validation, and testing sets.

4.9.2 Income by total number of households

As a first task, we will try to predict the first quartile of income for a census tract as a function of the number of households in the tract. Note that the number of households corresponds to the number of sampled households rather than the total number actually present in the region. As the boundaries between census tracts are somewhat randomly selected based on current populations, a relationship between these variables at first seems somewhat unlikely. There are two important things, however, that we can learn based on the number of households. First of all, more rural areas will tend to be farther away from the optimal census tract area as it is harder in rural areas to find a good split that does not cross state or county borders. Secondly, the Census Bureau only updates tract definitions every ten years and even then prefers to limit the modification of their boundaries. Therefore tracts with a large number of households have likely experienced recent growth and those with a small number of households have undergone a population shrinkage.

It is useful to begin by fitting a linear regression model predicting the income level as a function of the number of households in a census tract. The error rate from this model will serve as a good baseline for our linear smoothers. We will fit such as model on our training set and use the function `rmse` to evaluate it.

```
model <- lm(income_q1 ~ total_households, data = tract,
            subset = (flag == 1))
y_hat <- predict(model, tract)
rmse(y_hat, tract$income_q1, tract$flag)
```

```
       1        2        3
15056.84 14845.66 15245.68
```

We have not used the validation set (2) at all, and in theory it should be indistinguishable from the testing set. However, we see that the model does better on the validation set and worse on the testing set when compared to the training set. Let this serve as an empirical warning about overinterpreting minor differences in the performance of various models over different sets of data.

For this estimation task, we will use a cubic regression spline with knots evenly distributed between the 10th and 90th percentiles of the training data's range.

```
knots <- quantile(tract$total_households[tract$flag == 1],
```

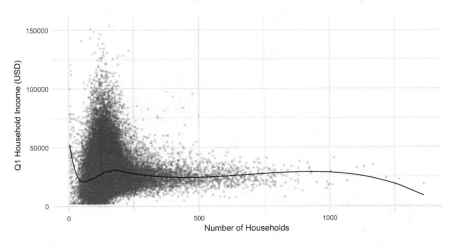

FIGURE 4.7: Scatter plot showing the number of households in each census tract and the first quartile of household income. The line shows the predicted values from a regression spline using cubic splines with 9 knots evenly spaced between the 10th and 90th percentiles of the training data.

```
seq(0.1, 0.9, by = 0.1))
```

From these knots, we can construct the training dataset X. As expected it has 13 columns: one for each knot (9), one for the order of the polynomial (3), and one for the intercept (1).

```
X <- splines::bs(tract$total_households,
                 knots = knots,
                 intercept = TRUE)
y <- tract$income_q1
dim(X)
```

```
[1] 66298    13
```

From this data matrix, we can fit a linear regression to the response variable, now saved as the vector y, over the training set. Because we already have data matrices, we will use `stats::lm.fit` to find the ordinary least squares fit rather than calling the wrapper `stats::lm`.

```
beta <- coef(lm.fit(X[tract$flag == 1,], y[tract$flag == 1]))
y_hat <- X %*% beta
rmse(y_hat, tract$income_q1, tract$flag)
```

```
        1         2         3
14835.16 14644.65 15054.47
```

The root mean squared error decreases across all three categories, showing the power of the cubic splines to capture non-linearities in the model. A plot of the fit is shown in Figure 4.7. Note that there is an obvious boundary problem in the model for tracts with a very low number of households, as the curve spikes up near zero.

4.9.3 Income by median age

Another demographic variable that has an interesting relationship to income is median age. The methods of this chapter are useful for studying how median age affects income, particularly at the first quartile, because it is likely that there is a non-linear relationship between these variables. Median ages that are very low indicate tracts that largely house students or other young populations, neither of which typically earn very much money. High median ages likely indicate retirement communities, which also have low earnings. The highest income is likely to correspond to median ages somewhere between mid-30s to late 50s.

As with total population, it is useful to see how well a linear model performs using only median age as a predictor variable.

```
model <- lm(income_q1 ~ median_age, data = tract,
            subset = (flag == 1))
y_hat <- predict(model, tract)
rmse(y_hat, tract$income_q1, tract$flag)
```

```
        1        2        3
14609.24 14382.60 14785.76
```

Despite our reasoning that this relationship should be relatively non-linear, this regression performs better than any of the models using total population. Clearly this is a useful variable to use in predicting first quartile household income. The fact that this new variable within a simple model outperforms fancier models using a different variable points to a general principal in statistical learning: models are only as good as the data you feed into them. More often than not, choices made in data collection and processing affect performance more than the specific algorithmic choices.

The linear model performs relatively well, but we have already argued that we expect a non-linear relationship between age and income so we will again try to fit regression splines to this problem. This time we will use natural cubic splines, which force the solution to be linear outside of the boundary knots, to illustrate how this affects the model.

```
knots <- quantile(tract$median_age[tract$flag == 1],
                  seq(0.1, 0.9, by = 0.1))
X <- splines::ns(tract$median_age,
                 knots = knots,
```

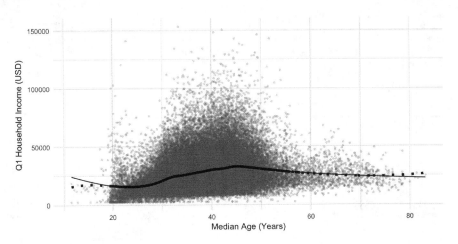

FIGURE 4.8: Scatter plot showing the median age of residents in each census tract and the first quartile of household income. The dashed line shows the predicted values from a regression spline using cubic splines with 9 knots evenly spaced between the 10th and 90th percentile of the training data. The solid line fits the same model using the natural cubic spline basis from the same knots.

```
                    intercept = TRUE)
y <- tract$income_q1
```

The performance of this model noticeably improves on the linear one.

```
beta <- coef(lm.fit(X[tract$flag == 1,], y[tract$flag == 1]))
y_hat <- X %*% beta
rmse(y_hat, tract$income_q1, tract$flag)
```

```
        1         2         3
14334.18 14085.15 14495.43
```

Figure 4.8 shows this fit as well as the fit using the cubic spline basis. Notice how the two models diverge for extreme median ages but agree almost everywhere else. In this case, it is not visually clear which model should be preferred. The plot verifies our initial theory that the income levels would drop off for extreme median ages; the maximum predicted income occurs around 46 years.

4.9.4   Median income by tract longitude

The two examples we have looked at so far illustrate the power of linear smoothers to predict univariate, non-linear relationships. These relationships have been relatively simple, however, being essentially monotonic over most of their ranges. Here we look at the significantly more complex relationship between longitude and median income. We would expect median incomes to jump up for longitudes corresponding to major cities and decrease rapidly for longitudes without a major metropolitan area. Over the entire United States, this change will likely occur several dozen times or more.

The following code will be easier to read if we first extract the independent and dependent variables of interest. Even though in this case our predictor variable is not a matrix, we will still use a capital X to represent it in keeping with the coding style used throughout this text.

```
X <- tract$lon
y <- tract$median_income
```

We believe that the relationship between these variables will have many critical points making regression splines less optimal as we would need to select ahead of time where all of these jumps may occur. Instead, we will use kernel regression here to estimate the model. The function `stats::ksmooth` is relatively straightforward, taking both a training set to fit the model with and data for which predictions are desired. The only catch is that the results are returned in a non-standard format. A list is given that includes an element with x-points and another element with the predicted y-values; the x-points have been rearranged in increasing order. We need to call the function `match` in order to extract the predictions corresponding with our original data.

```
res <- stats::ksmooth(X[tract$flag == 1], y[tract$flag == 1],
                      x.points = X, bandwidth = 2)
index <- match(X, res$x)
y_hat <- res$y[index]
rmse(y, y_hat, tract$flag)
```

```
        1        2        3
27835.07 27557.76 28054.40
```

We do not have a baseline model here to compare to in this case as linear regression seems wholly inappropriate for this relationship, but these results do at least indicate that we are not overfitting to the training set with a bandwidth of 2. Perhaps lowering this value would improve the model's performance.

The bandwidth parameter to `stats::ksmooth` is given in the units of the x-variable, so we should think of it in terms of degrees of longitude. Setting the bandwidth to 0.1 seems like a good place to start, but running the code above causes some difficulty.

```
res <- stats::ksmooth(X[tract$flag == 1], y[tract$flag == 1],
                      x.points = X, bandwidth = 0.1)
index <- match(X, res$x)
```

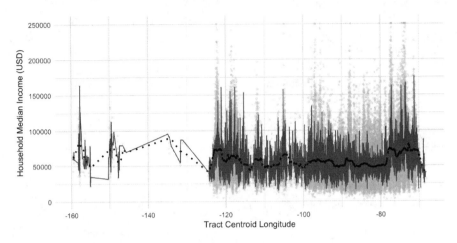

FIGURE 4.9: Scatter plot showing the longitude of each tract's centroid and the household median income. The lines show predictions from a kernel regression with a Gaussian kernel. The solid line uses a bandwidth of 0.01 and the dashed line uses a bandwidth of 2.

```
y_hat <- res$y[index]
rmse(y, y_hat, tract$flag)
```

```
         1              2              3
25941.34          NA             NA
```

The training set error decreases substantially, but we have introduced missing values in the validation and testing sets. What exactly is going wrong? Looking at which states the missing values come from reveals some of the difficulty.

```
table(tract[is.na(y_hat),"state"])
```

```
AK AZ CO HI ME MT NM NV SD TX UT WA WY
17  3  5 14  9  1  5  1  1  3  2  1  5
```

These states all contain relatively rural areas. The problem is that there are some census tracts that have a longitude that is not within 0.1 degrees of a tract in the training set despite our stratification to make sure each metropolitan area is split evenly throughout the sets. Of course, this never occurs within the training set because each tract is at a minimum within 0.1 degrees of itself.

There is a relatively simple fix to the problem of having missing values from a small bandwidth in the kernel smoothing function, however this currently requires writing our own implementation of the fix. Consider what happens when only one training sample is within a bandwidth of a testing point: the weighted mean sets the prediction exactly equal to that one point. Therefore, a reasonable thing to do once the bandwidth includes no training points is to find the closest training point and set the prediction equal to this value. In this way, the predictions will continue to be continuous in the bandwidth parameter (at least when using the Epanechnikov kernel). We can implement this using the k-nearest neighbors function `knn.reg` from the **FNN** package by setting k equal to one.

```
# One-dimensional kernel smoothing.
#
# Args:
#     X: Numeric matrix of training locations.
#     y: Numeric response vector of responses.
#     x_new: A vector of data values at which to estimate.
#     h: A numeric value giving the bandwidth of the kernel.
#
# Returns:
#     Numeric vector of predicted values for points in x_new.
ksmooth_smart <-
function(X, y, x_new, h) {
  res <- ksmooth(X, y, x.points=x_new, bandwidth=h)
  index <- match(x_new, res$x)
  y_hat <- res$y[index]

  if (any(index <- is.na(y_hat)))
  {
    y_hat[index] <- FNN::knn.reg(X, as.matrix(x_new[index],
                                              ncol=1L),
                                 y, k=1L)$pred
  }

  y_hat
}
```

Inside this function, we also wrap up the reordering of the inputs to further simplify our code. Testing the function we see that it successfully solves the problem with missing values for a small bandwidth.

```
y_hat <- ksmooth_smart(X[tract$flag == 1], y[tract$flag == 1],
                       x_new = X, h = 0.1)
rmse(y, y_hat, tract$flag)
```

```
        1        2        3
25941.34 26166.93 26661.44
```

Now that we have results for all three sets, it becomes clear that the smaller bandwidth 0.1 outperforms the larger one of 2.

In order to find the optimal bandwidth, we need to perform some type of validation. In this example, we will use simple validation by testing the performance of a number of bandwidths on the validation set.

```
bw_vals <- seq(0, 2, by = 0.01)
vals <- matrix(NA, nrow = length(bw_vals), ncol = 3)
for (i in seq_along(bw_vals)) {
  y_hat <- ksmooth_smart(X[tract$flag == 1],
                         y[tract$flag == 1],
                         x_new = X,
                         h = bw_vals[i])
  vals[i,] <- rmse(y, y_hat, tract$flag)
}
```

The best value occurs at 0.08, just slightly smaller than our prior guess of 0.1.

```
bw_vals[which.min(vals[,2])]
```

```
[1] 0.08
```

The performance on this bandwidth improves slightly on the values from our prior guess.

```
vals[which.min(vals[,2]),]
```

```
[1] 25834.14 26154.43 26667.42
```

Figure 4.9 shows the results of the kernel estimator for two different bandwidths. Notice how much more non-linear this fit is compared to our two previous applications of linear smoothers.

4.9.5 Public transit rates by tract latitude

If we can us longitude to predict a census tract's median income, perhaps we can also use latitude to predict other values that tend to be spatially correlated. Here we will try to build a predictive model for estimating the percentage of workers who commute via public transit. As before, we first extract the predictor and response variables from the data.

```
X <- tract$lat
y <- tract$public_transit
```

As with median income and longitude, it seems reasonable to expect latitude and public transit rates to have a very non-linear relationship. Here, we show how to use k-nearest neighbors to predict the relationship. For example, using k equal to 10 we have the following.

```
y_hat <- FNN::knn.reg(X[tract$flag == 1], as.matrix(X),
                      y[tract$flag == 1], k = 10)$pred
rmse(y, y_hat, tract$flag)
```

```
        1        2        3
8.782193 9.899090 9.650040
```

As with the bandwidth in kernel regression, we need to perform some sort of validation to figure out the optimal value of k. We will again do this using simple validation over the validation set.

```
k_vals <- seq(1, 500, by = 10)
vals <- matrix(NA, nrow = length(bw_vals), ncol = 3)
for (i in seq_along(k_vals)) {
  y_hat <- FNN::knn.reg(X[tract$flag == 1],
                        as.matrix(X),
                        y[tract$flag == 1],
                        k = k_vals[i])$pred
  vals[i,] <- rmse(y, y_hat, tract$flag)
}
```

The optimal value occurs at 111, quite a bit larger than our initial guess.

```
k_vals[which.min(vals[,2])]
```

```
[1] 111
```

```
vals[which.min(vals[,2]),]
```

```
[1] 9.228715 9.460985 9.268023
```

Notice that the performance has gone down on the training set, but up on the validation and testing sets. As we might expect, setting k too low leads to overfitting the model. In the extreme case of k equal to 1, the training error would be exactly equal to 0 (at least when there are no ties in the response vector). A plot of the k-nearest neighbors fit is shown in Figure 4.10, with a zoomed in version near the latitude of New York City in Figure 4.11. Particularly from the zoomed in version, we see that this fit has even more points of inflection that the longitude and median income model.

Another technique for validation of k-nearest neighbors is leave-one-out cross-validation. In this, we do cross-validation with the number of folds equal to the number of data points. The model needs to be refit for each data point in the training set. For some models this is quite time consuming, but for k-nearest neighbors it is fairly easy: we simply find the k closest training points to each input *not including the point itself* and use that as a prediction.

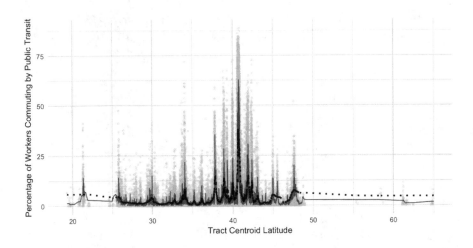

FIGURE 4.10: Scatter plot showing the latitude of each tract's centroid and the percentage of workers who commute by public transit. The lines show predictions from the k-nearest neighbors estimator. The solid line uses the 50 closest values and the dashed line uses the nearest 300.

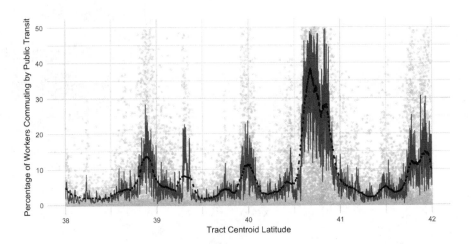

FIGURE 4.11: Scatter plot showing the latitude of each tract's centroid and the percentage of workers who commute by public transit, centered around New York City. The lines show predictions from the k-nearest neighbors estimator. The solid line uses the 50 closest values and the dashed line uses the nearest 300.

Running this with `knn.reg` occurs if we do not give a specific prediction set; helpfully the sum of squared residuals is even returned as the object `PRESS`.

```
k_vals <- seq(2, 500, by = 10)
vals <- rep(NA, length(k_vals))
for (i in seq_along(k_vals)) {
  PRESS <- FNN::knn.reg(X[tract$flag == 1],
                        y = y[tract$flag == 1],
                        k = k_vals[i])$PRESS
  vals[i] <- sqrt(PRESS / length(X))
}
```

Optimizing this quantity gives a value of 112, just one more than the technique using a separate validation set.

```
k_opt <- k_vals[which.min(vals)]
k_opt
```

```
[1] 112
```

The benefit of this form of cross-validation is that it allows us to eschew an independent validation set, thus effectively increasing the size of the training set, without incurring any noticeable increase in the computational time.

4.9.6 Housing costs by distance to city center

As a final application, we will try to predict the median housing costs, by household, as a function of the distance of the tract to its CBSA centroid. The center of the CBSA has been computed taking the center of mass for all of the census tracts, though this likely does not correspond directly to what might be economically or socially considered the center of the region.

```
X <- tract$center_dist
y <- tract$housing_costs
```

Unlike our other models, in this case we want to fit a separate model for each of the largest 20 CBSAs. For example, we will start by taking the data for the Seattle-Tacoma-Bellevue metropolitan area.

```
cbsa <- "Seattle-Tacoma-Bellevue, WA"
index <- which(tract[,"cbsa_name"] == cbsa)
X_cbsa <- X[index]
y_cbsa <- y[index]
train_cbsa <- tract$flag[index]
```

A good choice for this prediction problem is to use smoothing splines. They generally provide a smoother solution than kernel smoothing or k-nearest neighbors, and for this problem we do not expect the same number of non-linearities as there were in the examples that use longitude and latitude as independent variables. However, we also do not want to hand construct knots of each and every metropolitan area. Smoothing splines can be fit using the function `stats::smooth.spline`; as with kernel regression, the result is returned as a two-element list, but the input is (helpfully) not reordered.

```
res <- smooth.spline(X_cbsa[train_cbsa == 1],
                     y_cbsa[train_cbsa == 1],
                     spar = 0.4)
y_hat <- predict(res, X_cbsa)$y
rmse(y_cbsa, y_hat, train_cbsa)
```

```
        1         2         3
4225.530  5547.061  5270.790
```

Notice that we have set the tuning parameter `spar` rather than the λ value described in Section 4.6. The `spar` parameter is monotonically related to λ by the relationship but provides a much better parameter for tuning as it is scale-invariant (usually, the optimum occurs somewhere between 0 and 2).

We again need to use validation to set the tuning parameter for the smoothing spline. As with k-nearest neighbors, there is an efficient way of doing cross-validation that is supported by the `stats::smooth.spline` function. We will wrap up this particular cross-validation routine as a function accepting just a training set and testing set, and returning the smoothing spline object with the optimally learned parameter. This should serve as a useful template for many of the end-of-chapter exercises.

```
# Cross-validated one-dimensional smoothing splines.
#
# Args:
#     X: Numeric matrix of training locations.
#     y: Numeric response vector of responses.
#
# Returns:
#     A smooth.spline object using optimal spar value.
cv_smooth_spline <-
function(X, y) {

  spar_vals <- seq(0, 2, by = 0.01)
  vals <- rep(NA_real_, length(spar_vals))
  for (i in seq_along(spar_vals))
  {
    res <- smooth.spline(X, y, spar = spar_vals[i])
    vals[i] <- res$cv.crit
  }
```

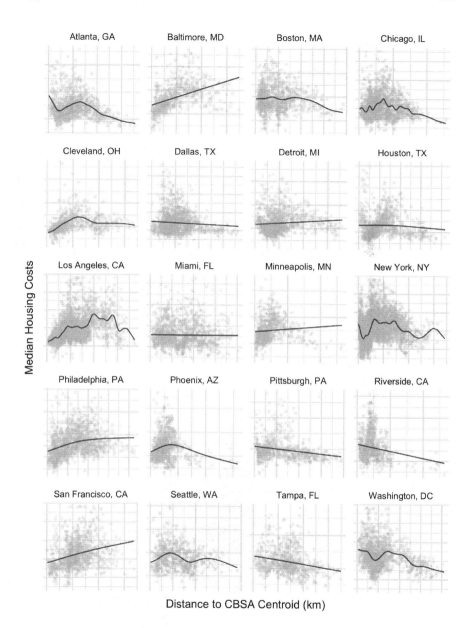

Distance to CBSA Centroid (km)

FIGURE 4.12: Scatter plots showing the distance to the city center and median housing costs for census tracts in the largest 20 community based statistical areas (CBSA). The line shows a cubic smoothing spline fit tuned using leave-one-out cross-validation. The axes are rescaled to match the ranges for each city.

```
spar <- spar_vals[which.min(vals)]
res <- smooth.spline(X, y, spar = spar)

res
}
```

Running this cross-validation routine on the data from Seattle-Tacoma-Bellevue, we see that our previous model was overfit to the training data. The new model performs worse on the training set but works significantly better on the validation and testing sets.

```
res <- cv_smooth_spline(X_cbsa[train_cbsa == 1],
                        y_cbsa[train_cbsa == 1])

y_hat <- predict(res, X_cbsa)$y
rmse(y_cbsa, y_hat, train_cbsa)
```

```
         1        2        3
  4952.104 4657.814 4718.134
```

Running this same routine for the largest 20 CBSA regions in the United States shows significantly different results across regions, as seen in Figure 4.12. In Baltimore, prices tend to increase steadily as the distance increases away from the city, likely as a result of (to the south) getting closer to the more expensive Washington, DC metropolitan area. San Francisco has the same relationship, partially due to the effect of Silicon Valley on the southern edge of the CBSA. Several of the largest cities (New York, Chicago, and LA) exhibit non-linear relationships, most likely as specific distances correspond to specific desirable or undesirable neighborhoods. Most importantly for our study, notice how well smoothing splines perform with no specific tuning needed for predicting a variety of relationships. For this reason, it is the most commonly used linear smoother to wrap within other algorithms. We will see a specific example of this when looking at additive models in Chapter 6.

4.10 Exercises

1. Write a function that uses gradient descent to solve for the optimal values of β_1 and β_2 to minimize the ordinary least squares loss given by Equation 4.1.

2. It is possible to use the `stats::poly` function directly in the formula interface to `stats::lm`. Write a test script to verify that our functions

`casl_nlm1d_poly` and `casl_nlm1d_poly_predict` work as expected compared to this formula interface.

3. Write a test script to verify that `casl_nlm1d_kernel` works as expected.

4. The formula in Equation 4.18 gives us a way to compute the effective degrees of freedom for a linear smoother given its hat matrix. Write functions `knn_edf`, `local_reg_edf`, and `kernel_reg_edf` that return the effective degrees of freedom given a training dataset X and the relevant tuning parameters.

5. Take advantage of banded, symmetric positive definite matrices Q and T in Section 4.6 to improve the performance of the `smspline` algorithm, and roughly estimate your performance savings in arithmetic operations for a given problem size. (Hint: consider using Cholesky factorization.)

6. Describe what happens to the `smspline` algorithm defined in Section 4.6 when the abscissae are not unique. Propose reasonable solutions to handling this problem. What does R's `smooth.spline` function do in this situation?

7. Using the testing dataset we used in this chapter and the functions you wrote in the previous question (`knn_edf`, `local_reg_edf`, and `kernel_reg_edf`), plot the testing error of all three estimators as a function of the effective degrees of freedom. What do you observe about their performance on this scale?

8. Set up a more formal test to evaluate the numerical error in using the truncated power basis compared to the B-spline basis. Use cubic splines but change the number of data points and the number of knots. At what limits does it appear that the noise degrades the predictive power of the method?

9. Kernels can also be used as density estimators. Specifically, we have

$$f_h(x) = \frac{1}{n} \sum_i K_h(x - x_i). \qquad (4.70)$$

In this setting we see again why it is important to have the integral of the kernel equal to 1. Write a function `kern_density` that accepts a training vector `x`, bandwidth `h`, and test set `x_new`, returning the kernel density estimate from the Epanechnikov kernel. Visually test how this performs for some hand-constructed datasets and bandwidths.

10. In Chapter 6 we will see how to extend linear smoothers to higher dimensions. Some smoothers, however, are easy to extend in a naïve way. For example, one could replace the absolute value in Equation 4.10 with the ℓ_2-distance between points. Write a function `kernel_reg_2d` that extends `kernel_reg` to two-dimensional input matrices `x` and `x_new`. Test your function on data distributed as $y = cos(x_1) + x_2^2$ under Gaussian noise.

11. We know that the matrix supplied by our function `casl_nlm1d_bspline_x` will be a sparse, banded matrix. Rewrite the function to return a sparse matrix making sure to never form the full dense matrix internally.

5

Generalized Linear Models

5.1 Classification with linear models

Many of the most interesting problems in statistical learning require the prediction of a response variable that takes on categorical values. Examples include spam detection, image recognition, predicting the outcome of a sporting event, and estimating the potential for a cataclysmic weather event. Our methods have so far focused only on the estimation of a continuous response variable. The flexibility of the linear model, fortunately, allows us to extend it in ways capable of estimating categorical responses.

To start, consider a task where the response variable takes on only two categories. In Section 5.5 we will see how to extend this setup to multiple classes. As a first step, we can create a numeric response vector Y the same way we constructed indicator variables for the matrix X. For example, in a classifier to predict the outcome of a sporting event we might define Y as

$$y_i = \begin{cases} 1, & \text{category}_i = \text{win} \\ 0, & \text{category}_i \neq \text{win} \end{cases}. \tag{5.1}$$

Assuming we also have a model matrix X there is no algorithmic reason that we cannot directly apply linear regression using this binary response Y. In fact, we could apply the regularization functions in Chapter 3 and the linear smoothers from Chapter 4 to this data as well. How can we use the output of such a model to predict the classes on a new dataset? The fitted values $\widehat{Y}$ will not be exactly equal to 0 or 1. We therefore discretize the fitted values according to

$$z_i = \begin{cases} 1, & \widehat{y}_i \geq 0.5 \\ 0, & \widehat{y}_i < 0.5 \end{cases} \tag{5.2}$$

and assign the predicted classes according the values z_i. Splitting by the value of 0.5 is reasonable being halfway between 0 and 1, though of course it is possible to pick a different cutoff value depending on the needs of a particular problem.

The strategy of converting the response to a numeric vector and discretizing the output is both important and powerful. It makes it possible to turn any generic algorithm for the prediction of a continuous response into an algorithm for predicting a categorical one. Not only is the algorithm unchanged, but the implementation also works without any modification. All that is needed are trivial pre- and post-processing steps. Empirically, it has been observed that this strategy produces reasonably good estimates for classification on new datasets.

Many of the theoretical properties of linear regression are broken if we have a binary response variable Y. To see why, consider the residual vector r. For a given value of i, the residual can take on only two values. These are defined explicitly as

$$r_i \in \left\{ 1 - x_i^t \beta, 0 - x_i^t \beta \right\}. \tag{5.3}$$

The dependency of r_i on the distribution of x_i breaks the assumption that the residuals are identically distributed. The variance of each residual is also dependent on the predictors according to the relationship

$$Var(y_i) = (x_i^t \beta) \cdot (1 - x_i^t \beta). \tag{5.4}$$

The variance is highest when the predicted probability is near 0.5 and lowest as it approaches 0 or 1. Put together, this means that the OLS procedure yields unbiased estimators but that these are no longer efficient. That is, there is a uniformly better way of computing the regression coefficients.

Another concern with applying linear models to classification tasks is that the linear model assumption

$$\mathbb{E}(y_i) = x_i^t \beta \tag{5.5}$$

can only hold if the data matrix X and regression vector β are constrained to yield predictions between 0 and 1. This is a not strictly an algorithmic problem; the discretization in Equation 5.2 can be easily applied to predictions outside of this range. However, if broken, the theoretical conception of the linear model no longer applies. In order to rectify this problem, we can modify the linear regression assumption by the use of a fixed real-valued function g,

$$\mathbb{E}(y_i) = g^{-1}(x_i^t \beta). \tag{5.6}$$

The function g is known as a *link function*. For classification we would like an inverse link function that takes any real number and maps it into the range $[0, 1]$. Why the whole range and not just the two values 0 and 1? While a specific observation must be one of these two integers, the expected value can fall anywhere in between these two extremes. The expected value of y_i is the probability that it will be equal to one and it is the probability that we want to model with our model. One choice of a function that maps to values in $[0, 1]$ is the inverse of a cumulative distribution function. Another option, by far the most popular, is the logit function. The inverse logit function is defined as

$$g^{-1}(z) = \frac{1}{1 + e^{-z}} = \frac{e^z}{1 + e^z}. \tag{5.7}$$

We will see in Section 5.2 where this particular choice comes from.

The relationship described in Equation 5.6 describes a class of estimators known as *generalized linear models*, or GLMs. A general approach to the solution of such models was first described by Fisher in 1935 [56]. The phrase "generalized linear model" as described here was defined by Nelder and Wedderburn in 1972 [124]. Our approach to these models generally follows the book by McCullagh and Nelder [118]. Recall that one way to motivate ordinary least squares is as the maximum likelihood estimator of a linear model with normally distributed errors. In a similar fashion, we can estimate the parameter β from a generalized linear model using maximum likelihood estimators. Here, we will derive the maximum likelihood equations when the response variable y is given by a Bernoulli distribution using the logit link function from Equation 5.7. The likelihood function is given by:

$$\mathcal{L}(y) = \prod_{i=1}^{n} p_i^{y_i} \cdot (1 - p_i)^{1-y_i} \tag{5.8}$$

where p_i is the probability that y_i is equal to one. The probability p_i is a function of both x_i and β. Taking logarithms and expanding the probabilities we have

$$l(y) = \sum_{i=1}^{n} y_i \cdot \log(p_i) + (1 - y_i) \cdot \log(1 - p_i) \tag{5.9}$$

$$= \sum_{i=1}^{n} \log(1 - p_i) + y_i \cdot \log\left(\frac{1 - p_i}{p_i}\right) \tag{5.10}$$

$$= \sum_{i=1}^{n} -\log(1 + e^{x_i^t \beta}) + y_i(x_i^t \beta). \tag{5.11}$$

Taking the derivative with respect to β_j gives

$$\frac{\partial l(y)}{\partial \beta_j} = -\sum_{i=1}^{n} \frac{e^{x_i^t \beta}}{1 + e^{x_i^t \beta}} \cdot x_{i,j} + y_i x_{i,j} \tag{5.12}$$

$$= \sum_{i=1}^{n} (y_i - p_i) x_{i,j}. \tag{5.13}$$

We can write this equation in a vector form as

$$\nabla_\beta l(y) = X^t(y - p). \tag{5.14}$$

The gradient with respect to β of the log-likelihood is known as the *score function*. To find the maximum likelihood estimators we need to find where the gradient is equal to zero. Notice that at the minimizer the residuals $y - p$ are perpendicular to the column space of X. This geometric property also defines the ordinary least squares solution through the normal equations.

Unlike solving the ordinary least squares optimization problem, there is no direct algorithm to find the maximum likelihood estimators for β in logistic regression. This can be seen by trying to set the gradient, Equation 5.14, equal to zero and noticing that the system of equations does not allow a sequence of algebraic manipulations in which the terms β_j are isolated on one side of the equation. Instead, an iterative algorithm must be used. The most common approach for solving GLMs is the Newton–Raphson method. The idea behind the method is to estimate the function of interest by a quadratic approximation at a given point. Because a quadratic function has a unique optimal value, the next iteration can jump to the optimal value implied by the approximation. Subsequent steps repeat this process until convergence is reached. Symbolically, the update steps for our task can be written as

$$\beta^{(k+1)} = \beta^{(k)} - H^{-1}(l)(\beta^{(k)}) \cdot \nabla_\beta(l)(\beta^{(k)}) \tag{5.15}$$

where $\beta^{(k)}$ is the estimate at the end of the kth iteration and H is the Hessian matrix of second derivatives. As this is a second order method, it generally converges very quickly to a stationary point. In order to use the Newton–Raphson method, we need to compute the Hessian matrix. The elements are given by

$$\frac{\partial^2 l(y)}{\partial \beta_k \partial \beta_j} = -\sum_i x_{i,j} \frac{\partial p_i}{\partial \beta_k} \tag{5.16}$$

$$= -\sum_i x_{i,j} \left(\frac{1}{1 + e^{-x^t \beta}} \right)^2 (-x_{i,k}) \cdot e^{-x_i^t \beta} \tag{5.17}$$

$$= -\sum_i x_{i,j} x_{i,k} p_i (1 - p_i). \tag{5.18}$$

The Hessian matrix can be written as the inner-product of the matrix X weighted by the variance of each observation

$$H(l) = X^t \cdot D \cdot X \tag{5.19}$$

where D is a diagonal matrix with elements

$$D_{i,i} = p_i \cdot (1 - p_i). \tag{5.20}$$

For now, we will assume that inverting this matrix is not an issue. In Section 5.3 we again take up the issue of numerical stability.

With the score function and Hessian matrix, we are now able to write an implementation of logistic regression. We will supply a maximum iteration number as well a basic stopping criterion when the updates seem to no longer be changing up to a tolerance.

```
# Logistic regression using the Newton-Ralphson method.
#
```

```
# Args:
#     X: A numeric data matrix.
#     y: Response vector.
#     maxit: Integer maximum number of iterations.
#     tol: Numeric tolerance parameter.
#
# Returns:
#     Regression vector beta of length ncol(X).
casl_glm_nr_logistic <-
function(X, y, maxit=25L, tol=1e-10)
{
  beta <- rep(0,ncol(X))
  for(j in seq(1L, maxit))
  {
    b_old <- beta
    p <- 1 / (1 + exp(- X %*% beta))
    W <- as.numeric(p * (1 - p))
    XtX <- crossprod(X, diag(W) %*% X)
    score <- t(X) %*% (y - p)
    delta <- solve(XtX, score)
    beta <- beta + delta
    if(sqrt(crossprod(beta - b_old)) < tol) break
  }
  beta
}
```

We now generate some random data to test this function with.

```
n <- 1000; p <- 3
beta <- c(0.2, 2, 1)
X <- cbind(1, matrix(rnorm(n * (p- 1)), ncol = p - 1))
mu <- 1 / (1 + exp(-X %*% beta))
y <- as.numeric(runif(n) > mu)
```

The function stats::glm provided in R implements the Newton–Raphson algorithm for solving generalized linear models. We can compare our implementation to the one in R.

```
beta <- casl_glm_nr_logistic(X, y)
beta_glm <- coef(glm(y ~ X[, -1], family = "binomial"))
cbind(beta, as.numeric(beta_glm))
```

```
            [,1]         [,2]
[1,] -0.27892466 -0.27892466
[2,] -2.00678693 -2.00678693
[3,] -1.01560168 -1.01560168
```

We see that our method, up to the precision of the printed values, gives exactly the same result as the method provided by R. It also is close to the value of β set in our simulation. We remark that the stopping critera used above, terminating the iterations whenever the change in the solution coefficients is small enough, was chosen for simplicity and differs from typical stopping criteria in production software like R. Generalized least squares problems in practice can be phrased using a *deviance residual* term that roughly corresponds to the ordinary least squares residual norm. The method is stopped when the deviance residual falls below a threshold; see the book by McCullagh and Nelder [118] for details.

5.2   Exponential families

An exponential family is a class of probability distributions that all share a specific form of density functions. The class is primarily defined for mathematical convenience, with general results allowing for the computation of properties such as sufficient statistics and specification of closed-form conjugate priors. Both discrete and continuous distributions are contained in the family. Specifically, single-parameter distributions in the exponential family are those that can be written as

$$f(z|\theta) = h(z) \cdot \exp\left\{\eta(\theta) \cdot T(z) - A(\theta)\right\} \tag{5.21}$$

for fixed functions h, T and A. Appropriate choices for these define many of the most well-known density functions, including those associated with the normal, exponential, Poisson, binomial, gamma, and beta distributions.

The structure of exponential families make them particularly suited for describing the distribution of the response variable in a generalized linear model. Notice that there are many equivalent ways of parametrizing a given distribution as an exponential family. If the functions η and T are equal to the identity function, the family is said to be in *canonical form*. In this section, we will assume that we are working with an exponential family in canonical form. Further, we will assume that the linear predictor $X\beta$ describes the canonical parameter η. The link function then describes the relationship between a specific value $\eta_i = \eta(\theta_i)$ and the mean of y_i,

$$\mathbb{E}y_i = \mu_i = g^{-1}(\eta_i) = g^{-1}(x_i^t\beta). \tag{5.22}$$

From here we can compute the score function for a canonical exponential family, starting with the likelihood function

$$\mathcal{L}(y) = \prod_{i=1}^{n} h(y_i) \cdot \exp\left\{x_i^t\beta \cdot y_i - A(x_i^t\beta)\right\}. \tag{5.23}$$

> ## Generalized Linear Model (GLM)
>
> Let f_θ denote an exponential family parametrized by the scalar parameter θ. Define a real-valued function g, the *link function*, and assume that
>
> $$\mathbb{E}y_i = g^{-1}(x_i^t \beta), \quad y_i \sim f_\theta$$
>
> for all i and some fixed regression vector β. This defines a generalized linear model. When not otherwise specified, estimates $\widehat{\beta}$ are computed by maximum likelihood.

The log-likelihood is then

$$l(y) = \sum_{i=1}^{n} x_i^t \beta \cdot y_i - A(x_i^t \beta) + log(h(y_i)), \tag{5.24}$$

with first derivatives given by

$$\frac{\partial l}{\partial \beta_j} = \sum_{i=1}^{n} x_{i,j} \cdot \left[y_i - A'(x_i^t \beta) \right] \tag{5.25}$$

where A' is the first derivative of the function A. The second derivatives are derived similarly as

$$\frac{\partial^2 l}{\partial \beta_k \partial \beta_j} = \sum_{i=1}^{n} x_{i,j} x_{i,k} \cdot \left[-A''(x_i^t \beta) \right]. \tag{5.26}$$

While we could directly plug these into the Newton–Raphson equations, it will be instructive to first understand exactly what derivatives of the function A mean in terms of the exponential family.

When a single-parameter exponential family is put into canonical form, the first derivative of A with respect to η gives the expected value of the random variable. Plugging this into the score function from Equation 5.25 yields

$$\nabla_\beta l = X^t(y - \mathbb{E}y), \tag{5.27}$$

which reduces to our score function for logistic regression when Y is distributed as a Bernoulli random variable. Likewise, the second derivative of A with respect to η yields the variance. The Hessian matrix then becomes

$$H = X^t \cdot D \cdot X, \tag{5.28}$$

where D is a diagonal matrix with elements given by

$$D_{i,i} = Var(y_i). \tag{5.29}$$

Formula 5.28 again simplifies to Formula 5.19 when applied to logistic regression.

We will now implement the Newton–Raphson steps for solving a generalized linear model with a response described by a canonical exponential family. This mostly follows the logistic regression function, with a few new input parameters.

```
# Solve generalized linear models with Newton-Ralphson method.
#
# Args:
#     X: A numeric data matrix.
#     y: Response vector.
#     mu_fun: Function from eta to the expected value.
#     var_fun: Function from mean to variance.
#     maxit: Integer maximum number of iterations.
#     tol: Numeric tolerance parameter.
#
# Returns:
#     Regression vector beta of length ncol(X).
casl_glm_nr <-
function(X, y, mu_fun, var_fun, maxit=25, tol=1e-10)
{
  beta <- rep(0,ncol(X))
  for(j in seq_len(maxit))
  {
    b_old <- beta
    eta    <- X %*% beta
    mu     <- mu_fun(eta)
    W      <- as.numeric(var_fun(mu))
    XtX    <- crossprod(X, diag(W) %*% X)
    score <- t(X) %*% (y - mu)
    delta <- solve(XtX, score)
    beta  <- beta + delta
    if(sqrt(crossprod(beta - b_old)) < tol) break
  }
  beta
}
```

Notice that this function takes functions as inputs. Users must specify the exponential family by supplying the mean and variance functions.

To test our new function, we generate some Poisson distributed data.

```
n <- 5000; p <- 3
beta <- c(-1, 0.2, 0.1)
X <- cbind(1, matrix(rnorm(n * (p- 1)), ncol = p - 1))
eta <- X %*% beta
```

```
lambda <- exp(eta)
y <- rpois(n, lambda = lambda)
```

We now run this data through our function and compare it to the results from
the `stats::glm` function

```
beta_hat <- casl_glm_nr(X, y,
                        mu_fun = function(eta) exp(eta),
                        var_fun = identity)
beta_glm <- coef(glm(y ~ X[,-1], family = "poisson"))
cbind(beta, beta_hat, as.numeric(beta_glm))
```

```
[1,] -1.0 -0.9823415 -0.9823415
[2,]  0.2  0.2069349  0.2069349
[3,]  0.1  0.1023110  0.1023110
```

Again, up to the accuracy of the print out, we see that our implementation
matches the output of the R function. The estimates are also quite close to
the β vector in the simulation.

5.3 Iteratively reweighted GLMs

Our implementation for solving generalized linear models matches the output
from R very well. However, if you unpack the core code in the `stats::glm`
function it will look quite different from our algorithm. In fact, the R implementation is applying a nearly identical algorithm re-written in a more
compact form. Specifically, the Newton–Raphson updates are re-written to
look like a weighted least squares problem. This compact form is commonly
presented in other texts and it is useful to derive it from our formulation.
Specifically, take the update function given by:

$$\beta^{(k+1)} = \beta^{(k)} - H^{-1}(l)(\beta^{(k)}) \cdot \nabla_\beta(l)(\beta^{(k)}) \tag{5.30}$$

Now plug in Formula 5.27 for the gradient and Formula 5.28 for the Hessian
as follows

$$\beta^{(k+1)} = \beta^{(k)} - H^{-1}(l)(\beta^{(k)}) \cdot \nabla_\beta(l)(\beta^{(k)}) \tag{5.31}$$

$$\beta^{(k+1)} = \beta^{(k)} + \left[X^t \cdot W \cdot X\right]^{-1} \cdot X^t(y - \mathbb{E}y^{(k)}) \tag{5.32}$$

where W is defined as a diagonal matrix of the variances of the responses

$$W_{i,i} = Var(y_i^{(k)}). \tag{5.33}$$

The update rule further simplifies if we define a vector a such that

$$a_i = \left(\frac{y_i - \mathbb{E}y_i^{(k)}}{Var(y_i^{(k)})} \right). \tag{5.34}$$

This yields

$$\beta^{(k+1)} = \beta^{(k)} + \left[X^t W X \right]^{-1} \cdot \left(X^t W^t \right) \cdot a \tag{5.35}$$

$$= \left[X^t W X \right]^{-1} X^t W^t \cdot \left\{ X\beta^{(k)} + a \right\} \tag{5.36}$$

$$= \left[X^t W X \right]^{-1} X^t W^t z, \tag{5.37}$$

where z has been defined as

$$z = X\beta^{(k)} + a. \tag{5.38}$$

The quantity z is called the *effective response*. Notice that if our estimates of y, $y^{(k)}$, are unbiased for y, then the term a in the definition of the effective response has zero mean. We can regard a as a noise term and therefore the signal component of z is a linear function of the X's.

The iterative updates described in Equation 5.37 lead to the iteratively reweighted least squares algorithm. It is algorithmically equivalent to the Newton-Ralphson steps we derived earlier, when using a canonical link function, but does yield a helpful way of thinking about exactly what the iterative algorithm is doing with the data. It is this form of the algorithm that is implemented in R and that is usually presented in textbooks. We feel that it is beneficial to see the precise algorithmic derivation in Section 5.2 before moving on to the method motivated by these heuristics.

It is occasionally useful to use an exponential family that is not in the canonical format. The iteratively re-weighted least squares algorithm can still be used in this case, however the equations for the weights and effective response need to be slightly updated. Namely, for all i from 1 to n, we have

$$W_{i,i} = \mu'(x_i^t \beta)^2 / Var(y_i^{(k)}) \tag{5.39}$$

$$z_i = X\beta + \frac{y_i - \mu(x_i^t \beta)}{\mu'(x_i^t \beta)}, \tag{5.40}$$

and the off-diagonal elements of $W_{i,i}$ remain equal to zero. When not in canonical form, the derivative of the mean function need not be equal to the variance function. When these are the same, we see that these equations reduce to the form given in Equation 5.37.

If we built a function to implement iteratively re-weighted least squares for a generic exponential family, we would now need three functions: the mean, the variance, and the derivative of the mean. Passing all of these parameters can become burdensome and error prone. R has a clever solution to this issue. Rather than passing all of these individual functions, R has objects describing

Iteratively Re-Weighted Least Squares (IRWLS)

Let y and X be the response and model matrix from a GLM with link function g over the exponential family f_θ. Denote an initial guess of the regression vector β as $\beta^{(0)}$. Also, for any non-negative integer k, denote $\widehat{y}_i^{(k)}$ to be a random variable distributed as f_θ where θ is chosen such that $\mathbb{E}\widehat{y}_i^{(k)}$ is equal to $g^{-1}(x_i^t \beta^{(0)})$. Then, recursively define for all $i = 1, \ldots, n$

$$\widehat{y}_i^{(k)} = g^{-1}(x_i^t \widehat{\beta}^{(k)})$$

$$W_{i,i}^{(k)} = (g^{-1})'(x_i^t \widehat{\beta}^{(k)})^2 / Var(\widehat{y}_i^{(k)})$$

$$z^{(k)} = x_i^t \widehat{\beta}^{(k)} + \frac{y_i - g^{-1}(x_i^t \widehat{\beta}^{(k)})}{(g^{-1})'(x_i^t \widehat{\beta}^{(k)})}$$

and as the next value $\widehat{\beta}^{(k+1)}$ as

$$\widehat{\beta}^{(k+1)} = (X^t W^{(k)} X)^{-1} X^t (W^{(k)})^t z^{(k)}.$$

Compute k^* as the number of iterations until either convergence or a maximum value has been hit. That is,

$$k^* = \arg\min\left\{\{k_{max}\} \cup \left\{k \,:\, |\widehat{\beta}^{(k+1)} - \widehat{\beta}^{(k)}| < \epsilon\right\}\right\}$$

for some pre-defined k_{max} and numerical tolerance ϵ. The estimator $\widehat{\beta}^{(k^*)}$ is called the *iteratively re-weighted least squares* (IRWLS) estimator.

all of the functions necessary for working with an exponential family. Here is an example of the parameters available from the binomial family using a logit link function:

```
names(binomial(link = "logit"))
```

```
[1] "family"     "link"        "linkfun"   "linkinv"
[5] "variance"   "dev.resids"  "aic"       "mu.eta"
[9] "initialize" "validmu"     "valideta"  "simulate"
```

We will now implement the iteratively re-weighted least squares algorithm, allowing users to pass a single family function.

```
# Solve generalized linear models with Newton-Ralphson method.
#
# Args:
#     X: A numeric data matrix.
#     y: Response vector.
#     family: Instance of an R 'family' object.
```

```
#      maxit: Integer maximum number of iterations.
#      tol: Numeric tolerance parameter.
#
# Returns:
#      Regression vector beta of length ncol(X).
casl_glm_irwls <-
function(X, y, family, maxit=25, tol=1e-10)
{
  beta <- rep(0,ncol(X))
  for(j in seq_len(maxit))
  {
    b_old <- beta
    eta <- X %*% beta
    mu <- family$linkinv(eta)
    mu_p <- family$mu.eta(eta)
    z <- eta + (y - mu) / mu_p
    W <- as.numeric(mu_p^2 / family$variance(mu))
    XtX <- crossprod(X, diag(W) %*% X)
    Xtz <- crossprod(X, W * z)
    beta <- solve(XtX, Xtz)
    if(sqrt(crossprod(beta - b_old)) < tol) break
  }
  beta
}
```

Our algorithm is now appears much more similar to the way it is implemented in R. The R function does a better job of dealing with numerical instability and error checking, both of which we will investigate in Section 5.4.

We can verify that the `casl_glm_irwls` function works for a non-standard exponential family. We will use the inverse Cauchy cumulative distribution function (CDF) as a link between the linear response $X\beta$ and the mean of the binomial. We generate some data that uses the inverse Cauchy CDF to compute probabilities so that the learned regression vector should be close to the vector used to simulate the data.

```
n <- 1000; p <- 3
beta <- c(0.2, 2, 1)
X <- cbind(1, matrix(rnorm(n * (p- 1)), ncol = p - 1))
mu <- 1 - pcauchy(X %*% beta)
y <- as.numeric(runif(n) > mu)
```

Next, we compute the regression coefficients from this model using both the `stats::glm` function and the `casl_glm_irwls`. For comparison, we compute the model using the standard link function with `stats::glm`.

```
beta <- casl_glm_irwls(X, y,
                       family = binomial(link = "cauchit"))
```

```
beta_glm_c <- coef(glm.fit(X, y,
                    family = binomial(link = "cauchit")))
beta_glm_l <- coef(glm.fit(X, y,
                    family = binomial(link = "logit")))
cbind(beta, beta_glm_c, beta_glm_l)
```

```
[1,] 0.2360408  0.2360436  0.2273324
[2,] 1.7464101  1.7464108  1.4726255
[3,] 0.9679743  0.9679725  0.8034749
```

We see that our solution matches the `stats::glm` model to the fifth digit (the difference is mostly due to the stopping criterion). The coefficients from using the Cauchy-based link function are close to those in the simulation. Also, we see that there is a fairly large difference between the output using the Cauchy-based link function and the logit link.

5.4   (⋆) Numerical issues

We have shown that finding the maximum likelihood estimators for a generalized linear regression reduces to solving a sequence of least squares problems. In Chapter 3 we discussed at length the numerical issues that arise in solving least squares problems. These concerns and associated solutions—using either the QR-decomposition or the SVD—all apply equally as well to generalized linear models. Numerical problems do become much more of a concern with iteratively re-weighted least squares.

One difficulty is that the weights in the matrix W can become very small, even reducing to a zero value at double precision. When the variance terms become very small, the effective residuals grow prohibitively large. There are two possible quick fixes for this problem, both of which involve setting a tolerance level and checking whether weights fall below this threshold. We can either remove rows with small weights or manually convert the weights to our lower threshold. In practice, both seem to perform similarly over a fixed set of iterations. R chooses to remove small weights because that makes it easier to check for convergence. When fitting binomial models, R will warn users when probabilities become very close to zero or one. We can see this by running a simulation with extreme probabilities:

```
n <- 1000; p <- 5
beta <- rep(20, p)
X <- cbind(1, matrix(rnorm(n * (p- 1)), ncol = p - 1))
mu <- 1 / (1 + exp(-X %*% beta))
y <- as.numeric(runif(n) > mu)
beta_glm <- coef(glm(y ~ X[, -1], family = "binomial"))
```

```
Warning message:
glm.fit: fitted probabilities numerically 0 or 1 occurred
```

The binomial model is, however, fairly well behaved. Running our original code produces very similar results compared to the `stats::glm` function that is checking for and removing small weights.

```
beta <- casl_glm_irwls(X, y, family = binomial())
cbind(beta, as.numeric(beta_glm))
```

```
          [,1]      [,2]
[1,] -18.45671 -18.45671
[2,] -18.69986 -18.69986
[3,] -18.59445 -18.59445
[4,] -18.75675 -18.75675
[5,] -18.47664 -18.47664
```

Notice that our implementation closely matches the `stats::glm` function even though many of the predicted probabilities are extremely close to zero or one, which often causes numeric issues.

```
probs <- 1 / (1 + exp(-X %*% beta))
quantile(probs, c(0.1, 0.9))
```

```
          20%          80%
2.220446e-16 9.999937e-01
```

In practice, the warning message is perhaps even more beneficial than the numerical safe-guards in the algorithm, at least for the binomial model. While our naïve approach yields nearly the exact same coefficients, outside of a contrived simulation, predicted probabilities near zero or one usually indicate there is some overfitting in the model and the predicted probabilities cannot be trusted.

In each iteration of the IRWLS algorithm, we solve a least squares problem with the same matrix X. Solving the weighted normal equations is the most computationally intensive step in the algorithm. If we are clever, we can compute the QR decomposition of the model matrix X once and use this throughout each cycle of the algorithm. Here, we implement the algorithm of [127], while also checking for small weights.

```
# Generalized linear models with Newton-Ralphson and QR.
#
# Args:
#     X: A numeric data matrix.
#     y: Response vector.
#     family: Instance of an R 'family' object.
```

```
#      maxit: Integer maximum number of iterations.
#      tol: Numeric tolerance parameter.
#
# Returns:
#      Regression vector beta of length ncol(X).
casl_glm_irwls_qr <-
function(X, y, family, maxit=25, tol=1e-10)
{
  s <- eta <- 0
  QR <- qr(X)
  Q  <- qr.Q(QR)
  R  <- qr.R(QR)

  for(j in seq_len(maxit))
  {
    s_old <- s
    mu    <- family$linkinv(eta)
    mu_p  <- family$mu.eta(eta)
    z     <- eta + (y - mu) / mu_p
    W     <- as.numeric(mu_p^2 / family$variance(mu))
    wmin  <- min(W)
    if(wmin < sqrt(.Machine$double.eps))
      warning("Tiny weights encountered")
    C     <- chol(crossprod(Q, W*Q))
    s     <- forwardsolve(t(C), crossprod(Q, W*z))
    s     <- backsolve(C, s)
    eta   <- Q %*% s
    if(sqrt(crossprod(s - s_old)) < tol) break
  }
  beta <- backsolve(R, crossprod(Q, eta))
  beta
}
```

We will compute the time it takes to run this new function and compare it to our original implementation and the `stats::glm.fit` function.

```
n <- 10000; p <- 25
beta <- rep(1, p)
X <- cbind(1, matrix(rnorm(n * (p- 1)), ncol = p - 1))
mu <- 1 / (1 + exp(-X %*% beta))
y <- as.numeric(runif(n) > mu)
t1 <- system.time({casl_glm_irwls(X, y, family = binomial())})
t2 <- system.time({casl_glm_irwls_qr(X, y, family = binomial())})
t3 <- system.time({glm.fit(X, y, family = binomial()) })
cbind(a = t1, t2, t3)[1:3,]
```

	t1	t2	t3
user.self	22.081	0.081	0.086
sys.self	2.427	0.006	0.004
elapsed	24.531	0.088	0.090

We see that the new technique is much faster than our initial approach. In this example, in fact, it is slightly faster that the R implementation.

5.5 (⋆) Multi-Class regression

We started this chapter with a motivating example of building models for classification. The generalized linear regression framework offers a solution to the binary classification task while also giving linear regression solutions for other data types such as count data. Yet, we still do not have a way of doing classification with more than two classes.

In Section 5.1 we saw a way of converting any prediction algorithm capable of estimating a continuous response into an algorithm for predicting classes. There are two similarly generic strategies for converting any binary classifier into a K-class algorithm. The *one-vs-all* approach fits K binary models, one for each class. Using the entire dataset, each model predicts whether an observation is in the kth class or not. To predict new values, apply all of the models and assign whichever class has the largest probability. The *one-vs-one* technique fits a model for each pair of classes, a total of $K(K-1)/2$ total models. When predicting a class on a new sample, each model is fit and assigned a vote for whichever class the sample most likely belongs to. The class with the most votes is assigned as the prediction. Both of these techniques usually work well when the number of samples is evenly distributed and the number of classes is not too large.

There are also methods for directly performing multi-class classification. The multinomial model, for example, extends logistic regression to deal with multiple classes and provides predicted probabilities for each class. Rather than deriving it here, however, we will delay discussion of the multinomial until it comes up again in the context of neural networks (see Section 8.7). The standard implementation in R in fact comes from the `multinom` function in the **nnet** (Neural Network) package. Trying to work out the equations for the multinomial directly takes a bit of work but comes along 'for free' when solving neural networks. For these models, dealing with multiple classes requires no special machinery but is a natural consequence of the model structure.

5.6   Implementation and notes

R provides a fast algorithm for running generalized linear models with the `stats::glm` function. Family functions provide access to the most popular exponential families and link functions. Using improper link functions, such as a binomial with a log link, can be tricky out of the box. The starting guess used by R often yields invalid models and there is no mechanism for determining a valid set. Otherwise, the function performs very well for datasets that fit comfortably in memory. Almost all of the input options such as sample weights and offsets provided by `stats::lm` are also included in `stats::glm`. The methods available for `stats::lm` models (e.g., `coef`, `predict`, `confint`) have also been extended to work with generalized linear models. An implementation of iteratively re-weighted least squares with sparse matrices is given by the `glm4` function in the **MatrixModels** package.

Our discussion here focused on exponential families parametrized by a single unknown value θ. In many applications one may wish to use a larger class that includes more than one parameter. The most straightforward way to manage this is to assume that all but one of the parameters are fixed (but unknown) constants. These constants can be learned by maximum likelihood. In fact, this is exactly what is done with linear regression. We assume the errors are normal with a mean given as a linear function of X and a fixed, unknown noise variance σ^2. The `stats::glm` function supports this approach for the normal and Gamma families. The package **MASS** provides `glm.nb` to extend this to the negative binomial with a fixed, unknown dispersion. Further extensions are provided by contributed packages; notably **gamlss** provides the ability to model up to four unknown parameters as either constants or as a function of their own data matrices. Packages such as **arm**, **bayesm**, and **MCMCglmm** provide functions to running Bayesian generalized linear regression models. Finally, **glm2** refines the base `stats::glm` function to help with convergence issues, **brglm** includes a bias reduction method for binary regression, and **safeBinaryRegression** provides a function that overloads `stats::glm` to test for the existence of the MLE estimators.

In Chapters 6 and 7 we will study additive models and the elastic net. The packages associated with these models are **mgcv** and **glmnet**, respectively. These extend the regularization methods of Chapter 3 and the smoothing methods of Chapter 4. The 'g' in both package names stands for generalized in the same sense as a generalized linear model. Both support a wide variety of exponential families and corresponding link functions. Going forward, these will allow us to apply techniques to both continuous and categorical responses.

5.7   Application: Chicago crime prediction

5.7.1   Data and setup

The City of Chicago data portal provides open access to many datasets produced by the city. One of the largest and most well-known is the Chicago Crime dataset. This contains one row for any crime reported, by either the police or through public tips, within the city limits. Variables include the address of the reported crime, a description of the crime type, the neighborhood name, a timestamp, and whether the crime resulted in an arrest. Data are available over any timespan over the past ten years. Visitors to the website can download the data as a plain-text csv file. Here, we have grabbed a subset of the data from the first six months of 2017. Further filtering and summarizing of the data was done within each of the following applications as needed.

5.7.2   Binomial: Predicting arrests

Our first task is to use the covariates to determine whether or not a crime resulted in an arrest. Each row of the dataset here consists of a single reported crime. We have filtered it to only include burglaries, robberies, assaults, and narcotics violations. The data has also been reduced in size by taking only crimes listed as occurring in a residence, on the street, or in a restaurant. Additional variables include whether this was a domestic crime (committed by someone in their own home), the month, and the hour of the crime bucketed into four groups. Here are the first few rows of the dataset:

```
chi_crimes <- read.csv("data/chi_crimes.csv", as.is=TRUE)
head(chi_crimes[, c("arrest_flag", "domestic", "hour", "loc",
                    "month", "type")])
```

	arrest_flag	domestic	hour	loc	month	type
1	0	false	(3,17]	Residence	1	Burglary
2	0	false	(3,17]	Street	1	Robbery
3	0	true	(3,17]	Residence	1	Assault
4	1	false	(3,17]	Street	1	Narcotics
5	0	true	(3,17]	Street	1	Assault
6	0	false	(21,24]	Street	1	Robbery

A table of the crime type and arrest flag shows a strong relationship between these two variables.

```
table(chi_crimes$arrest_flag, chi_crimes$type)
```

```
    Assault Burglary Narcotics Robbery
0    2858     2110         1    1672
1     566       80      1745      92
```

All but one of the narcotics violations resulted in an arrest compared to only 3.6% of burglaries.

We start by fitting a binomial model with the standard canonical logit link function on this dataset. We will use all of the available variables, converting month into a categorical covariate.

```
model <- glm(arrest_flag ~ domestic + hour + loc +
                          factor(month) + type,
             data = chi_crimes,
             family = binomial)
chi_crimes$glm_pred <- predict(model,
                              newdata = chi_crimes,
                              type = "response")


summary(model)
```

```
Coefficients:
                 Estimate Std. Error z value Pr(>|z|)
(Intercept)      -1.92770    0.15170 -12.708  < 2e-16 ***
domestictrue      0.16654    0.10623   1.568   0.1170
hour(21,24]       0.27310    0.14330   1.906   0.0567 .
hour(3,17]       -0.16579    0.10090  -1.643   0.1003
hour[0,3]        -0.01052    0.13737  -0.077   0.9390
locRestaurant     0.91385    0.14114   6.475 9.49e-11 ***
locStreet         0.17549    0.10255   1.711   0.0870 .
factor(month)2    0.28542    0.14765   1.933   0.0532 .
factor(month)3    0.30909    0.14335   2.156   0.0311 *
factor(month)4    0.02621    0.14600   0.180   0.8575
factor(month)5    0.19841    0.13904   1.427   0.1536
factor(month)6    0.05895    0.14033   0.420   0.6744
typeBurglary     -1.49858    0.14146 -10.594  < 2e-16 ***
typeNarcotics     9.13464    1.00039   9.131  < 2e-16 ***
typeRobbery      -1.31703    0.12443 -10.585  < 2e-16 ***
```

The regression table looks very similar to the regression tables we built for linear models in Chapter 2. For each component of β we have an estimate and standard error. There is also a test statistic and p-value corresponding to a hypothesis test with a null hypothesis that β_j is equal to zero. An exact interpretation of these values is more difficult, however, because there is not a clear one-to-one relationship between these factors and how they affect the predicted probabilities. We can make some inferences based on the signs and magnitudes of the variables. For example, we see that crimes occurring in restaurants are much more likely to lead to an arrest relative to those on the street. As we saw in the table, a narcotics violation has the largest regression coefficient. All of the variables are categorical, so scaling of the input variables does not matter, so we can infer from this that a narcotics violation is the strongest single predictor of an arrest. The month of March, for some reason, had a higher arrest rate than all of the other months.

As mentioned in Section 5.1, it is possible to fit a linear model to the binary response variable. We will run a linear regression here on the same variables and compare the results to the logistic regression.

```
model <- lm(arrest_flag ~ domestic + hour + loc +
                    factor(month) + type,
            data = chi_crimes)
chi_crimes$lm_pred <- predict(model,
                        newdata = chi_crimes)
summary(model)
```

```
Coefficients:
                 Estimate Std. Error t value Pr(>|t|)
(Intercept)     0.1432452  0.0112023  12.787  < 2e-16 ***
domestictrue    0.0150026  0.0097908   1.532   0.1255
hour(21,24]     0.0179071  0.0103515   1.730   0.0837 .
hour(3,17]     -0.0110010  0.0070357  -1.564   0.1179
hour[0,3]      -0.0024178  0.0096894  -0.250   0.8030
locRestaurant   0.0940146  0.0127730   7.360 1.99e-13 ***
locStreet       0.0122636  0.0081290   1.509   0.1314
factor(month)2  0.0177739  0.0097782   1.818   0.0691 .
factor(month)3  0.0215002  0.0096786   2.221   0.0263 *
factor(month)4  0.0007064  0.0095588   0.074   0.9411
factor(month)5  0.0127663  0.0092906   1.374   0.1694
factor(month)6  0.0026389  0.0093095   0.283   0.7768
typeBurglary   -0.1152318  0.0093131 -12.373  < 2e-16 ***
typeNarcotics   0.8373250  0.0086281  97.046  < 2e-16 ***
typeRobbery    -0.1133421  0.0085709 -13.224  < 2e-16 ***
```

Notice that the test statistic values are very similar to those in the logistic regression. Strongly positive terms here are strongly positive in the other model; the same holds for strongly negative terms as well. The specific coefficients, on the other hand, are much different. This should not be surprising as the connection between the coefficients in the linear model and the predicted values is different. We can loosely interpret the predicted values from the linear model as probabilities of a crime resulting in arrest. Although the model assumptions are broken by this model, it is in some ways easier to reason about the meaning of these coefficients than in the logistic regression case. For example,

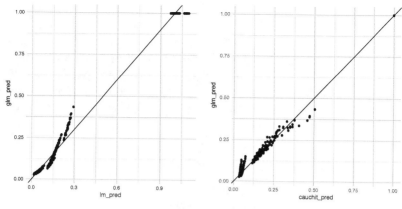

(a) Linear model vs. logistic model   (b) Cauchy-link vs. logit-link

FIGURE 5.1: Scatter plot of predicted probabilities of a crime leading to arrest from three different models. Each model has the same covariates but uses a different link function or exponential family to describe the output.

we see that a crime occurring in a restaurant is about 9.4% more likely to result in arrest than one occurring in a residence, all other details being held constant. Burglaries are 11% less likely to lead to an arrest than an assault.

To illustrate the similarities and differences, we fit a second binomial model with a Cauchy link function.

```
model <- glm(arrest_flag ~ domestic + hour + loc +
                          factor(month) + type,
             data = chi_crimes,
             family = binomial(link = "cauchit"))
chi_crimes$cauchit_pred <- predict(model,
                                   newdata = chi_crimes,
                                   type = "response")
```

A comparison of the predicted values from our three models is shown in Figure 5.1. Generally, there is a high correlation between the predicted values produced by all three models. The linear model differs most distinctly from the logistic function for points with a very high probability of arrest. These are all likely narcotics violations. The logistic functions pushes these all to a number very close to 1 while the linear model spreads these out from about 95% to 105%. Here, the form of the logistic model is helping create a better model that does not give illogical predictions greater than 100% and reasonably assigns all narcotics crimes to a percentage of about 99%. This structure is impossible for the linear model to replicate. The logistic regression and Cauchy-based binomial model visually agree on the points with a very high probability of arrest. They differ most strongly on points with a low probability of arrest. The heavy tails in the Cauchy link function lead to more extreme probabilities on the lower end (the predictions near 1 are also more extreme, but this is not visible at the scale of the plot).

In this example we see that the largest departures between the models occur for extreme values of the probabilities. We have purposely chosen an example here that has predicted values over a wide range of probabilities. If the probabilities are all concentrated over a small range, the three models will usually provide very similar results.

5.7.3   Poisson: Predicting crime counts

In the 1920s, the City of Chicago was partitioned into 77 divisions known as 'community areas.' These have no political jurisdiction, but were established by the presence of physical boundaries and census data. They have been used for urban planning and are officially recognized by the city government. For this analysis, we once again filtered crimes to only include burglaries, robberies, assaults, and narcotics violations. We then grouped all reported crimes by crime type, hour, month and community area. Here are the first few rows of data.

```
ca <- read.csv("data/ca.csv", as.is=TRUE)
head(ca)
```

	community_area	hour	month	type	n	hourb
1	44	5	1	Robbery	3	(3,17]
2	71	5	1	Robbery	0	(3,17]
3	1	5	1	Robbery	0	(3,17]
4	25	5	1	Robbery	1	(3,17]
5	24	5	1	Robbery	0	(3,17]
6	48	5	1	Robbery	0	(3,17]

For each group we have counted the number of crimes that occurred in a given grouping. Our goal is to model the number of crimes as a function of the month, bucketed hour, and crime type.

The variable of interest in this task is a count value. It seems reasonable then to fit a generalized linear model using a Poisson distribution.

```
model <- glm(n ~ factor(hour) + factor(month) + type,
             data = ca,
             family = poisson())
ca$pred_pois <- predict(model,
                        data = ca,
                        type = "response")
summary(model)
```

```
Coefficients:
                Estimate Std. Error z value Pr(>|z|)
(Intercept)      0.18689    0.02062   9.064  < 2e-16 ***
hourb(21,24]    -0.26371    0.02464 -10.703  < 2e-16 ***
hourb(3,17]     -0.37929    0.01545 -24.545  < 2e-16 ***
hourb[0,3]      -0.63586    0.02206 -28.824  < 2e-16 ***
factor(month)2  -0.15597    0.02221  -7.023 2.17e-12 ***
factor(month)3  -0.08319    0.02178  -3.819 0.000134 ***
factor(month)4  -0.05613    0.02163  -2.595 0.009470 **
factor(month)5   0.05809    0.02102   2.763 0.005725 **
factor(month)6   0.03530    0.02114   1.670 0.094950 .
typeBurglary    -0.41975    0.01646 -25.500  < 2e-16 ***
typeNarcotics   -0.65127    0.01771 -36.779  < 2e-16 ***
typeRobbery     -0.56239    0.01721 -32.686  < 2e-16 ***
```

As was the case with logistic regression, the results roughly resemble those from a linear regression but the coefficients are harder to directly interpret. We see that there is a higher rate of crimes in May and June, and the lowest in February. The highest rate of crimes occur in the baseline hour bucket, 17:00 to 21:00, corresponding to the evening hours.

Is the Poisson distribution a reasonable model here? With binary data this is not something we need to worry about because the Bernoulli distribution provides a complete parametrization over all possible random variables distributed over two values. With count data this is not the case. One way to test our model assumption comes from realizing that we can convert count data into binary data by asking whether or not *any* crimes occurred. We can get this binary response from the observed values as well as the implied probabilities from the fitted Poisson model:

```
ca$n_one <- as.numeric(ca$n > 0)
ca$pred_pois1 <- 1 - dpois(0, lambda = ca$pred_pois)
```

With a binary response, we can fit a binomial regression using the same co-variates.

```
model <- glm(n_one ~ factor(hour) + factor(month) + type,
             data = ca,
             family = binomial())
ca$pred_binom <- predict(model, data = ca, type = "response")
```

Figure 5.2 shows a scatter plot of the predicted probabilities from the binomial model compared to the implied probabilities from the Poisson model. The implied probabilities are systematically higher in the Poisson model compared to the binomial regression.

By construction, the Poisson regression is producing fitted values $\hat{y}$ that correctly captures the mean of the response y. However, the model implied by

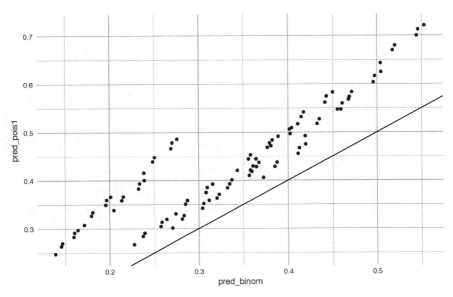

FIGURE 5.2: Predicted probabilities for whether a crimes will be reported for a particular crime type during a given hour in a one of the 77 community areas. Comparison of the logistic model and implied probabilities from a generalized linear model using the Poisson distribution showing a high degree of under-dispersion.

this is overpredicting the number of buckets where any crimes are detected. This indicates that the Poisson is not a very accurate model for this data. All of the counts are concentrated in a smaller set of buckets than is predicted by the Poisson distribution. In the context of crime counts this seems reasonable. Some community areas are more heavily policed, more heavily populated, and more prone to a particular type of crime.

5.7.4   Negative binomial: Underdispersion

In Figure 5.2, we saw the discrepancy between the logistic regression and the implied zero counts from the Poisson distribution. The phenomenon observed here is known as *underdispersion*; the counts are not as spread out as we would expect given the model, leading to more zeros than predicted by the model. The graph roughly shows two lines of data: one corresponding to narcotics violations and the second to the other crime types. It will be useful to separate these into two datasets and handle them differently because each has a different degree of underdispersion.

```
narco <- ca[ca$type == "Narcotics",]
other <- ca[ca$type != "Narcotics",]
```

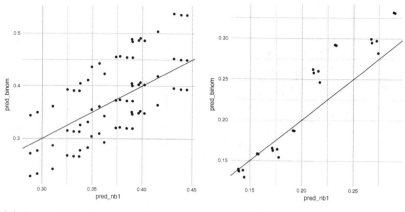

(a) Assault, Burglary, and Robberies   (b) Narcotics Violations

FIGURE 5.3: Predicted probabilities for whether crimes will be reported for a particular crime type during a given hour in a one of the 77 community areas. Comparison of the logistic model and implied probabilities from a generalized linear model using the negative binomial distribution.

We can now fit a model to each subset with an extra fixed, but unknown, term to capture the dispersion.

The negative binomial is a two-parameter exponential family that generalizes the concept of a Poisson distribution. It also has support on the non-negative integers and can be used to model count data. The extra term is used to describe the degree of dispersion. To fit a negative binomial regression, we use the glm.nb function from the **MASS** package. Here, we use it to fit the narcotics data.

```
library(MASS)
model <- glm.nb(n ~ hourb + factor(month),
                data = narco)
narco$pred_nb <- predict(model, data = narco,
                type = "response")
narco$pred_nb1 <- 1 - dnbinom(0, size = model$theta,
                mu = narco$pred_nb)
model$theta
```

```
[1] 0.1878204
```

The parameter θ reports the degree of dispersion. A value of 1 corresponds to the Poisson family, and values less than 1 indicate underdispersion. The value of 0.187 here is quite low, showing that narcotics violations are concentrated in a small number of buckets.

We can repeat this analysis with the three other crime types grouped together.

```
model <- glm.nb(n ~ hourb + factor(month),
                data = other)
other$pred_nb <- predict(model, data = other,
                         type = "response")
other$pred_nb1 <- 1 - dnbinom(0, size = model$theta,
                              mu = other$pred_nb)
model$theta
```

```
[1] 0.8099156
```

The underdispersion here is less extreme, only 0.8, but still very much present in the data.

As with the Poisson model, we can run a binomial regression on each of these datasets and then compare the probabilities implied by the negative binomial regression.

```
model <- glm(n_one ~ hourb + factor(month),
             data = narco, family = binomial())
narco$pred_binom <- predict(model, data = narco,
                            type = "response")
model <- glm(n_one ~ hourb + factor(month) + type,
             data = other, family = binomial())
other$pred_binom <- predict(model, data = other,
                            type = "response")
```

The results are shown in Figure 5.3. The non-narcotics crimes now seem to be well described, at least at 0, by the negative binomial regression. The narcotics violations perform much better than the Poisson model, though do still have a slight degree of underdispersion for values with a higher predicted probability. This could be fixed by splitting the data again, or by using a package such as **gamlss** that allows for describing variable dispersion terms as a linear function of the predictor variables.

5.8 Exercises

1. In Equation 5.2, we mentioned that other cutoffs could be used 'when appropriate.' Describe a specific situation when a low cutoff might make sense and another where a high value would be more appropriate.

2. We mentioned that the Hessian matrix in Equation 5.19 can be more ill-conditioned than the matrix $X^t X$ itself. Generate a matrix X and propa-

bilities p such that the linear Hessian $(X^t X)$ is well-conditioned but the logistic variation is not.

3. Outline and implement a basic first-order solution method for the GLM maximum likelihood problem that uses only gradient information, avoiding the Hessian matrix entirely. Describe the pros and cons of such a method compared to the Newton–Rhapson solution method.

4. In Section 5.7.4 we saw that the dispersion of the negative binomial is a function of the crime type. Use the **gamlss** package to fit a single negative binomial model with the dispersion given by the categorical variable describing crime type. How does this model compare to our crime-specific models?

5. It is possible to incorporate a ridge penalty into the maximum likelihood estimator. Modify the function `casl_glm_irwls` to include an ℓ_2-norm penalty on the regression vector for a fixed tuning parameter λ.

6

Additive Models

6.1 Multivariate linear smoothers

Our treatment of linear smoothers in Chapter 4 focused on the prediction of a response variable y by a scalar quantity x. We investigated several techniques for estimating non-linear relationships between two variables and studied the computational properties of these estimators. Most applications of statistical learning, however, are concerned with predicting a response as a function of many variables. At this point we have only shown how to compute linear models in the multivariate case. In this chapter, we extend unidimensional linear smoothers to predictions using many variables.

Several of the non-parametric methods we have studied can be extended easily to multivariate data. Both kernel regression, Section 4.3, and local regression, Section 4.4, predict new values by fitting linear models with sample weights. In theory, a new model is fit for any new input data value x_{new} by assigning weights to training points x_i based on their proximity to x_{new}, for some definition of distance. These techniques work in higher dimensions by fitting a multivariate model with sample weights that are also defined by some distance metric. For example, kernel regression can be defined as

$$d_i = ||x_{new} - x_i||_2 \tag{6.1}$$

$$w_i = K_h \left(\frac{d_i}{h} \right) \tag{6.2}$$

$$W = diag(w_1, \ldots, w_n) \tag{6.3}$$

$$\widehat{y}_{new} = x_{new}^t (X^t W X)^t X^t W y \tag{6.4}$$

$$= x_{new}^t \widehat{\beta}_{x_{new}} \tag{6.5}$$

for an given choice of the kernel K and bandwidth $h > 0$. Local regression works similarly with a different choice of the function describing the weights w_i. To illustrate this idea, we implement multivariate kernel regression.

```
# Nonlinear regression using kernel regression.
#
# Args:
#     X: A numeric data matrix.
```

```
#       y: Response vector.
#       X_new: Numeric data matrix of prediction locations.
#       h: The kernel bandwidth.
#
# Returns:
#       A vector of predictions; one for each row in X_new.
casl_nlm_kernel <-
function(X, y, X_new, h)
{
  apply(X_new, 1, function(v)
  {
    dists <- apply((t(X) - v)^2, 2, sum)
    W <- diag(casl::casl_util_kernel_epan(dists, h = h))
    beta <- Matrix::solve(Matrix::crossprod(X, W %*% X),
                          Matrix::crossprod(W %*% X, y))
    v %*% beta
  })
}
```

The Epanechnikov kernel is described in Equation 4.12. Here we are using the function `casl_util_kernel_epan` as implemented in Section 4.3.

To test our multidimensional linear smoother, we need to generate some non-linear data. Here, we generate a data matrix with three columns with all of the values randomly chosen on the unit interval. The mean of the response y is high when X is close to either the origin or to the unit vector.

```
n <- 1000; p <- 3
X <- matrix(runif(n * p), ncol = 3)
y_mean <- dnorm(apply(X^2, 1, sum) * 2) +
          dnorm(apply((X - 1)^2, 1, sum) * 2)
y <- y_mean + rnorm(n, sd = 0.1)
X_test <- matrix(runif(n * p), ncol = 3)
y_mean <- dnorm(apply(X_test^2, 1, sum) * 2) +
          dnorm(apply((X_test - 1)^2, 1, sum) * 2)
y_test <- y_mean + rnorm(n, sd = 0.1)
```

In order to test for overfitting, we will also construct an independent testing set. While relatively simple, this is an interesting data example for non-parametric regression because we would expect methods that do not find interactions between the variables to not perform very well. For example, if X_2 and X_3 are both near zero, there should be a negative relationship between X_1 and y as the largest values occur when X is near the origin. If X_2 and X_3 are close to 1 then the relationship is reversed.

Using this test data, we fit a linear regression on the training data and find predicted values on the test set. We need to include an explicit intercept term in the model.

```
beta_lm <- coef(lm.fit(cbind(1, X), y))
y_pred_lm <- cbind(1, X_test) %*% beta_lm
lm_rmse <- sqrt(mean((y_pred_lm - y_test)^2))
```

Next, we fit the kernel regression for across a set of bandwidth values. For each value we compute the root mean squared error on the training set.

```
h_vals <- c(5, 4, 3, 2, 1.5, 1, 0.75, 0.5, 0.4,
            0.2, 0.1, 0.05)
rmse_kr <- rep(NA_real_, length(h_vals))
for (i in seq_along(h_vals)) {
  y_pred_kr <- casl_nlm_kernel(cbind(1, X), y,
                               cbind(1, X_test),
                               h = h_vals[i])
  rmse_kr[i] <- sqrt(mean((y_pred_kr - y_test)^2))
}
y_pred_kr <- casl_nlm_kernel(cbind(1, X), y,
                             cbind(1, X_test), h=.1)
```

Figure 6.1 shows the predictive power of the linear model compared to the kernel regressions. We see that the kernel regression significantly outperforms the linear model, particularly as the bandwidth decreases. The optimal value occurs somewhere between bandwidths of 0.1 and 0.2. Note that past 0.05, in addition to overfitting the data, we run into numerical issues. For at least one observation in the test set, the corresponding weights on the training set are non-zero on fewer than 3 inputs. Fixing this would require implementing a multivariate version of the `ksmooth_smart` function from Section 4.9.4.

Extending basis expansion estimators to multivariate data takes only slightly more work compared to the distance based smoothers. Consider a polynomial basis expansion. In p dimensions we need to consider not only powers of each component of the data but also interactions between individual dimensions. One method for producing such polynomials is to include terms for all combinations of the variables where the sum of the exponents is bounded by some value k. That is, we have terms of the form

$$\prod_{j=1}^{p} x_{i,j}^{\alpha_j}, \quad \sum_j \alpha_j \leq k. \tag{6.6}$$

We will refer to this set of basis functions as a kth order polynomial expansion. The most subtle part of implementing this is determining the valid set of components for α. Thankfully, the R function `stats::poly` will handle that aspect for us.

```
# Nonlinear regression using polynomial regression.
#
# Args:
```

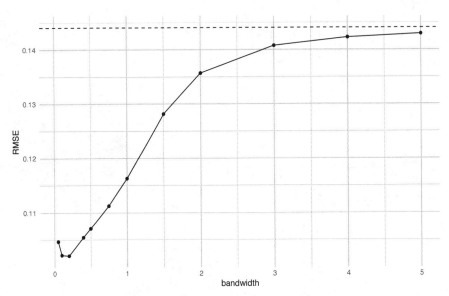

FIGURE 6.1: Root-mean squared error from simulated data with three columns and a continuous response variable. The dashed line is the linear regression solution while the solid lines shows a kernel regression predictor for various choices of the bandwidth parameter.

```
#     X: A numeric data matrix.
#     y: Response vector.
#     X_new: Numeric data matrix of prediction locations.
#     k: Order of the polynomial
#
# Returns:
#     A vector of predictions; one for each row in X_new.
casl_nlm_poly <-
function(X, y, X_new, k)
{
  Z <- poly(X, degree = k)
  Z_new <- cbind(1, predict(Z, X_new))
  Z <- cbind(1, Z)
  beta <- coef(lm.fit(Z, y))
  y_hat <- Z_new %*% beta
  y_hat
}
```

We will fit this to the same simulated data used with the kernel regression function.

```
y_pred_poly <- casl_nlm_poly(X, y, X_test, k = 3)
```

```
sqrt(mean((y_pred_poly - y_test)^2))
```

```
[1] 0.104289
```

The result is almost as good as the best kernel estimator and is certainly more predictive than the straightforward linear regression.

A similar procedure exists for extending any basis expansion technique, including smoothing splines, to multidimensional data. If we have a set of basis functions $\{\phi_j\}_{j=1}^{k}$ over the real line, we can construct the tensor product basis by considering products formed by all combinations of these bases for each column of the data. We can write this explicitly as

$$f(x_1, \ldots, x_p) = \sum_{i_1=1}^{k} \cdots \sum_{i_p=1}^{k} c_{i_1,\ldots,i_p} \prod_{j=1}^{p} \phi_{i_j}(x_j). \tag{6.7}$$

Our polynomial basis is slightly different than this tensor product. They would be equivalent if we replaced the condition from Equation 6.6, where the sum of the α's is less than k, with the less-strict requirement that each α must be less than k.

6.2 Curse of dimensionality

We have seen that it is generally straightforward, in an algorithmic sense, to extend linear smoothers in one dimension to prediction tasks with many predictor variables. Unfortunately the performance of these generalizations deteriorates quickly as the number of dimensions grows. Estimating nonparametric regression functions becomes all but impossible as the dimensions grow beyond more than 4 or 5 without additional constraints. The required constraints go beyond the simple (Lipschitz) continuity requirements needed for reasonably predictive estimators in lower dimensions. We will discuss various constraints from several approaches throughout the remainder of this text. In this section, we first explore the nature of what becomes difficult in higher dimensions.

The tensor product method of combining univariate basis functions is a great trick due to its simplicity to compute and ability to work with any set of basis functions. Consider a set of p basis functions $\{\phi_j\}_{j=1}^{p}$. What is the resulting dimensionality of the tensor product in d-dimensional space? This is a simple combinatorics question, resulting in a total set of p^d basis functions. If we want to fit a truncated power basis of order 2 (see Section 4.5) with 16 knots, for a total of $1 + 3 + 16 = 20$ terms, with two-dimensional data results in a total of 400 coefficients that must be estimated. This is already a significant increase in model complexity relative to the one-dimensional case.

Using the truncated power basis with only 3 knots, for a total of 7 terms, with 12 variables yields over 13.8 billion coefficients to estimate in the tensor product. Three knots each will only capture a very weak form of non-linearity and 12 variables is fairly small compared to many problems in statistical learning. Clearly, tensor products will not be a viable solution to fitting non-parametric regression in more than a few dimensions.

The *curse of dimensionality* is a collection of related concepts that, among other things, show that the explosion of terms needed for a tensor product approach to multivariate non-parametric regression is an unavoidable problem. Consider the k-nearest neighbors estimator. We have already seen that this can be easily extended to multivariate data. There is no exploding set of terms that needs to be estimated with the nearest neighbors algorithm. It seems equally complex regardless of the number of dimensions it is applied within. We will use a simple simulation to explain the difficulty with nearest neighbors as the dimensionality of the space increases. Let us generate data randomly distributed over the unit cube $[-1, 1]^d$ for dimensions ranging from 1 to 18 and compute the proportion of points that are within a distance of 1 from the origin. For context, in two dimensions this should be roughly the ratio between the size of a circle of radius 1 and the unit cube: $\pi/4 \approx 0.79$.

```
n <- 1e6
d_vals <- 1:18
props <- rep(NA_real_, length(d_vals))
for (i in seq_along(d_vals)) {
  d <- d_vals[i]
  X <- matrix(runif(n * d, min = -1, max = 1), ncol = d)
  dist <- apply(X^2, 1, sum)
  props[i] <- mean(dist <= 1)
}
```

The output in the variable **props** decays so quickly that it is difficult to even plot it very well. Here, instead, is a print out of the 18 proportions from the simulation:

```
options(width = 58)
sprintf("d=%02d, prop=%01.6f", d_vals, props)
```

```
 [1] "d=01, prop=1.000000" "d=02, prop=0.785359"
 [3] "d=03, prop=0.523600" "d=04, prop=0.308163"
 [5] "d=05, prop=0.164188" "d=06, prop=0.080972"
 [7] "d=07, prop=0.037154" "d=08, prop=0.015808"
 [9] "d=09, prop=0.006371" "d=10, prop=0.002455"
[11] "d=11, prop=0.000905" "d=12, prop=0.000301"
[13] "d=13, prop=0.000089" "d=14, prop=0.000036"
[15] "d=15, prop=0.000008" "d=16, prop=0.000003"
[17] "d=17, prop=0.000001" "d=18, prop=0.000000"
```

In only 18 dimensions, not a single one of the million points sampled in the simulation were within a distance of 1 of the origin. The lesson here is that for data randomly distributed in high dimensions, the nearest neighbors to any observation will still be fairly far away. These 'neighbors' will therefore be poor estimates of how the function behaves at a point of interest.

Our simulation of points in the unit cube illustrates that the typical distances between points grow larger in higher dimensions. While true, this is not fundamentally the problem that occurs when applying distance-based estimators to multidimensional data. If the scale of the distances was simple scaling upwards this might not be an issue. Kernel estimators would have larger bandwidths and the collection of nearest neighbors would just have numerically larger distances to their reference point. Yes, distances are growing, but not very quickly. The variance of a single uniform random variable distributed between -1 and 1 is $1/3$. Using Jenson's inequality for concave functions, we have an easy upper bound on how fast the average distance from the origin should be for points in our simulation:

$$\mathbb{E}\sqrt{X_1^2 + \cdots + X_d^2} \leq \sqrt{\mathbb{E}[X_1^2 + \cdots + \mathbb{E}X_d^2]} \tag{6.8}$$

$$= \sqrt{[\mathbb{E}X_1^2 + \cdots + \mathbb{E}X_d^2]} \tag{6.9}$$

$$= \sqrt{\frac{1}{3} + \cdots + \frac{1}{3}} = \sqrt{\frac{d}{3}}. \tag{6.10}$$

Also, the maximum distance between the origin and a point in the farthest corner of the unit cube is only $\sqrt{d}$. Therefore, the scale of distance should only be growing at a rate proportional to the square-root of the dimension d. The rate at which we saw points pushed away from the origin appears, in contrast, to be decaying at an exponential rate.

At its core, the curse of dimensionality says that our intuition about space built from living in only three dimensions does not extend well to geometric properties in higher dimensions. Our simulation shows that points are moving away from the origin at an exponential rate while our derivation illustrates that the average distances should only increase at a rate proportional to the square root of the dimension. To rectify these apparently conflicting statements, we now generate data from a 100-dimensional unit cube and look at the whole distribution of distances from the origin.

```
d <- 100
X <- matrix(runif(n * d, min = -1, max = 1), ncol = d)
dist <- sqrt(apply(X^2, 1, sum))
```

A histogram of these distances is given in Figure 6.2. The figure shows that the distances are clustering around a particular value. The concentration around this mean is what causes the proportion of points close to the origin to shrink so rapidly. The mean itself grows only at a rate proportional to the square root of d. To summarize, all of the sampled points appear to be approximately the

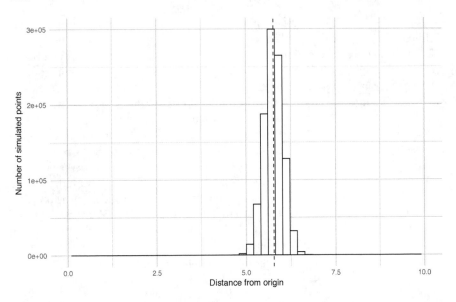

FIGURE 6.2: Random samples were uniformly drawn from a 100-dimensional unit cube centered on the origin. This histogram shows the density of the distances of these points from the origin. The scale on the x-axis shows the theoretical limits of distance, from 0 (at the origin) to $\sqrt{100}$ at a corner of the unit cube. The dashed line indicates an upper bound on the expected mean of the distances.

same distance away from the origin. This phenomenon is a specific example of the *concentration of measure*, a general result showing that well-behaved functions of independent random variables approximate a constant as the number of variables becomes large.

The concentration of distances around a single value illustrates the actual problem with distance-based prediction functions in high dimensions: if all points look approximately the same distance away from a reference point, there is no sensible set of sample weights to place on the training data for prediction at a point x_{new}.

6.3   Additive models

One way to circumvent the curse of dimensionality is to restrict the class of non-parametric relationships between X and y that are under consideration. Linear models are a particularly strict form of restriction where each predictor variable is assumed to have a fixed partial derivative with respect to the response. Additive models relax this assumption while remaining clear of the curse of dimensionality. Additive models assume that the mean of y can be decomposed into a sum of univariate functions over the predictors,

$$\mathbb{E}\left[y_i|x_i\right] = \alpha + \sum_j f_j(x_{i,j}). \tag{6.11}$$

This allows models such as the saddle function, $x_1^2 - x_2^2$, but not those with interactions between the variables. The effect of changing any of predictor variables is independent of the values of the other predictors. Additive models, therefore, capture non-linearities but disallow for interactions.

Notice that the functions in Equation 6.11 are generally non-unique. We could, for example, add a fixed constant to f_1 and subtract the same constant from f_2. The resulting relationship with y remains unchanged. To remove this degree of freedom, we force the sample mean of each function f_j to be zero over the training data. The intercept parameter α serves to capture the mean offset of y. The additive model can also be extended to the class of generalized additive models (GAMs). For a link function g, the relationship between the expected value of y and the predictors X is

$$g\left(\mathbb{E}\left[y_i|x_i\right]\right) = \alpha + \sum_j f_j(x_{i,j}), \tag{6.12}$$

for y defined over an exponential family. In both the linear and generalized additive models, the functions f_j are typically estimated by (possibly penalized) maximum likelihood estimators where each f_j is restricted to some class of functions. This class is defined over a finite set of basis functions.

How do additive models avoid the curse of dimensionality? For distance-based approaches, to estimate f_j we only need to find observations that are close to a reference point in the dimension x_j. The additive model assumption that the effect of x_j does not depend on the other coordinates means that those points close in the jth dimension are the only data needed to estimate f_j. When using a basis expansion, the full tensor product is no longer needed. For a set of basis functions $\{\phi_k\}_{k=1}^p$, an additive model must only consider the basis expansion

$$\bigcup_{j=1}^p \{\phi_k(x_j)\}_{k=1}^p. \tag{6.13}$$

because the inclusion of any cross terms would violate the additive assumption. Counting these, we only need $d \cdot p$ basis functions to estimate a linear model with d dimensions and a basis with p terms. Compare this to the exponential scaling for unconstrained non-parametric regression.

The *backfitting* algorithm is a common approach to estimating the functions f_j from training data. It can be used for generalized linear models, but we describe it here in the linear case. We start by defining scalar quantities representing the value of the f_j's applied to each of the elements of the data matrix X,

$$\widehat{f}_{i,j} = \widehat{f}_j(x_{i,j}). \tag{6.14}$$

Setting $\widehat{\alpha}$ to the mean of y and initializing all of the $\widehat{f}_{i,j}$ to zero, we then cycle through j by

$$\widehat{f}_{i,j} \leftarrow \mathcal{S}\left[y_i - \widehat{\alpha} - \sum_{k \neq j} \widehat{f}_{i,j} \right] \tag{6.15}$$

$$\widehat{f}_{i,j} \leftarrow \widehat{f}_{i,j} - \frac{1}{n}\sum_i \widehat{f}_{i,j}, \tag{6.16}$$

where $\mathcal{S}$ is some pre-determined smoothing function. This could be a smoothing spline, polynomial regression, or kernel regression. These updates are then repeated again over all values j until convergence of the predicted values $\widehat{f}$. To predict values on new data points, we can save the output of the smoothers as real-valued functions, thus getting functions $\widehat{f}_j$ that can be applied to any inputs.

Next, we will implement the backfitting algorithm for a linear additive model. Here, we use a smoothing spline as the smoothing operator and make sure to save the models as well as their predicted values.

```
# Fit linear additive model using backfit algorithm.
#
# Args:
#     X: A numeric data matrix.
#     y: Response vector.
#     maxit: Integer maximum number of iterations.
#
# Returns:
#     A list of smoothing spline objects; one for each column
#     in X.
casl_am_backfit <-
function(X, y, maxit=10L)
{
  p <- ncol(X)
  id <- seq_len(nrow(X))
  alpha <- mean(y)
  f <- matrix(0, ncol = p, nrow = nrow(X))
  models <- vector("list", p + 1L)

  for (i in seq_len(maxit))
  {
    for (j in seq_len(p))
```

```
    {
        p_resid <- y - alpha - apply(f[, -j], 1L, sum)
        id <- order(X[,j])
        models[[j]] <- smooth.spline(X[id,j], p_resid[id])
        f[,j] <- predict(models[[j]], X[,j])$y
    }
    alpha <- mean(y - apply(f, 1L, sum))
}


models[[p + 1L]] <- alpha
return(models)
}
```

Notice that the algorithm does not require storing the old iteration values because the backfitting algorithm uses only the current values of $\widehat{f}_{i,j}$, even inside the innermost loop.

To predict the output of the additive model at new observations, the function `casl_am_predict` applies the component-wise models and adds together their results.

```
# Predict values from linear additive model.
#
# Args:
#     models: A list of smoothing spline objects; one for each
#             column in the training data matrix.
#     X_new: Numeric data matrix of prediction locations.
#
# Returns:
#     A list of smoothing spline objects; one for each column
#     in X.
casl_am_predict <-
function(models, X_new)
{
    p <- ncol(X_new)
    f <- matrix(0, ncol = p, nrow = nrow(X_new))
    for (j in seq_len(p))
    {
        f[,j] <- predict(models[[j]], X_new[,j])$y
    }

    y <- apply(f, 1L, sum) + models[[p + 1L]]

    list(y=y, f=f)
}
```

The prediction returns the fitted values y as well as the individual quantities

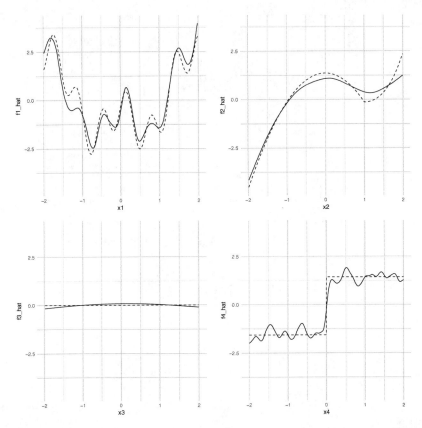

FIGURE 6.3: Predicted component values from an additive model simulation. Solid curves show solutions using smoothing splines and successive backfitting learned from noisy data y. Dashed curves give the true means.

$\widehat{f}_{i,j}$. The latter will be useful for plotting the individual components of each univariate function.

We now test our additive model implementation on simulated data. The simulation has 4 predictor variables and conforms to the additive model assumptions. The value of f_1 is a noisy, periodic function; f_2 is defined as a continuous piece-wise polynomial; f_3 is equal to zero; and f_4 is a scaled version of the `sign` function.

```
n <- 500; p <- 4
X <- matrix(runif(n * p, min = -2, max = 2), ncol = p)
f1 <- cos(X[,1] * 4) + sin(X[,1] * 10) + X[,1]^(2)
f2 <- -1.5 * X[,2]^2 + (X[,2] > 1) * (X[,2]^3 - 1)
f3 <- 0
f4 <- sign(X[,4]) * 1.5
```

```
f1 <- f1 - mean(f1); f2 <- f2 - mean(f2)
f3 <- f3 - mean(f3); f4 <- f4 - mean(f4)


y <- 10 + f1 + f2 + f3 + f4 + rnorm(n, sd = 1.2)
```

The `casl_am_backfit` function can then be used to predict the functions f_j.

```
models <- casl_am_backfit(X, y, maxit = 1)
pred <- casl_am_predict(models,X)
```

The predicted values $\widehat{f_j}$ are shown relative to the true f_j in Figure 6.3. We included sufficient noise in the data to prevent perfect reconstruction of the functions, but across each dimension the model generally captures the correct patterns. The function f_4 causes the most difficulty due to the discontinuity at zero, which smoothing splines have trouble approximating.

The additive model formulation in Equations 6.11 and 6.12 can be extended to include known interactions. Examples include interactions between latitude and longitude or variables that measure the same quantity at different time points or using different summary statistics (e.g., mean income and median income). We can define each f_j to come from a different class of allowed functions. In the extreme this could include forcing some variables to only have linear effects, a sensible assumption in many cases. Putting these together, we could have models such as

$$\mathbb{E}\left[y_i|x_i\right] = \alpha + \beta_1 \cdot x_{i,1} + \beta_2 \cdot x_{i,1} \cdot x_{i,2} + f_1(x_{i,3}, x_{i,4}) + f_2(x_{i,5}). \quad (6.17)$$

The backfitting algorithm can be modified to learn functions of this form as well. Where there are parametric terms, the smoother used in the backfitting algorithm is just the least squares smoother. For functions of more than one variable, the techniques from Section 6.1 can be employed as smoothers. Some sources describe the class of additive models as consisting of this more general formulation.

6.4   (⋆) Additive models as linear models

The backfitting algorithm is convenient because it can easily be applied using any linear smoother. The downside is that the iterative nature of the algorithm can take a long time to converge. Several of the smoothers we saw in Chapter 4 can be written as a linear model over a set of expanded basis functions. When using these smoothers to estimate each individual f_j, the class of allowed models over all the variables is once again just a linear regression. Take for example an additive model with two predictors

$$\mathbb{E}\left[y_i|x_i\right] = \alpha + f_1(x_{i,1}) + f_2(x_{i,2}). \quad (6.18)$$

If a two-dimensional polynomial basis expansion is used as a smoothing function, the class of all models can be written as

$$\mathbb{E}\left[y_i | x_i\right] = \alpha + x_{i,1}\beta_1 + x_{i,1}^2\beta_2 + x_{i,2}\beta_3 + x_{i,2}^2\beta_4. \tag{6.19}$$

The unknown parameters can then be solved directly using ordinary least squares normal equations. This can be extended to generalized additive models, where iteratively re-weighted least squares can be used to solve the linear model in the expanded basis. While iterative, the IRWLS algorithm is a second-order method and usually converges very quickly.

Writing additive models as linear models over a small set of basis functions is a useful technique for working with datasets that have weakly non-linear relationships with the response variable. Detecting intricate non-linearities requires using adaptive methods such as smoothing splines. Smoothing splines can be written as penalized linear regressions over a set of basis functions, with a penalty given by a term proportional to $||\Omega\beta||_2^2$ for a properly defined matrix Ω (see Section 4.6). It is possible to write the optimization formulation of fitting smoothing splines to an additive model as a single penalized regression problem. This can then be solved directly using penalized weighted least squares.

Let $Z^{(j)}$ be the basis expansion defined by the truncated power basis applied to the variable x_j, $\beta^{(j)}$ be the regression coefficients over the truncated basis functions, and $\Omega^{(j)}$ be the associated matrix of penalties. Recall that the penalty matrix will be non-zero only on a small number of bands around the diagonal. The additive model can be parametrized as

$$\mathbb{E}\left[y | X\right] = \alpha + \sum_j Z^{(j)}\beta^{(j)}. \tag{6.20}$$

Maximizing the penalized likelihood yields the least squares optimization problem

$$\underset{\alpha,\beta}{\arg\max}\left\{||y - \alpha - \sum_j Z^{(j)}\beta^{(j)}||_2^2 + \sum_j \lambda^{(j)}||\Omega^{(j)}\beta^{(j)}||_2^2\right\}. \tag{6.21}$$

Putting terms together,

$$Z = \begin{pmatrix} 1 & Z^{(1)} & Z^{(2)} & \cdots Z^{(p)} \end{pmatrix} \tag{6.22}$$

$$\beta = \begin{pmatrix} \alpha \\ \beta^{(1)} \\ \vdots \\ \beta^{(p)} \end{pmatrix} \tag{6.23}$$

$$\Omega = \begin{pmatrix} 0 & \sqrt{\lambda^{(1)}} \cdot \Omega^{(1)} & 0 & \cdots & 0 \\ 0 & 0 & \sqrt{\lambda^{(2)}} \cdot \Omega^{(2)} & \cdots & 0 \\ \vdots & \vdots & & \ddots & \vdots \\ 0 & 0 & \cdots & 0 & \sqrt{\lambda^{(p)}} \cdot \Omega^{(p)} \end{pmatrix} \tag{6.24}$$

The penalty terms $\lambda^{(j)}$ can be different for each component of the model. Otherwise we would be assuming that each function f_j has an equal degree of smoothness. To simplify notation, the λ terms have been subsumed into the Ω matrix. Putting this notation together, we can write Equation 6.21 as a single penalized regression task:

$$\arg \max_{\beta} \left\{ ||y - Z\beta||_2^2 + ||\Omega\beta||_2^2 \right\}. \qquad (6.25)$$

with solutions given analytically by

$$\widehat{\beta} = \left[Z^t Z + \Omega^t \Omega \right]^{-1} Z^t y. \qquad (6.26)$$

As with linear regression problems, this can be solved numerically in an efficient way through the use of matrix decompositions in order to compute the inverse matrix product. The generalized additive model can be solved similarly by replacing the penalized least squares with penalized iteratively reweighed least squares.

Representing generalized additive models as linear models has several benefits. We have a direct solution in the linear case and a second-order solution in the generalized case. Features such as the inclusion of mixed models, robust techniques, and inference-based metrics built for linear regression can often be directly extended to additive models. It can also simplify the implementation. Once the penalty and model matrices are constructed, these can be passed to general purpose linear model routines. One potential downside is the need to store a large design matrix Z entirely in memory. With backfitting only a n-by-d matrix (for d basis functions) is needed at any given point. Here we need the entire n-by-$(d \cdot p)$ matrix.

Notice that solving the linear regression task only finds the solution for a specific value of the tuning parameters λ_j. With backfitting we can use a line search at each step to select a good tuning parameter λ_j via cross-validation or some model selection metric (e.g., AIC, BIC, Mallow's C_p). In the regression formulation of additive models, optimal values for each λ_j need to be found simultaneously. For more than a few predictor variables a complete grid search becomes infeasible. A commonly used metric for model selection is the generalized cross-validation (GCV). It approximates leave-one-out cross-validation with a simple formula that can be applied to any linear smoother. For the set-up here, the formula becomes

$$GCV = \frac{1}{n} \sum_i \left(\frac{y_i z_i^t \widehat{\beta}}{1 - df/n} \right)^2 \qquad (6.27)$$

$$df = \text{tr} \left[Z(Z^t Z + \Omega^t \Omega)^{-1} Z^t \right]. \qquad (6.28)$$

It is possible to find the gradient of the GCV here with respect to the vector λ. This can be used to apply Newton's method or some other optimization algorithm. For each value tested with a given λ, the penalized least squares algorithm must be reconstructed though the design matrix Z remains constant.

When fitting generalized additive models, it is possible to update the λ terms at the same time as the sample weights and effective response are changing in the iteratively re-weighted least squares. Otherwise, the λ optimization can be done as an additional outside loop to the algorithm. The first choice is often faster though the second has known convergence results.

6.5 (⋆) Standard errors in additive models

One useful feature of additive models is the ability to easily visualize the fitted model with a plot such as the one in Figure 6.3. This is possible because each term is a univariate function that can be shown on a two-dimensional plot. The extended form of additive models (Equation 6.17) can be plotted similarly using contour plots for the two-way non-parametric fits. Plots of fitted values from additive models can be made more useful with the inclusion of standard errors. An additional benefit of the penalized regression formulation of additive models is that there is a straightforward way of computing the standard errors.

Consider the fitted values $\widehat{y}$ as a function of the fitted regression vectors $\widehat{\beta}^{(j)}$:

$$\widehat{y} = \widehat{\alpha} + \sum_j \widehat{f}^{(j)} \tag{6.29}$$

$$= \widehat{\alpha} + \sum_j Z^{(j)} \cdot \widehat{\beta}^{(j)} \tag{6.30}$$

For each plot in the visualization of an additive model, we want to see how much variation there is in that particular component. The standard errors we want to compute, therefore, are given by the square root of the variance of the terms $\widehat{f}^{(j)}$. This is given by

$$Var(\widehat{f}^{(j)}) = Var(Z^{(j)} \cdot \widehat{\beta}^{(j)}) \tag{6.31}$$

$$= Z^{(j)} \cdot Var(\widehat{\beta}^{(j)})(Z^{(j)})^t. \tag{6.32}$$

The variance matrix of $\widehat{\beta}^{(j)}$ must be derived by taking the corresponding rows and columns of the variance matrix for the entire $\widehat{\beta}$ vector. This is given by

$$Var(\widehat{\beta}) = \widehat{\sigma}^2 \cdot \left[Z^t Z + \Omega^t \Omega\right]^{-1} Z^t Z \left[Z^t Z + \Omega^t \Omega\right]^{-1}. \tag{6.33}$$

Plugging this into Equation 6.32, the process of calculating the variance matrix is a straightforward set of matrix operations. The resulting matrix will be a large n-by-n matrix. However, only the diagonal elements of the variance are needed. We can be clever in our implementation to only compute the diagonal elements that are needed.

We now implement the technique for solving additive models as linear models derived in Section 6.4. Standard errors are also computed and returned to the user. There are a number of steps in the process so we split the algorithm into several sub-functions. First, we need a function to return the entire matrix Z given the input X. We will produce a truncated power basis of a fixed order using evenly spaced knots.

```
# Compute additive model design matrix.
#
# Args:
#     X: A numeric data matrix.
#     order: Order of the polynomial.
#     nknots: Number of knots in the power basis.
#
# Returns:
#     A list of smoothing spline objects; one for each column
#     in X.
casl_am_ls_basis <-
function(X, order, nknots)
{
  Z <- matrix(rep(1, nrow(X)), ncol = 1L)
  for (j in seq_len(ncol(X)))
  {
    knots <- seq(min(X[, j]), max(X[, j]),
                 length.out = nknots + 2L)
    knots <- knots[-c(1, length(knots))]
    Z2 <- casl::casl_nlm1d_trunc_power_x(X[, j], knots,
                                        order = order)[, -1L]
    Z <- cbind(Z, Z2)
  }
  Z
}
```

The function `casl_nlm1d_trunc_power_x` was defined in Section 4.5.

Next, we need a function to return the matrix Ω. Here we will use a simple diagonal matrix that defines ridge regression over the basis functions.

```
# Compute penalty matrix for penalized additive models.
#
# Args:
#     Z: A numeric data matrix.
#     lambda: Vector of penalty terms, with one value per
#             column in the original data matrix.
#
# Returns:
#     The penalty matrix for the additive model.
casl_am_ls_omega <-
```

```
function(Z, lambda)
{
  d <- (ncol(Z) - 1L) / length(lambda)
  omega <- Matrix::Matrix(0, ncol = ncol(Z), nrow = ncol(Z))
  Matrix::diag(omega)[-1L] <- rep(sqrt(lambda), each = d)
  omega
}
```

It is straightforward to compute $\widehat{\beta}$ as using the matrices Z. We will compute this as well as the fitted values $\widehat{y}$ and the predicted noise variance $\widehat{\sigma}^2$ inline in our final function.

Once we have the entire regression vector, it will be helpful to compute the values $\widehat{f}^{(j)}$ and store these individually. The `casl_am_ls_f` function here does exactly that given the matrix Z, the learned regression vector, and the number of columns in the original dataset X.

```
# Compute predicted values for penalized additive models.
#
# Args:
#     Z: A numeric data matrix.
#     beta_hat: Estimated regression vector.
#     p: Number of columns in original data matrix.
#
# Returns:
#     Predicted values, f_j, from the additive model.
casl_am_ls_f <-
function(Z, beta_hat, p)
{
  d <- (ncol(Z) - 1L) / p
  f <- matrix(NA_real_, ncol = p + 1L, nrow = nrow(Z))
  for (j in seq_len(p))
  {
    id <- 1L + seq_len(d) + d * (j - 1L)
    f[, j] <- Z[, id] %*% beta_hat[id]
  }
  f[, p + 1L] <- beta_hat[1L]
  f
}
```

Similarly, we also need to compute the standard errors for each of the values in the matrix `f`.

```
# Compute standard errors for penalized additive models.
#
# Args:
#     Z: A numeric data matrix.
#     omega: A penalty matrix.
```

```
#       beta_hat: Estimated regression vector.
#       s_hat: Estimated noise variance.
#       p: Number of columns in original data matrix.
#
# Returns:
#       A matrix of standard errors for the predicted values,
#       f_j, from the additive model.
casl_am_ls_f_se <-
function(Z, omega, beta_hat, s_hat, p)
{
  d <- (ncol(Z) - 1L) / p
  f_se <- matrix(NA_real_, ncol = p + 1L,
                           nrow = nrow(Z))
  ZtZ <- Matrix::crossprod(Z, Z)
  B <- solve(ZtZ + Matrix::crossprod(omega, omega),
             Matrix::t(Z))
  beta_var <- Matrix::tcrossprod(B) * s_hat
  A <- Z %*% beta_var
  f_se <- apply((Z %*% beta_var) * Z, 1, sum)
  f_se
}
```

We are careful to never construct the entire n-by-n variance matrix.

Using all of these helper functions, it is relatively straightforward to construct the function `casl_am_ls` that takes the raw data and returns the fitted values from the additive model. Note that it requires the user to specify a vector of penalty terms λ, one for each column of X.

```
# Additive model using truncated power basis.
#
# Args:
#       X: A numeric data matrix.
#       y: Response vector.
#       lambda: Vector of penalty terms, with one value per
#               column in the original data matrix.
#       order: Order of the polynomial.
#       nknots: Number of knots in the fit.
#
# Returns:
#       A list of the fitted values.
casl_am_ls <-
function(X, y, lambda, order=3L, nknots=5L)
{
  # get basis and penalty matricies
  Z <- casl_am_ls_basis(X, order=order, nknots=nknots)
  omega <- casl_am_ls_omega(Z, lambda)
```

```
# compute predicted beta and f_j values
beta_hat <- Matrix::solve(Matrix::crossprod(Z, Z) +
                          Matrix::crossprod(omega, omega),
                          Matrix::crossprod(Z, y))
f <- casl_am_ls_f(Z, beta_hat, p = ncol(X))

# fitted values for y and sigma^2
y_hat <- apply(f, 1L, sum)
s_hat <- sum((y - y_hat)^2) / (nrow(Z) - ncol(Z))

# get standard errors for f_j
f_se <- casl_am_ls_f_se(Z, omega, beta_hat, s_hat,
                        p = ncol(X))

# return list of values
return(list(y_hat=y_hat, f=f, f_se=f_se))
}
```

The output returns the fitted values $\widehat{y}$ and $\widehat{f}^{(j)}$ along with the standard errors for the latter.

Using the linear model based implementation, we fit an additive model to the simulated data from Section 6.3. We manually adjusted the lambda terms to improve the fit.

```
model <- casl_am_ls(X, y, lambda = rep(0.01, 4))
```

A visualization of the fitted value as well as the standard errors is given in Figure 6.4. Notice that the standard errors do a relatively good job of indicating where the true mean is likely to differ from the predicted mean function. Overall, the performance of this model is not quite as accurate compared to the one shown in Figure 6.3. This is not due to a difference between backfitting and penalized least squares but rather that our backfitting algorithm used an adaptive smoothing spline instead of the fixed penalty terms used here.

6.6   Implementation and notes

Simple additive models can be fit in R by directly calling the function `stats::lm` or `stats::glm` with the appropriate basis expansion functions. For example, we can fit a three-parameter additive logistic regression as:

```
model <- glm(y ~ poly(x1) + splines::bs(x2, df = 4) +
                 poly(x3),
             family = binomial())
```

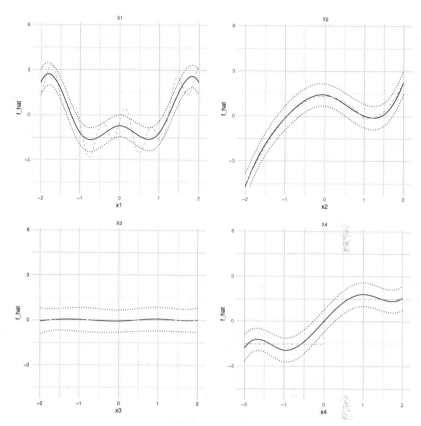

FIGURE 6.4: Predicted component values from an additive model simulation. Solid curves show solutions using smoothing splines and penalized least squares. Dashed curves give the true means and dotted lines give standard errors.

This approach makes it easy to accommodate weak forms of non-linearity into any model constructed using the formula interface in R. The downside is that it does not allow for adaptive smoothing functions nor does it yield coefficients or standard errors for each term (coefficients and standard errors are returned separately for each term in the basis expansion, which cannot be trivially combined).

There are three packages of particular importance that implement additive models: **gam** [70], **mgcv** [177], and **gamlss** [137]. All three of these implement generalized additive models and allow users to describe these models using a formula interface. New functions, such as **s** in **gam** and **te** in **mgcv**, are given that describe what type of smoother to use for each term. This makes it easy to include two-way interactions as well as standard linear terms. The **gam** package uses backfitting whereas **mgcv** implements the penalized iteratively re-weighted leasts squares algorithm. The latter is typically faster and preferred. The **mgcv** package also has functions for extending additive models to include mixed effects. The **gamlss** package allows for generalized additive models where up to four parameters of the underlying distribution of y are allowed to vary with the data.

6.7 Application: NYC flights data

6.7.1 Data and setup

The data here consists of commercial domestic flights departing from one of the three major airports serving the New York City metropolitan area: JFK, LaGuardia (LGA), and Newark (EWR). The data includes all flights from the year 2013. The raw data comes courtesy of the US Bureau of Transportation Statistics and has been conveniently published in a stand-alone R package **nycflights13**. We will look at the dataset from two different levels. First, counts are aggregated into hour buckets at each airport. A Poisson regression model is used to predict the number of flights departing within a given hour. The second and third models look at features of individual flights. Information is available about the weather, airplane, and the arrival airport.

6.7.2 Estimating number of flights arrived

The dataset `cnt` contains the number of flights that took off in a given hour at a specific airport. The data contains the day of the year (starting on 1 January), the hour, the day of the week, the airport, and the temperature at the start of the hour. Here are the first few rows of the data:

```
cnt <- read.csv("data/cnt.csv")
head(cnt)
```

```
  n origin wday doy hour  temp
1  5    LGA    1   1    5 39.92
2 20    LGA    1   1    6 39.92
3 13    LGA    1   1    7 39.92
4 19    LGA    1   1    8 39.92
5 18    LGA    1   1    9 39.92
6 14    LGA    1   1   10 39.92
```

We will use the `gam` function from the **mgcv** package to fit an additive model to predict the number of flights departing in a given time window. We include a linear offset for the origin airport and linear smoothers for the other terms. The

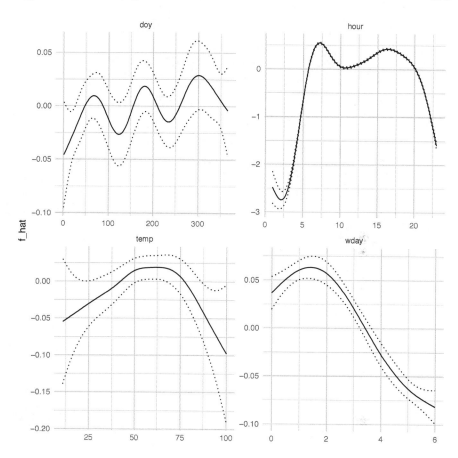

FIGURE 6.5: Estimated means and standard deviations of a Poisson GAM model for predicting the number of departures that will occur in a given hour at a specific New York City area airport.

weekday term, which has only 7 unique values, requires that we manually set a number of knots as the default is too large. We set the family to `poisson()` to indicate that the response variable should be modelled as a count.

```
model <- mgcv::gam(n ~ origin + s(wday, k = 4) + s(doy) +
                       s(hour) + s(temp),
                   data = cnt,
                   family = poisson())
```

The summary of the model, among other model statistics, shows the linear offsets for the `origin` variable.

```
summary(model)
```

```
Family: poisson
Link function: log

Formula:
n ~ origin + s(wday, k = 4) + s(doy) + s(hour) + s(temp)

Parametric coefficients:
             Estimate Std. Error z value Pr(>|z|)
(Intercept)  2.830813   0.006342  446.37   <2e-16 ***
originJFK   -0.103929   0.009579  -10.85   <2e-16 ***
originLGA   -0.136508   0.008811  -15.49   <2e-16 ***
---
Signif. codes:
0 '***' 0.001 '**' 0.01 '*' 0.05 '.' 0.1 ' ' 1

Approximate significance of smooth terms:
          edf Ref.df   Chi.sq  p-value
s(wday) 2.948  2.998  164.522  < 2e-16 ***
s(doy)  8.729  8.973   69.009 2.49e-11 ***
s(hour) 8.640  8.911 4366.309  < 2e-16 ***
s(temp) 3.725  4.707    7.345     0.17
---
Signif. codes:
0 '***' 0.001 '**' 0.01 '*' 0.05 '.' 0.1 ' ' 1

R-sq.(adj) =  0.504   Deviance explained = 52.9%
UBRE = 0.43089  Scale est. = 1        n = 4296
```

All things being equal, the highest number of flights depart Newark and the fewest depart LaGuardia. The day of the week, day of the year, and hour of the day are all significant effects on the number of departures. However, the temperature does not have a statistically significant effect after other effects are accounted for.

The fitted models for the other variables are shown in Figure 6.5. Standard error curves are included along with the means. The strongest predictor is the hour of the day, which peaks around 07:00 and drops off sharply before 05:00 and after 20:00. The number of flights decreases through the week (day 0 is Monday). More flights arrive when the temperature is between 50 and 75 degrees Fahrenheit. There is a periodic relationship with the day of the year, with a general increase throughout the year.

6.7.3 Arrival delay prediction

Now, we will turn to data on the flights themselves. Here we have coded a flight as delayed if it arrives more than 30 minutes late. Covariates include the

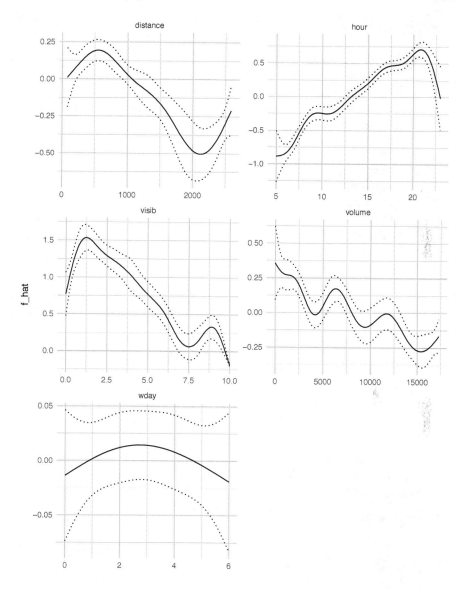

FIGURE 6.6: Estimated means and standard deviations of a binomial GAM model for predicting whether a flight will be more than 30 minutes late on arrival.

visibility, distance of the flight, and the volume of flights landing each year from NYC to a particular destination airport.

```
flights <- read.csv("data/flights.csv")
```

```
head(flights)
```

```
   delayed origin wday visib hour distance seats volume
1        0    EWR    1    10    5     1400   149   7198
2        1    LGA    1    10    5     1416   149   7198
3        1    JFK    1    10    5     1089   178  11728
4        0    LGA    1    10    6      762   178  17215
5        0    EWR    1    10    5      719   191  17283
6        1    EWR    1    10    6     1065   200  12055
  air_time         lon       lat
1      227   -95.34144  29.98443
2      227   -95.34144  29.98443
3      160   -80.29056  25.79325
4      116   -84.42807  33.63672
5      150   -87.90484  41.97860
6      158   -80.15275  26.07258
```

For this data, we will fit an additive model for whether a flight is delayed using a binomial exponential family. We include a linear term for the origin and again put all of the other covariates in as linear smoothers.

```
model <- mgcv::gam(delayed ~ origin + s(wday, k = 4) +
                   s(visib) + s(hour) + s(distance) +
                   s(seats, k = 5) + s(volume),
                   data = flights,
                   family = binomial())
```

Both the weekday and number of seats need to have manually set values of k given the small number of unique values.

The model summary shows that the worst airport to depart from in terms of delays is Newark, followed by JFK.

```
summary(model)
```

```
Family: binomial
Link function: logit

Formula:
delayed ~ origin + s(wday, k = 4) + s(visib) + s(hour) +
    s(distance) + s(seats, k = 5) + s(volume)

Parametric coefficients:
             Estimate Std. Error z value Pr(>|z|)
(Intercept) -1.194353   0.007706 -155.00   <2e-16 ***
originJFK   -0.218520   0.012351  -17.69   <2e-16 ***
```

```
originLGA   -0.134665   0.012601   -10.69   <2e-16 ***
---
Signif. codes:
0 '***' 0.001 '**' 0.01 '*' 0.05 '.' 0.1 ' ' 1

Approximate significance of smooth terms:
             edf  Ref.df   Chi.sq  p-value
s(wday)     2.968   2.999    502.9  <2e-16 ***
s(visib)    8.726   8.967   4314.8  <2e-16 ***
s(hour)     8.813   8.988  11446.6  <2e-16 ***
s(distance) 8.747   8.962    162.0  <2e-16 ***
s(seats)    3.983   4.000    866.2  <2e-16 ***
s(volume)   8.637   8.950    239.3  <2e-16 ***
---
Signif. codes:
0 '***' 0.001 '**' 0.01 '*' 0.05 '.' 0.1 ' ' 1

R-sq.(adj) =  0.0714   Deviance explained = 6.41%
UBRE = 0.01701  Scale est. = 1         n = 265222
```

Figure 6.6 shows the means and standard deviations of the remaining terms. Once again, the hour of the day is the most predictive variable. Later flights have a higher probability of being delayed, likely because of cascading issues occurring earlier in the day. These drop-off in the later hours as the volume of flights decreases. The visibility conditions also have a significant impact on flight delays, with poor visibility conditions causing much higher rates of large delays. Flights to farther away locations are less likely to be delayed, possibly due to the fact that they can make up lost time in the air and because longer flights have more built-in room within the published arrival times. Destination airports with higher volumes have fewer delays, and the day of the week seems to have very little effect on whether a flight suffers a large arrival delay.

6.7.4   Predicting time spent in the air

As a final task, we will use the location of a destination airport to predict how long the plane will spend in the air. Likely this will be directly correlated to the distance between the destination and arrival airports. We use this here simply to illustrate the use of a two-dimensional linear smoother. This is accomplished by giving both latitude and longitude to a single call of the function s.

```
model <- mgcv::gam(air_time ~ s(lon, lat),
                   data = flights)
```

Contour lines for this fit, plotted over a map of the United States, are shown in Figure 6.7. As expected, the fitted values approximate the distance of points from New York City.

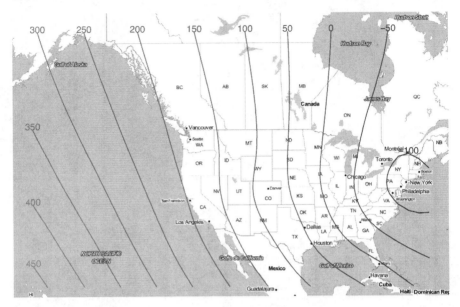

FIGURE 6.7: Contours showing predicted flight times as a function of latitude and longitude plotted over a map of the United States. All flights originated in New York City.

6.8 Exercises

1. In the example from Section 6.7.4 we used the smoother function `s`. Change this to `te` to use the tensor product smoother. How do the resulting contours change, if it all?

2. Write a test function to compare the `mgcv` function with the implementation `casl_am_backfit`. How well does our reference implementation match the **mgcv** package's function?

3. Write a test function to compare the `mgcv` function with the **gam** package's gam function. How well do the models match one another? Include a timing element as well. Which model runs faster?

4. Modify the function `casl_am_backfit` to include iteratively re-weighted least squares for logistic regression.

5. The **mgcv** package includes the function `bam` that fits generalized additive models over larger datasets by reducing the required memory for the fitting procedure. Run a simulation on a dataset with 15 columns and 150,000 rows. Compare the speed and memory footprint of `mgcv::gam` and `mgcv::bam`. Is there a noticeable difference in either?

7

Penalized Regression Models

7.1 Variable selection

The task of figuring out what variables to use, *variable selection*, is an important aspect in the construction of both predictive and inferential models. In many cases this step depends heavily on data availability, deep domain knowledge, and project-specific constraints. These external considerations, while important, are outside of the scope of our treatment. Instead, data-driven approaches to variable selection are the focus here. That is, we want methods that stwart with a large set of potential variables (perhaps everything available) in our data matrix X and use the data itself to determine which variables should be included in the final output. Such an approach is particularly important when there is a large number of variables that could be used in a model. This is common in applications such as gene expression studies and text classification tasks, both of which typically have tens of thousands of available variables [107].

Classical methods for computing data-driven variable selection typically fall under a class of algorithms known as *stepwise regression* [21]. In backward stepwise regression, we start with a standard regression model using all of the variables. Variables are then removed from the model due to some selection criterion, such as removing variables with small T-statistics. The new model is then refit with the smaller set of variables and the selection criterion is applied once again. Eventually this leads to a standard regression model over a reduced set of variables. The forward selection variation uses the same approach, but starts with an empty model and iteratively determines the next most significant term to add, stopping when some metric (i.e., AIC or Mallow's C_p) is locally optimized. Even when not used formally, the ideas behind stepwise regression are frequently applied in the medical and social sciences. Econometrics texts, for example, often promote a process by which an initial model is built, the fit and residuals are evaluated, and variables and interaction terms are iteratively added and removed as needed [178]. Hand-constructed stepwise regression can lead to erroneous conclusions if hypothesis tests from the final model are presented without correcting for the post hoc variable selection step [120]. Done maliciously—pejoratively known as *p-hacking*—this can have serious effects for the advancement of research and is currently a hotly debated topic of interest within statistics [64, 75, 123]. When used for purely predictive purposes, such concerns are avoided and stepwise regression is a perfectly valid approach. However, empirical results suggest that it often leads to non-optimal results and its ad hoc description makes it difficult to provide a solid theoretical treatment [153].

When performing data-driven variable selection it is not always necessary to reconstruct a new data matrix X with a limited number of columns. In the context of a regression model, for example, we could instead work with the entire dataset X but restrict ourselves to regression vectors β that have some components exactly equal to zero. Those variables associated with the zero components are effectively left out of the model. Using this approach, we can write regression models that perform variable selection and model estimation simultaneously by (optionally) returning some components of β exactly equal to zero. Any such model is said to have a *parsimonious* property. Models that combine selection and estimation in a single step avoid many of the pitfalls of iterative techniques and typically outperform them on out-of-sample prediction tasks. It is this class of models that we focus our attention on in the remainder of this chapter.

7.2   Penalized regression with the ℓ_0- and ℓ_1-norms

We can derive a combined approach to variable selection and model estimation by formalizing our goal as an optimization task. Abusing terminology, we will define the ℓ_0-"norm" of a vector as a count of the number of non-zero components it contains

$$||v||_0 = \# \left\{ i : v_i \neq 0 \right\}. \tag{7.1}$$

While this does not satisfy the conditions of a proper norm (see Section A.1), Equation 7.1 can be understood as a measurement of the size of the vector v. The goal of model selection within the regression framework can be seen as an attempt to balance the fit of the model with the size of the ℓ_0-norm of the regression vector. This task now appears similar to the ridge regression framework from Section 3.2. In ridge regression we minimize a linear combination of the residual sum of squares and the ℓ_2-norm of the regression vector. A model selection regression can be defined similarly as

$$\widehat{\beta}^{\ell_0}(\lambda) \in \arg\min_b \left\{ ||y - Xb||_2^2 + \lambda \cdot ||b||_0 \right\}. \tag{7.2}$$

The tuning parameter λ, as with ridge regression, can be selected using some form of validation or cross-validation. We now have an optimization problem whose solution should produce a model where variables that do not significantly improve the model fit are excluded entirely. Empirical research has shown that Equation 7.2 typically outperforms models from stepwise regression and includes the most important variables in the final model [31].

Unfortunately, the model described by Equation 7.2 is nearly impossible to compute for any reasonably large set of starting variables. The optimization problem is not continuous in b, let alone convex, making most general purpose algorithms unusable. The only sure way to minimize the quantity is to exhaustively check all possible sets of variables that could be included in the model; a total of 2^p total calculations. As an alternative, we need to approximate the explicit model selection optimization problem with a computationally tractable approach that still returns model vectors with some components set exactly to zero.

The ℓ_1-norm of a vector is given by adding together the absolute values of each of its components

$$||v||_1 = \sum_j |v_j|. \tag{7.3}$$

Unlike the ℓ_0-norm, this is a proper mathematical norm and a continuous function from $\mathbb{R}^p$ into $\mathbb{R}$. Replacing the norm in Equation 7.2 with the ℓ_1-norm yields

$$\widehat{\beta}^{\ell_1}(\lambda) \in \arg\min_b \left\{ ||y - Xb||_2^2 + \lambda \cdot ||b||_1 \right\}. \tag{7.4}$$

This estimator is commonly called the LASSO (least absolute shrinkage and selection operator) [154]. Remarkably, this variation is able to approximate the behavior of the model selection estimator yet yields a solvable optimization task. For sufficiently large values of λ some components of $\widehat{\beta}^{\ell_1}$ will be set exactly to zero. The optimization problem is convex and therefore can be approximately solved by a number of general purpose algorithms; in Section 7.5 we derive a specific approach that is particularly adept at solving this particular task.

There are several approaches to understanding why ℓ_1-penalized regression leads to a parsimonious regression vector. The most straightforward explanation is to realize that the derivative of the objective function in Equation 7.4 with respect to b_j will be undefined at zero due to the absolute value function. The value $b_j = 0$ is a critical point and therefore a possible solution to the minimization task. A similar geometric argument comes from re-writing the lasso estimator in its primal form as a constrained optimization task parametrized by a value $s > 0$

$$\widehat{\beta}^{\ell_1}(s) \in \arg\min_b \left\{ ||y - Xb||_2^2 \quad \text{s.t.} \quad ||b||_1 \leq s \right\}. \tag{7.5}$$

It is left as an exercise to show that this is equivalent to Equation 7.4 for some mapping between λ and s. Figure 7.1 illustrates, in two dimensions, how the sharp corners of an ℓ_1 constraint can lead an estimator with coefficients equal to zero. In the following section we show a direct derivation of how the ℓ_1-penalized estimator approximates the ℓ_0-penalized estimator.

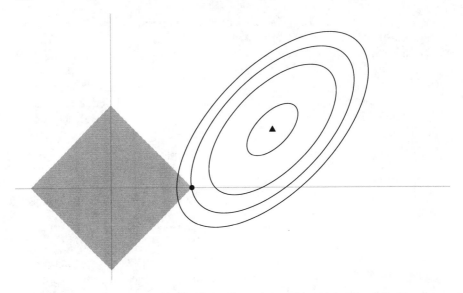

FIGURE 7.1: The contour lines show the values of a quadratic objective function with a minimal point centered on the black triangle. The minimal value of the objective function constrained to an ℓ_1-disk (the grey region) is given by the solid black dot. This leads to an optimal value where the y-coordinate is set exactly equal to zero.

7.3   Orthogonal data matrix

Another instructive method for understanding the relationship between the ℓ_0- and ℓ_1-penalized regression models is to consider the simple example where the columns of X are uncorrelated. Specifically, assume that each variable in X has also been standardized such that

$$X^t X = \mathbb{1}_p. \tag{7.6}$$

In this case, we can derive an analytic solution to both penalized regression models for any given value of λ.

The ℓ_0-penalized objective model with an orthogonal data matrix can be written as

$$g(b) = y^t y + b^t X^t X b - 2b^t X^t y + \lambda \cdot ||b||_0 \tag{7.7}$$

$$= y^t y + \sum_j \left\{ b_j^2 - 2b_j \cdot X_j^t y + 1_{b_j \neq 0} \cdot \lambda \right\} \tag{7.8}$$

$$= y^t y + \sum_j g_j(b_j) \tag{7.9}$$

The problem is now decoupled over the components b_j and we can optimize each function g_j independently. It is this property that allows us to provide an analytic solution. To minimize g_j, first consider the case when b_j is not equal to zero. Then, the optimal value comes from taking the derivative

$$g_j'(b_j) = 2b_j - 2 \cdot X_j^t y \tag{7.10}$$

and setting it equal to zero

$$b_j = X_j^t y, \tag{7.11}$$

which yields an objective value of

$$g_j(X_j^t y) = b_j^2 - 2b_j \cdot X_j^t y + \lambda \tag{7.12}$$

$$= \lambda - \left(X_j^t y \right)^2. \tag{7.13}$$

This will be smaller than the value $g_j(0)$ if and only if $|X_j^t y| > \lambda^{1/2}$. Putting these two conditions together, we can write the ℓ_0-penalized regression solution as

$$\widehat{\beta}_j^{\ell_0} = \begin{cases} X_j^t y, & |X_j^t y| > \lambda^{1/2} \\ 0, & \text{else} \end{cases} \tag{7.14}$$

By selecting a sufficiently large value of λ, we explicitly see how the ℓ_0-penalty performs automatic variable selection.

A similar approach can be applied to find an analytic solution to ℓ_1-penalized linear regression. The objective function can be decoupled over each variable as

$$f(b) = y^t y + b^t X^t X b - 2b^t X^t y + \lambda \cdot ||b||_1 \tag{7.15}$$

$$= y^t y + \sum_j \left\{ b_j^2 - 2b_j \cdot X_j^t y + \lambda \cdot |b_j| \right\} \tag{7.16}$$

$$= y^t y + \sum_j f_j(b_j) \tag{7.17}$$

We can minimize each quantity f_j inside the sum in Equation 7.17 independently. To work with the absolute value in Equation 7.17, we start by considering what happens when b_j is non-negative. We can remove the absolute value and we are left with

$$f_j(b_j) = b_j^2 + b_j \cdot \left(\lambda - 2X_j^t y \right). \tag{7.18}$$

To minimize this, take the derivative and set it equal to zero. This yields

$$b_j = X_j^t y - \frac{\lambda}{2}. \tag{7.19}$$

From our assumption, this is only valid if $b_j = X_j^t y \geq \lambda/2$. Assuming instead that b_j is non-positive yields a similar expression with the sign of λ flipped,

$$b_j = X_j^t y + \frac{\lambda}{2}, \tag{7.20}$$

which is valid if $X_j^t y \leq -\lambda/2$. Putting these conditions together, and extracting the equality case out for clarity, we have the analytic solution

$$\widehat{\beta}_j^{\ell_1} = \begin{cases} X_j^t y - \frac{\lambda}{2} & X_j^t y \geq \frac{\lambda}{2} \\ 0 & X_j^t y \in \left(-\frac{\lambda}{2}, \frac{\lambda}{2}\right) \\ X_j^t y + \frac{\lambda}{2} & X_j^t y \leq \frac{-\lambda}{2} \end{cases} \tag{7.21}$$

for ℓ_1-penalized regression under an orthogonal design matrix X. Specifically, the ℓ_1-penalty shrinks each coordinate towards zero by a linear factor of $\lambda/2$ and sets the coordinate to zero if it is within a factor of $\lambda/2$. By setting a sufficiently large value of λ, we see an explicit derivation showing that the ℓ_1-penalty term performs data-driven variable selection simultaneously while fitting the regression model.

The solutions for both penalized estimators can be visualized as a form of thresholding on the ordinary least squares estimator. We will write the ℓ_0-penalized solution from Equation 7.14 concisely using the hard-threshold function H as

$$\widehat{\beta}_j^{\ell_0} = H(X_j^t y, \lambda^{1/2}). \tag{7.22}$$

where $H(a, b)$ pushes any value of a within a distance of b to the origin to zero and leaves other values untouched. Similarly, the soft-threshold function $S(a, b)$ shrinks the first argument towards zero by up to a linear factor of the second, and sets it equal to zero if it is within a distance of b to the origin. This can be written directly as

$$S(a, b) = \text{sign}(a) \left(|a| - b\right)_+. \tag{7.23}$$

We can compactly write the ℓ_1-penalized estimator from Equation 7.21 using the soft-threshold function as

$$\widehat{\beta}_j^{\ell_1} = S(X_j^t y, \lambda/2). \tag{7.24}$$

These equations illustrate the similarities and differences between the ℓ_0- and ℓ_1-penalized estimators. Both use the tuning parameter λ to determine which variables to remove from the model; a sufficiently large value will lead to aggressive variable selection. However, the hard-threshold produces a discontinuous solution whereas the soft-threshold is continuous in each parameter $\widehat{\beta}_j$ as a function of λ.

To illustrate the effects of hard- and soft-thresholding, we can look at the solution paths to the ℓ_0- and ℓ_1-penalized regression models applied to a simulated orthonormal data matrix X. A straightforward way to produce such a data matrix is to construct a random matrix and extract its left singular vectors.

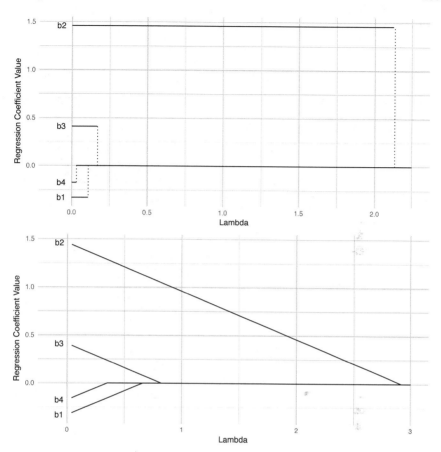

FIGURE 7.2: Solution paths of an ℓ_0-penalized regression (top) and ℓ_1-penalized regression (bottom) as a function of the penalty term λ.

```
Z <- matrix(rnorm(1000 * 4), ncol = 4)
X <- svd(Z)$u
round(crossprod(X), 10)
```

We will generate a response vector y by adding the first two components of X and adding some Gaussian noise.

```
y <- X %*% c(1, 1, 0, 0) + rnorm(1000, sd = 0.7)
Xty <- crossprod(X, y)
```

A graphical depiction of the solutions to Equations 7.4 and 7.2 are given in Figure 7.2. We see that for the penalty equal to zero, both models yield the ordinary least squares regression model. For larger values of λ, some of the co-efficients are set exactly to zero. The ℓ_0-penalized regression paths have hard, discontinuous jumps whereas the ℓ_1-penalized paths are continuous piecewise linear curves.

7.4   Convex optimization and the elastic net

The objective function of the ℓ_1-penalty, unlike the ℓ_0-penalty, is a continuous function in the regression vector. Not only is the objective function continuous, it is also a convex. We say that a function $f : \mathbb{R}^p \to \mathbb{R}$ convex if for any values $b_1, b_2 \in \mathbb{R}^p$ and quantity $t \in [0, 1]$ we have

$$f(tb_1 + (1 - t)b_2) \leq tf(b_1) + (1 - t)f(b_2). \tag{7.25}$$

Replacing the inequality in Equation 7.25 with a strict inequality for $t \in (0, 1)$ yields a definition of a strict convexity. If p is equal to 1 and f is twice differentiable, convexity is equivalent to requiring that the second derivative is always non-negative. We leave as a series of exercises several properties of convex functions that can be used to show that the ℓ_1-penalized objective function is in fact convex. When the matrix X is full-rank, the objective function will be strictly convex for all values of λ.

A convex function does not have any local minima that are not also global minima. In other words, if the value b_0 minimizes f over a neighborhood of b_0, it must also minimize f over its entire domain. To see this, assume that b_1 is any point that is not a global minima but set b_2 equal to a global minima of f. Then, for any $t \in [0, 1)$, we have

$$tf(b_1) + (1 - t)f(b_2) < tf(b_1) + (1 - t)f(b_1) = f(b_1), \tag{7.26}$$

and by convexity this implies that

$$f(tb_1 + (1 - t)b_2) < f(b_1). \tag{7.27}$$

For any neighborhood around b_1 we can find a t close enough to 1 such that $tb_1 + (1 - t)b_2$ is in that neighborhood, and therefore b_1 cannot be a local minimum. A similar derivation shows that a strictly convex function has (at most) a single global minimizer.

The lack of local optima makes it possible to optimize convex objective functions using general purpose algorithms. The study of these methods, *convex optimization*, is in fact a large sub-domain within the field of optimization. Examples of popular techniques include L-BFGS [117], the Nelder–Mead simplex method [65], conjugate gradients [57], gradient descent [30], bundle methods [105], interior points [5], and subgradient projection [20]. In theory any of these can be applied to solving the ℓ_1-penalized regression problem, but most general purpose algorithms tend to be non-optimal. One reason is that many convex optimization techniques cannot reasonably handle optimization

> ### Elastic Net
>
> Given a data matrix X and response vector y, the elastic net is defined as the solution to the optimization task
>
> $$\arg\min_b \left\{ \frac{1}{2n} ||y - Xb||_2^2 + \lambda \left((1-\alpha)\frac{1}{2}||b||_2^2 + \alpha||b||_1 \right) \right\}$$
>
> for any values $\lambda > 0$ and $\alpha \in [0, 1]$.

problems with a large number of variables, which is the case when applying penalized regression to a model matrix with thousands of variables. Secondly, most iterative general-purpose algorithms will never set components of $\widehat{\beta}$ exactly to zero. Instead, these components will be set to a small value that must be manually truncated. This introduces a potential source of error and fails to make use of the useful structure of the ℓ_1-penalty.

The least angle regression (LARS) algorithm is an algorithm specifically designed to solve the ℓ_1-penalized regression problem [54]. In Section 7.3 we saw that with an orthogonal data matrix the solution $\widehat{\beta}_j^{\ell_1}$ as a function of λ was continuous and piecewise linear. It can be shown that this property extends to an arbitrary data matrix. The LARS algorithm makes use of this fact to derive equations that compute the slopes and change points of these piecewise linear paths. By providing a fast and direct solution, the LARS algorithm contributed significantly to the adoption of ℓ_1-penalized regression within many fields of statistics. The explicit formulas in the model also led to many theoretical results on its convergence and allowed for many popular extensions [31]. Unfortunately, the LARS algorithm is not able to handle generalized linear models. It also suffers from a problem known as *cascading errors* when applied to large datasets. In place of the LARS algorithm, in Section 7.5 we instead derive a coordinate descent algorithm that builds off of the LARS algorithm but is able to compute a solution at any value of λ without having to construct an entire piecewise linear model. Also, as we show in Section 7.7, coordinate descent can be seamlessly extended to work with generalized linear models.

Prior to deriving the coordinate descent algorithm, it will be useful to define a slight generalization and re-parametrization of the model defined in Equation 7.4. We define the objective function f for some $\lambda > 0$ and $\alpha \in [0, 1]$ as

$$f(b; \lambda, \alpha) = \frac{1}{2n} ||y - Xb||_2^2 + \lambda \left((1-\alpha)\frac{1}{2}||b||_2^2 + \alpha||b||_1 \right) \qquad (7.28)$$

and corresponding *elastic net* [189] estimator as

$$\widehat{\beta}^{enet}(\lambda, \alpha) \in \arg\min_b f(b; \lambda, \alpha). \qquad (7.29)$$

Setting α to 1 yields a (re-parameterized) version of the LASSO regression and setting it to 0 reconstructs the ridge regression estimator. Adding a small ℓ_2-penalty preserves the variable selection and convexity properties of the ℓ_1-penalized regression, while reducing the variance of the model when X contains sets of highly correlated variables.

7.5 Coordinate descent

Coordinate descent is a general purpose convex optimization algorithm particularly well-suited to solving the elastic net equation. Coordinate descent successively minimizes the objective function along each variable. In every step all but one variable is held constant and a value for the variable of interest is chosen to minimize the constrained problem. This process is applied iteratively over all of the variables until the algorithm converges. As we will see, each univariate subproblem closely resembles the orthogonal solution derived in Section 7.3.

We begin by writing Equation 7.28 in terms of the individual values of b as

$$f(b) = \frac{1}{2n} \sum_{i=1}^{n} \left(y_i - \sum_{j=1}^{p} x_{ij} b_j \right)^2 + \lambda \sum_{j=1}^{p} \frac{1}{2} (1 - \alpha) b_j^2 + \alpha |b_j|. \tag{7.30}$$

Let $\tilde{b}$ be a vector of candidate values of the regression vector b and assume $b_l > 0$. Then the derivative of this function with respect to b_l at $b = \tilde{b}$ is

$$\frac{\partial f}{\partial b_l} \Big|_{b = \tilde{b}} = -\frac{1}{n} \sum_{i=1}^{n} x_{il} \left(y_i - \sum_{j \neq l}^{p} x_{ij} \tilde{b}_j + x_{il} \tilde{b}_l \right) + \lambda \left((1 - \alpha) \tilde{b}_l + \alpha \right)$$

$$= -\frac{1}{n} \sum_{i=1}^{n} x_{il} \left(y_i - \tilde{y}_i^{(l)} \right) + \sum_{i=1}^{n} x_{il}^2 \tilde{b}_l + \lambda (1 - \alpha) \tilde{b}_l + \lambda \alpha \tag{7.31}$$

where

$$\tilde{y}_i^{(l)} = \sum_{i=1}^{n} x_{ij} \tilde{b}_j$$

is the contribution of all regressors in the model _except_ the lth.

There are two challenges to using this approach to find the optimal value for b. First, as mentioned before there is a discontinuity when any element of b is zero. If this discontinuity did not exist then any $\tilde{b}_j$ could be optimized by setting the function to zero and solving for $\tilde{b}_j$ resulting in

$$\tilde{b}_l = \frac{\frac{1}{n} \sum_{i=1}^{n} x_{il} \left(y_i - \tilde{y}^{(l)} \right) - \lambda \alpha}{1 + \lambda (1 - \alpha)}. \tag{7.32}$$

Second, finding $\tilde{b}_j$ requires knowing all other values of $\tilde{b}$ except for the jth element. If we already knew how to get optimized values for all of the other $\tilde{b}_j$ then we could probably get $\tilde{b}_j$ and we would not be in this situation to begin with. To find the optimized values of $\tilde{b}$, we will rely on coordinate descent, which starts with an initial "guess" for each value of $\tilde{b}$ and updates each element over and over until the updated values do not change (or only change a little bit).

Suppose we have just started the algorithm on a dataset with $\alpha = 1$. Our initial guess for the values of $\tilde{b}$ is a $p \times 1$ vector of zeros. The numerator of 7.32 is

$$\frac{1}{n} \sum_{i=1}^{n} x_{il} y_i - \lambda. \tag{7.33}$$

We will assume that $\sum_{i=1}^{n} x_{il} y_i$ is positive since that is how we arrived at Equation 7.31. If it is not, then you need to find the derivative assuming $b_l < 0$. If $\frac{1}{n} \sum_{i=1}^{n} x_{il} y_i - \lambda \geq 0$, then there is no problem, we simply update according to 7.32. However, if this value is less than zero, then we increase the penalty for b_j in Equation 7.30. We incur a smaller penalty by setting it to zero.

We can then generalize Equation 7.33 using the soft-thresholding function as

$$\tilde{b}_l \leftarrow \frac{S\left(\frac{1}{n}\sum_{i=1}^{n} x_{il}\left(y_i - \tilde{y}_i^{(l)}\right), \lambda\alpha\right)}{1 + \lambda(1-\alpha)}. \tag{7.34}$$

The algorithm updates each of the values of $\tilde{b}$ based on the initial value. When it has completed, we have an updated estimate of $\tilde{b}$. If the penalty is large, then more of these values will be zero. The algorithm continues, this time using the new values of $\tilde{b}$ as the initial value. The algorithm stops when the updated values of $\tilde{b}$ are not different than the values in the previous iteration.

To implement, we start with the soft threshold function.

```
# Soft threshold function.
#
# Args:
#     a: Numeric vector of values to threshold.
#     b: The soft thresholded value.
#
# Returns:
#     Numeric vector of the soft-thresholded values of a.
casl_util_soft_thresh <-
function(a, b)
{
  a[abs(a) <= b] <- 0
  a[a > 0] <- a[a > 0] - b
```

```
  a[a < 0] <- a[a < 0] + b
  a
}
```

Next, we implement a function for updating $\tilde{b}$. The function will take the data, parameters, and previous estimates for the slope coefficients and return the updated slope coefficients. Our implementation will normalize the residuals with a weight parameter W, similar to the iteratively re-weighted least squares model from Section 5.3.

```
# Update beta vector using coordinate descent.
#
# Args:
#     X: A numeric data matrix.
#     y: Response vector.
#     lambda: The penalty term.
#     alpha: Value from 0 and 1; balance between l1/l2 penalty.
#     b: A vector of warm start coefficients for the algorithm.
#     W: A vector of sample weights.
#
# Returns:
#     A matrix of regression vectors with ncol(X) columns
#     and length(lambda_vals) rows.
casl_lenet_update_beta <-
function(X, y, lambda, alpha, b, W)
{
  WX <- W * X
  WX2 <- W * X^2
  Xb <- X %*% b

  for (i in seq_along(b))
  {
    Xb <- Xb - X[, i] * b[i]
    b[i] <- casl_util_soft_thresh(sum(WX[,i, drop=FALSE] *
                                      (y - Xb)),
                                  lambda*alpha)
    b[i] <- b[i] / (sum(WX2[, i]) + lambda * (1 - alpha))
    Xb <- Xb + X[, i] * b[i]
  }
  b
}
```

Fitting the elastic net model requires calling `casl_lenet_update_beta` iteratively, until a stopping condition is reached. We define a tolerance such that if the maximum difference between slope coefficients in successive iterations is small, then we are close enough to the optimal value. Alternatively, we also terminate the algorithm if we have reached a maximum number of iterations. If the underlying search space has a small gradient, then it can be difficult to find the optimal slope coefficients.

```
# Compute linear elastic net using coordinate descent.
#
# Args:
#     X: A numeric data matrix.
#     y: Response vector.
#     lambda: The penalty term.
#     alpha: Value from 0 and 1; balance between l1/l2 penalty.
#     maxit: Integer maximum number of iterations.
#     tol: Numeric tolerance parameter.
#
# Returns:
#     Regression vector beta of length ncol(X).
casl_lenet <-
function(X, y, lambda, alpha = 1,
         b = matrix(0, nrow=ncol(X), ncol=1),
         tol = 1e-5, maxit = 50, W = rep(1, length(y))/length(y))
{
  for (j in seq_along(lambda))
  {
    if (j > 1)
    {
      b[,j] <- b[, j-1, drop = FALSE]
    }

    # Update the slope coefficients until they converge.
    for (i in seq(1, maxit))
    {
      b_old <- b[, j]
      b[, j] <- casl_lenet_update_beta(X, y, lambda[j], alpha,
                                       b[, j], W)
      if (all(abs(b[, j] - b_old) < tol)) {
        break
      }
    }
    if (i == maxit)
    {
      warning("Function lenet did not converge.")
    }
  }
  b
}
```

The implementation also contains a maximum number of iterations to prevent the algorithm from running indefinitely in case of numeric issues.

We can now test the function `casl_lenet` using a small simulated dataset. One powerful feature of the elastic net is the ability to use a *high-dimensional* data matrix X, that is, a data matrix containing more columns than rows. To test this specific feature, we generate a data matrix with 1000 rows and 5000 columns. The regression vector will, however, contain only 3 non-zero terms.

```
n <- 1000
p <- 5000
X <- matrix(rnorm(n * p), ncol = p)
beta <- c(3, 2, 1, rep(0, p - 3))
y <- X %*% beta + rnorm(n = n, sd = 0.1)
```

We will fit a linear elastic net on this data for the penalty tuning parameter λ equal to 0.1. In the output, we display only those components that are non-zero.

```
bhat <- casl_lenet(X, y, lambda = 0.1)
names(bhat) <- paste0("v", seq_len(n))
bhat[bhat != 0]
```

```
        v1        v2        v3
2.9123985 1.9134943 0.9126593
```

We see that at the value 0.1 the estimated model correctly extracts only the first three coefficients. The values of these are close, though all slightly smaller due to the shrinkage of the ℓ_1-penlaty, to the simulated values of 3, 2, and 1. To compare, we also find the solution with the penalty tuning parameter set to the larger value of 2.

```
bhat <- casl_lenet(X, y, lambda = 2)
names(bhat) <- paste0("v", seq_len(n))
bhat[bhat != 0]
```

```
        v1        v2
1.1930363 0.1868619
```

Here, only the two largest coefficients are included in the model and the coefficients are much smaller than the simulated values. This tuning parameter would be good for identifying the most interesting features in the data matrix X, but is unlikely to produce predictions that are as accurate as the smaller tuning parameter value. Finally, we include an additional ℓ_2-penalty by setting α equal to 0.5, keeping the larger value of λ.

```
bhat <- casl_lenet(X, y, lambda = 2, alpha = 0.5)
names(bhat) <- paste0("v", seq_len(p))
bhat[bhat != 0]
```

v1	v2	v3
1.09603735	0.57688867	0.09540428

As was the case with ridge regression, the ℓ_2-penalty here spreads out coefficient values more evenly over the variables. Now, the third variable is again included in the model and the coefficient value for the first term is reduced.

7.6   (⋆) Active set screening using the KKT conditions

To fit a regression model, we need to decide on an appropriate value for λ. If it is too big, then its corresponding terms will dominate the penalty in Equation 7.28 and all of the slope coefficients will be estimated as zero. If it is too small, then we will overfit to the training data. This would be like including non-significant variables in ordinary least squares regression. A good approach, much like the ridge regression penalty, is to use validation or cross-validation to find the best tuning parameter for prediction purposes.

What values of λ should we use when performing cross-validation? At zero, the entire ordinary least squares solution is returned, and at some sufficiently large λ all of the coordinates are set to zero. Interesting solutions occur between these extremes. Consider the case that α is equal to 1. From the coordinate descent updates in Equation 7.32

$$\tilde{b}_j = \frac{\frac{1}{n}X_j^t y - \lambda}{1 + \lambda} \tag{7.35}$$

It follows that if $|X_j^t y| \leq \lambda$ then $\tilde{b}_j = 0$. The smallest value of λ for which all slope coefficients are zero is therefore:

$$\lambda_{max} = \underset{j \in \{1,2,\dots,p\}}{\mathrm{argmax}} \left| \frac{X_j^t y}{n} \right|. \tag{7.36}$$

Therefore, a common technique is to use a sequence of tuning parameters distributed between 0 and λ_{max} for cross-validation. Typically these are selected so that the tuning parameters are evenly distributed on the log-scale.

Now, suppose we have found the slope coefficients for a given penalty λ_1, denoted $\widehat{b}(\lambda_1)$, such that some of the coeffients are zero and some are not. If we are given a new penalty, λ_2 with $\lambda_2 < \lambda_1$, then we could refit, using the new penalty parameter and $\widehat{b}(\lambda_1)$ as the starting point. This would surely converge more quickly than an initial guess of all zero values. However, for high-dimensional problems our algorithm would still spend considerable time in the `casl_lenet_update_beta` function fitting zero slope coefficients. If $\lambda_1 - \lambda_2$ is small, then we know that most of the zeros in $\widehat{b}(\lambda_1)$ will be the same as those in $\widehat{b}(\lambda_2)$. So we might like to do a better job using this information to fit $\widehat{b}(\lambda_2)$ more efficiently.

The set of non-zero coefficients in $\widehat{b}(\lambda_2)$ is probably a subset of those in $\widehat{b}(\lambda_1)$. So, we could fit $\widehat{b}(\lambda_1)$, identify the coefficients that are non-zero and only fit the non-zero coefficients for λ_2. We will call the set of non-zero coefficients the *active set*. If the difference $\lambda_1 - \lambda_2$ is small enough then the active sets for the two penalties are the same and we can quickly fit $\widehat{b}(\lambda_2)$ by updating the non-zero coefficients of $\widehat{b}(\lambda_2)$ with the new penalty.

If the difference is not small then we still need to add the new non-zero coefficients to the active set. To do this, we will test to see if our our estimate of the coefficients is optimal. If it is not, then we can detect at which coefficients the solution is not optimal, add them to the active set, and refit. First, we will show how to detect that slope coefficients are optimal. From Equation (7.31) it follows that, if $\alpha = 1$, the optimal fitted values of b, denoted $\widehat{b}$ would be such that

$$\sum_{i=1}^{n} x_{il} \left(y_i - \sum_{j=1}^{p} x_{ij}\widehat{b}_j \right) = \lambda s_j \qquad (7.37)$$

for $j = 1, 2, ..., p$ where

$$s_j \in \begin{cases} 1 & \text{if} \quad \widehat{b}_j > 0 \\ -1 & \text{if} \quad \widehat{b}_j < 0 \\ [-1, 1] & \text{if} \quad \widehat{b}_j = 0 \end{cases} \qquad (7.38)$$

However, we can say something stronger than this. This formulation fits into the Karush–Kuhn–Tucker (KKT) conditions [93, 97] which means that if these conditions have not been met, then our slope coefficient vector cannot be optimal. Furthermore, the conditions are in terms of the individual slope coefficients. This means we can identify which coefficients violate the conditions. These will correspond to coefficients that are not in the active set but should be. Checking to see which coefficients are optimal and which are not can be implemented as follows.

```
# Check current KKT conditions for regression vector.
#
# Args:
#     X: A numeric data matrix.
#     y: Response vector.
#     b: Current value of the regression vector.
#     lambda: The penalty term.
#
# Returns:
#     A logical vector indicating where the KKT conditions have
#     been violated by the variables that are currently zero.
```

```
casl_lenet_check_kkt <-
function(X, y, b, lambda)
{
  resids <- y - X %*% b
  s <- apply(X, 2, function(xj) crossprod(xj, resids)) /
           lambda / nrow(X)

  # Return a vector indicating where the KKT conditions have
  # been violated by the variables that are currently zero.
  (b == 0) & (abs(s) >= 1)
}
```

To fit coefficients corresponding to the next penalty parameter the current coefficients will be used as the starting point and we will fit only on the active set. After fitting, the KKT conditions will be checked. If there are no violations then we are guaranteed to have found the optimal solution for the given penalty. If there are violations then the corresponding slope coefficients are added to the active set and the model is refit. This continues until there are no KKT violations at which point the algorithm can proceed to the next penalty parameter.

```
# Update beta vector using KKT conditions.
#
# Args:
#     X: A numeric data matrix.
#     y: Response vector.
#     b: Current value of the regression vector.
#     lambda: The penalty term.
#     active_set: Logical index of the active set of variables.
#     maxit: Integer maximum number of iterations.
#
# Returns:
#     A list indicating the new regression vector and active set.
casl_lenet_update_beta_kkt <-
function(X, y, b, lambda, active_set, maxit=1000L)
{
  if (any(active_set)) {
    b[active_set, ] <- casl_lenet(X[, active_set, drop = FALSE],
                                  y, lambda, 1,
                                  b[active_set, , drop = FALSE],
                                  maxit = maxit)
  }

  kkt_violations <- casl_lenet_check_kkt(X, y, b, lambda)

  while(any(kkt_violations))
```

```
{
  active_set <- active_set | kkt_violations
  b[active_set, ] <- casl_lenet(X[, active_set, drop = FALSE],
                                y, lambda, 1,
                                b[active_set, , drop = FALSE],
                                maxit = maxit)
  kkt_violations <- casl_lenet_check_kkt(X, y, b, lambda)
}

list(b=b, active_set=active_set)
}
```

The estimated slope coefficients and the current active set are returned by the function.

To see how the screening rules work in practice, we will again generate some random data. Here the generation process puts non-zero weights on only the first 10 components of the regression vector, with the size of the components decreasing from 1 to 0.1

```
n <- 1000L
p <- 5000L
X <- matrix(rnorm(n * p), ncol = p)
beta <- c(seq(1, 0.1, length.out=(10L)), rep(0, p - 10L))
y <- X %*% beta + rnorm(n = n, sd = 0.15)
```

If we compute the lasso regression vector at $\lambda = 0.5$, only the first five coefficients are selected by the model.

```
bhat <- casl_lenet(X, y, lambda=0.5)
which(bhat != 0)
```

```
[1] 1 2 3 4 5
```

Suppose we now want to compute the solution to the lasso regression at some smaller value of λ, such as 0.3. We can use our new function, `casl_lenet_check_kkt`, to determine what new variables may need to be included in the model.

```
which(casl_lenet_check_kkt(X, y, bhat, lambda=0.3))
```

```
[1] 6 7
```

We see that the KKT conditions are violated for the 6th and 7th coefficients. This matches our expectation that a smaller value of λ should include more of the non-zero elements of the regression vector β. We can then fit a lasso regression on just the first seven coefficients of the regression vector.

```
active_set <- seq(1L, 7L)
bhat <- rep(0, p)
bhat[active_set] <- casl_lenet(X[, active_set], y, lambda=0.3)
```

Using the new regression vector, the KKT conditions can again be checked to see if they are violated anywhere.

```
which(casl_lenet_check_kkt(X, y, bhat, lambda=0.3))
```

```
integer(0)
```

As the KKT conditions hold on each coordinate, the current value of **bhat** is a valid solution to the lasso regression problem at $\lambda = 0.3$.

We can wrap up the above logic as a new function to more efficiently compute the linear elastic net. The function will standardize the design matrix and center the response. Columns with constant values will be removed.

```
# Apply coordinate descent screening rules.
#
# Args:
#     X: A numeric data matrix.
#     y: Response vector.
#     lambda: The penalty term.
#     b: A matrix of warm start coefficients for the algorithm.
#     maxit: Integer maximum number of iterations.
#
# Returns:
#     Named list of parameters for use in the lenet algorithm.
casl_lenet_screen <-
function(X, y, lambda, b=matrix(0, nrow=ncol(X),
                                ncol=length(lambda)),
         maxit = 10000L)
{
  a0 <- mean(y)
  y <- y - mean(y)

  X <- scale(X)
  center <- attributes(X)[['scaledcenter']]
  scale <- attributes(X)[['scaledscale']]

  keep_cols <- which(scale > 1e-10)
  X <- X[, keep_cols]
  center <- center[keep_cols]
  scale <- scale[keep_cols]

  active_set <- b[, 1] != 0
```

```
lsu <- casl_lenet_update_beta_kkt(X, y, b[, 1, drop=FALSE],
                                  lambda[1], active_set,
                                  maxit)
b[, 1] <- lsu$b

for (i in seq_along(lambda)[-1])
{
  lsu <- casl_lenet_update_beta_kkt(X, y, b[, i - 1L,
                                    drop=FALSE],
                                    lambda[i], lsu$active_set,
                                    maxit)

  b[, i] <- lsu$b
}

list(b=b, a0=a0, center=center, scale=scale,
     keep_cols=keep_cols)
}
```

Notice that in this implementation the initial active set will be empty. It proceeds by fitting $\hat{b}$ for each of the values of λ based on the previous coefficients and active set, if they exist.

7.7 (⋆) The generalized elastic net model

The linear elastic net can be generalized in the same way ordinary least squares was generalized to the GLM in Section 5.3 [61]. The penalized aximum likelihood estimate is determined by iteratively transforming the fitted values, response, and weights and then fitting the linear model. In linear regression the model was fit by solving the normal equations. For the elastic net, this is done by minimizing the elastic net penalty, which we have done via coordinate descent. In other words, the implementation of the generalized elastic net looks similar to the linear elastic net but includes a sample weighting that depends on the link function and exponential family used in the model.

Here is the complete implementation of the generalized elastic net function.

```
# Compute generalized linear elastic net with coordinate descent.
#
# Args:
#     X: A numeric data matrix.
#     y: Response vector.
#     lambda: The penalty term.
#     alpha: Value from 0 and 1; balance between l1/l2 penalty.
#     family: Instance of an R 'family' object.
```

```
#      maxit: Integer maximum number of iterations.
#      tol: Numeric tolerance parameter.
#
# Returns:
#      Regression vector beta of length ncol(X).
casl_glenet <-
function(X, y, lambda, alpha=1, family=binomial(),
         maxit=10000L, tol=1e-5)
{
  b <- matrix(0, nrow=ncol(X), ncol=length(lambda))

  if (!is.null(colnames(X)))
  {
    rownames(b) <- colnames(X)
  }

  for (j in seq_along(lambda))
  {
    if (j > 1L)
    {
      b[, j] <- b[, j - 1L]
    }
    for (i in seq_len(maxit))
    {
      eta <- X %*% b[, j]
      g <- family$linkinv(eta)
      gprime <- family$mu.eta(eta)
      z <- eta + (y - g) / gprime
      W <- as.vector(gprime^2 / family$variance(g)) / nrow(X)

      old_b <- b[,j]
      b[, j] <- casl_lenet_update_beta(X, z, lambda[j], alpha,
                                       b[, j, drop = FALSE], W)

      if (max(abs(b[, j] - old_b)) < tol)
      {
        break
      }
    }
    if (i == maxit)
    {
      warning("Function casl_glenet did not converge.")
    }
  }
  b
```

```
}
```

Notice here that we make use of the sample weight value that we originally incorporated into the function `casl_lenet_update_beta` for just this purpose.

To test the generalized version of the elastic net function, we will again simulate a dataset with a large number of variables and true regression vector with a small set of non-zero components. Here we generate data such that the response is a binary variable related through the logit link function to the projected values $X\beta$.

```
n <- 1000L
p <- 5000L
X <- matrix(rnorm(n * p), ncol = p)
beta <- c(seq(1, 0.1, length.out=(10L)), rep(0, p - 10L))
p <- 1 / (1 + exp(- X %*% beta))
y <- as.numeric(runif(n) < p)
```

Using the function `casl_glenet`, the solution at a particular value of λ can be computed.

```
bhat <- casl_glenet(X, y, lambda=0.05, family=binomial())
active_set <- which(bhat != 0)
cbind(active_set, beta=beta[active_set], bhat=bhat[active_set])
```

```
     active_set beta       bhat
[1,]          1  1.0 0.4206053
[2,]          2  0.9 0.4070979
[3,]          3  0.8 0.3860851
[4,]          4  0.7 0.1694066
[5,]          5  0.6 0.2465033
[6,]          6  0.5 0.1506650
```

The logistic lasso function detects the 6 largest components of the regression vector. While the specific values of the estimated regression vector are not a good match from the true regression vector, given the amount of noise in the data, the overall values are all within the correct order of magnitude and exhibit a general decay from the largest to smallest coordinates.

7.8 Implementation and notes

There are many excellent R packages for fitting ℓ_1-penalized models and their variants. The **lars** package implements the classical LARS path algorithm for solving the linear regression LASSO [71]. An extension to proportional hazards models is provided by the **penalized** package [68]. A path solution applied to an extended class of penalized models is given by **genlasso** [12]. Extensions of LARS to large, potentially out-of-memory datasets, are given in **biglasso** [186]. An implementation of the coordinate descent algorithm is provided by the **glmnet** package [61]. It supports generalized linear models, elastic net penalties, multinomial regression, and cross-validation. The package also allows users to input sparse data matrices. For general purpose applications, **glmnet** is currently our recommended starting place for fitting ℓ_1-penalized models in R.

7.9 Application: Amazon product reviews

7.9.1 Data and setup

In order to demonstrate the powerful benefits of ℓ_1-penalized regression, we will look at several predictive modeling tasks using a dataset of product reviews from Amazon. Our dataset consists of a random sample of one-hundred thousand reviews of books and an equal number of reviews from films/television series. Our goal is to build models that predict (1) how helpful others considered the review to be, (2) how many stars the review gave to the product, and (3) whether the review is of a book or of a film/television show. We make use of word counts as features in all three models.

The data for this application are stored in two files. The first is a CSV file containing one row, a total of 200,000, for each Amazon product review along with relevant metadata.

```
amazon <- read.csv("data/amazon-meta.csv", as.is=TRUE)
head(amazon)
```

	overall	helpful	total	class
1	5	0	0	book
2	5	0	2	book
3	5	0	0	book
4	5	0	0	book
5	5	7	9	book
6	5	0	0	book

We also have a data matrix X, an object known as the term frequency matrix, that counts how often various words are used in the text. Specifically, the element $X_{i,j}$ counts the number of times the word w_j appears in the ith review in the dataset **amazon**, for some pre-specified set $\{w_j\}_{j=1}^p$ containing the words of interest. We produced this matrix using the package **cleanNLP** [10], and read it an as given below.

```
X <- readRDS("data/amazon-tf.Rds")
```

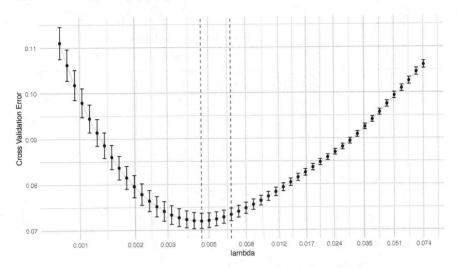

FIGURE 7.3: Cross-validated prediction errors using the elastic net model to predict review helpfulness. Error bars give the standard deviation of the average error metrics across ten folds.

We included any words that occurred at least 0.1% and no more than 50% of the reviews. We will be able to make use of the same matrix X throughout the examples. A more extensive treatment of term frequency matrices, related computational issues, and applications is given in Sections 10.5 and 10.8.

7.9.2 Review helpfulness

For any review on Amazon, customers may vote on whether the review is or is not helpful to them. Our first model will try to predict the proportion of users that vote on a review as being helpful. We will use reviews from films and television shows and only include those that received at least 10 votes. Using this dataset, we will use the function `glmnet::cv.glmnet` to fit a cross-validated elastic net model with α equal to 0.9.

```
y <- amazon$helpful / amazon$total
index <- which(amazon$total > 10 & amazon$class != "book")
out <- glmnet::cv.glmnet(X[index, ] , y[index],
                         family = "gaussian",
                         nfolds = 10L, alpha = 0.9)
```

The model fits 50 values of the tuning parameter λ across 10 cross-validation splits. Figure 7.3 shows the out-of-sample error estimates from the cross-validated linear elastic net model. The "U" shape in the plot is common. When the penalty is small, a large number of regressors are included in the model that are not associated with the outcome. When the model is applied to the test data, the error is high. As the penalty increases the model removes more of the unassociated variables thereby reducing the error. Eventually, the penalty is high enough that predictive variables are removed from the model and the error increases.

At the optimal cross-validated value there are a total of 871 non-zero components from a total of 10,000 possible variables.

```
z <- coef(out, s = out$lambda.min)[-1]
sum(z != 0)
```

```
[1] 871
```

In the code here, we remove the first coefficient, which corresponds to an unpenalized intercept term. Notice that there are two dashed lines in Figure 7.3. The left-most line corresponds to the minimizer of the cross-validated error. The second line shows the largest λ for which the cross-validated error is within one standard error (vertical bars in the plot) of the minimal value. In practice using this choice often leads to smaller models containing significantly fewer variables.

```
z <- coef(out, s = out$lambda.1se)[-1]
sum(z != 0)
```

```
[1] 424
```

Here, we see that the number of non-zero terms has decreased by approximately half by using the one-standard deviation rule.

If we want to investigate which variables are included in the model, it would be difficult to look at hundreds of variables (at least, it is difficult to show such a large set of variables in a textbook). Instead, we will look at the non-zero terms using a larger value of λ. The lambda values supplied by **glmnet** are given from largest to smallest; here we select the 11th largest value.

```
z <- coef(out, s = out$lambda[11])[-1]
names(z) <- rownames(coef(out, s = out$lambda[11]))[-1]
round(sort(z[z != 0]), 5)
```

boring	waste	stupid	her	quot
-0.06803	-0.03161	-0.02753	0.00029	0.00101
young	his	most	from	both
0.00237	0.00247	0.00357	0.00464	0.00553
one	well	superb	who	performances
0.00580	0.00747	0.00750	0.00785	0.00791
has	also	best	great	perfect
0.00805	0.00874	0.01110	0.01196	0.01457

cast	excellent	dvd	wonderful
0.01798	0.01847	0.02005	0.02388

The words "boring," "waste," and "stupid" all contribute negatively to whether a review of a film or television show is judged to be helpful. These are all strong negative words that would likely appear in short reviews without much nuance or context. The other terms are all either positive or neutral. Nouns such as "cast" and "performances" point to longer, more thoughtful reviews that may be most helpful to others interested in buying a movie or television show through Amazon.

7.9.3   Review stars

A rating from 1 to 5 stars is attached to each review on Amazon indicating how much a reviewer recommends the product to others. Using the same term frequency data, we can fit a penalized regression model to predict the number of stars a review will be given based on the text in the review. Here, we use all one-hundred thousand book reviews in our corpus and apply a standard LASSO regression with $\alpha = 1$ using `glmnet::cv.glmnet`. While we could treat each of the five ratings as categorical variables, we will instead consider the number of stars as a continuous response.

```
y <- amazon$overall
index <- which(amazon$class == "book")
out <- glmnet::cv.glmnet(X[index,], y[index],
                         family = "gaussian",
                         nfolds = 10L, alpha = 1)
```

In order to understand the resulting model, we select a relatively large value of λ and look at the non-zero coefficients.

```
z <- coef(out, s = out$lambda[7])[-1]
names(z) <- rownames(coef(out, s = out$lambda[7]))[-1]
round(sort(z[z != 0]), 5)
```

waste	boring	disappointing	disappointment
-0.42804	-0.33760	-0.20917	-0.11783
didn't	finish	disappointed	just
-0.06713	-0.05859	-0.05490	-0.03882
plot	too	bad	nothing
-0.03382	-0.03337	-0.03261	-0.02430
don't	would	author	i
-0.00850	-0.00630	-0.00103	-0.00041

Notice that all of these values are negative. The most frequent star rating in our dataset is a 5, so it does seem reasonable that the model assumes a high value as a default and only decreases this when terms particularly associated with negative reviews are used. We see that the selected terms fit into two broad groups. The first consists of words with a negative connotation ("waste," "bad," "disappointing") and the second includes neutral terms ("i," "just," and "would"). The second set is using stylistic features to capture indicators of more critical reviews. Critical reviews of a book likely contain phrases such as "I would have ..." or "it just needed ...". Words common across these phrases show up in the resulting model.

7.9.4 Product classification

For our final model we will use the entire dataset and try to predict whether a review is of a book or a review of a film/television show. In order to do this we will use a binomial regression model, which is implemented in **glmnet** using the approach shown in Section 7.7. To run a binomial elastic net model with `glmnet::cv.glmnet`, we need to convert our response variable into a vector of 0s and 1s and select the appropriate family option ("binomial") in the function call.

```
y <- as.numeric(amazon$class != "book")
index <- which(amazon$helpful > 100)
out <- glmnet::cv.glmnet(X, y, family = "binomial",
                         nfolds = 3L, alpha = 0.9)
```

The structure of the output from the function call is exactly the same as in the linear regression cases. Once again, we will look at the coefficients using a slightly higher value for λ to investigate which variables are included in the model.

```
z <- coef(out, s = out$lambda[30])[-1]
names(z) <- rownames(coef(out, s = out$lambda[30]))[-1]
round(sort(z[z != 0]), 5)
```

book	read	reading	books	author	written
-1.03617	-0.98012	-0.51980	-0.49322	-0.28205	-0.16839
reader	novel	story	seen	classic	comedy
-0.12649	-0.11378	-0.02769	0.00188	0.01775	0.01865
films	actors	watched	cast	watching	acting
0.07897	0.15902	0.16162	0.18930	0.21566	0.40285
dvd	watch	movies	film	movie	
0.52786	0.60892	0.64073	0.70981	1.16311	

Here we have a good mix of positive and negative terms. The negative words all directly relate to the writing and books (such as "read," "author," and "novel") and the positive terms relative to films and television ("dvd," "watch," and "actors"). The elastic net has done an excellent job of picking a set of terms that distinguish between these two categories and illustrates the power of ℓ_1-penalized regression to datasets with a large number of variables.

7.10   Exercises

1. Show that for any $\lambda > 0$, there exists a value s such that Equations 7.4 and 7.5 are equivalent.

2. Assuming an orthogonal design matrix as given in Equation 7.6, find an analytic solution to the elastic net penalty.

3. Show that if f and g are both convex functions, then their sum must also be convex.

4. Illustrate that the absolute value function is convex. Using the result from the previous exercise, show that the ℓ_1-norm is also convex.

5. Prove that the elastic net objective function is convex using the results from the previous two exercises.

6. Find the KKT conditions for glmnet when $0 < \alpha \leq 1$ and implement a `lasso_reg_with_screening` function that takes an α parameter.

8

Neural Networks

8.1 Dense neural network architecture

Neural networks are a broad class of predictive models that have enjoyed considerably popularity over the past decade. Neural networks consist of a collection of objects, known as neurons, organized into an ordered set of layers. Directed connections pass signals between neurons in adjacent layers. Prediction proceeds by passing an observation X_i to the first layer; the output of the final layer gives the predicted value $\widehat{y}_i$. Training a neural network involves updating the parameters describing the connections in order to minimize some loss function over the training data. Typically, the number of neurons and their connections are held fixed during this process. The concepts behind neural networks have existed since the early 1940s with the work of Warren McCulloch and Walter Pitts [119]. The motivation for their work, as well as the method's name, drew from the way that neurons use activation potentials to send signals throughout the central nervous system. These ideas were later revised and implemented throughout the 1950s [55, 139]. Neural networks, however, failed to become a general purpose learning technique until the early-2000s, due to the fact that they require large datasets and extensive computing power.

Our goal in this chapter is to de-mystify neural networks by showing that they can be understood as a natural extension of linear models. Specifically, they can be described as a collection of inter-woven linear models. This shows how these techniques fit naturally into the sequence of other techniques we have studied to this point. Neural networks should not be understood as a completely separate approach to predictive modeling. Rather, they are an extension of the linear approaches we have studied applied to the problem of detecting non-linear interactions in high-dimensional data.

A good place to begin is the construction of a very simple neural network. Assume that we have data where the goal is to predict a scalar response y from a scalar input x. Consider applying two independent linear models to this dataset (for now, we will ignore exactly how the slopes and intercepts will be determined). This will yield two sets of predicted values for each input,

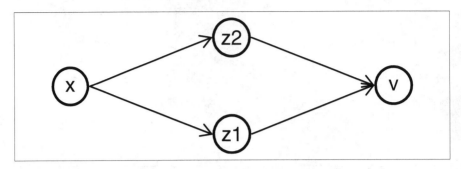

FIGURE 8.1: A diagram describing a simple neural network with one input (x), two hidden nodes (z_1 and z_2), and one output (v). Arrows describe linear relationships between the inputs and the outputs. Each node is also associated with an independent bias term, which is not pictured.

which we will denote by z_1 and z_2

$$z_1 = b_1 + x \cdot w_1 \tag{8.1}$$

$$z_2 = b_2 + x \cdot w_2 \tag{8.2}$$

This requires four parameters: two intercepts and two slopes. Notice that we can consider this as either a scalar equation for a single observation x_i or a vector equation for the entire vector of x.

Now, we will construct another linear model with the outputs z_j as inputs. We will name the output of this next regression model z_3:

$$z_3 = b_3 + z_1 \cdot u_1 + z_2 \cdot u_2. \tag{8.3}$$

Here, we can consider z_3 as being the predicted value $\widehat{y}$. For a visual description of the relationship between these variables, see Figure 8.1. What is the relationship between z_3 and x? In fact, this is nothing but a very complex way of describing a linear relationship between z_3 and x using 7 parameters instead of 2. We can see this by simplifying:

$$z_3 = b_3 + z_1 \cdot u_1 + z_2 \cdot u_2 \tag{8.4}$$

$$= b_3 + (b_1 + x \cdot w_1)u_1 + (b_2 + x \cdot w_2)u_2 \tag{8.5}$$

$$= (b_3 + u_1 \cdot b_1 + u_2 \cdot b_2) + (w_1 \cdot u_1 + w_2 \cdot u_2) \cdot x \tag{8.6}$$

$$= (\text{intercept}) + (\text{slope}) \cdot x \tag{8.7}$$

Figure 8.2 shows a visual demonstration of how the first two regression models reduce to form a linear relationship between x and z_3.

What we have just described is a very simple neural network with four neurons: x, z_1, z_2, and z_3. As shown in Figure 8.1, these neurons are organized into three layers with each neuron in a given layer connected to every neuron in the following layer. The first layer, containing just x is known as the *input*

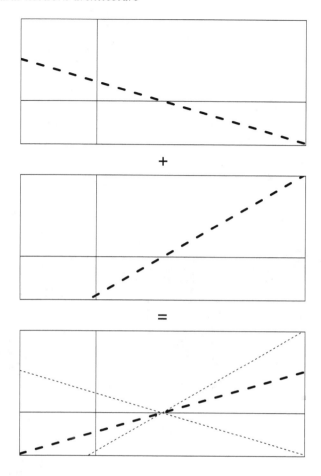

FIGURE 8.2: The top two plots show a linear function mapping the x-axis to the y-axis. Adding these two functions together yields a third linear relationship. This illustrates a simple neural network with two hidden nodes and no activation functions.

layer and the one containing just z_3 known as the *output layer*. Middle layers are called *hidden layers*. The neurons in the hidden layers are known as *hidden nodes*. As written, there is no obvious way of determining the unknown slopes and intercepts. Using the mean squared error as a loss function is reasonable but the model is over-parametrized; there are infinitely many ways to describe the same model. More importantly, there is seemingly no benefit to the neural network architecture here compared to that of a simple linear regression. Both describe the exact same set of relationships between x and y but the latter

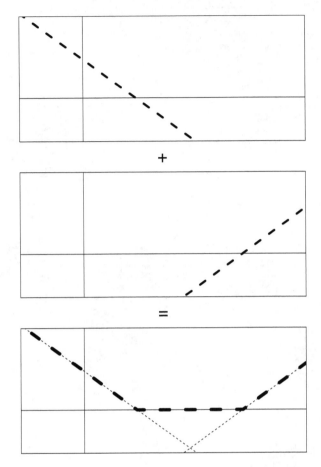

FIGURE 8.3: The top two plots show a linear function mapping the x-axis to the y-axis. The third diagram shows what happens when we add the first two lines together after applying the ReLU activation. In other words, we take the positive part of each function and add it to the other. The resulting function in the third diagram now gives a non-linear relationship between the x-axis and y-axis, illustrating how the use of activation functions is essential to the functioning of neural networks.

has a fast algorithm for computing the unknown parameters and usually has a unique solution.

One small change to our neural network will allow it to capture non-linear relationships between x and y. Instead of writing z_3 as a linear function of z_1

and z_2, we first transform the inputs by a function σ. Namely,

$$z_3 = b_3 + \sigma(z_1) \cdot u_1 + \sigma(z_2) \cdot u_2 \tag{8.8}$$

The function σ is not a learned function. It is just a fixed mapping that takes any real number and returns another real number. In the neural network literature it is called an *activation function*. A popular choice is known as a *rectified linear unit (ReLU)*, which pushes negative values to 0 but returns positive values unmodified:

$$ReLU(x) = \begin{cases} x, & \text{if } x \geq 0 \\ 0, & \text{otherwise} \end{cases} \tag{8.9}$$

The addition of this activation function greatly increases the set of possible relationships between x and z_3 that are described by the neural network. Figure 8.3 visually shows how we can now create non-linear relationships by combining two linear functions after applying the ReLU function. The output now looks similar to a quadratic term. By including more hidden units into the model, and more input values into the first layer, we can create very interesting non-linear interactions in the output. In fact, as shown in the work of Barron, nearly any relationship can be approximated by such a model [16].

Allowing the neural network to describe non-linear relationships does come at a cost. We no longer have an analytic formula of finding the intercepts and slopes that minimizes the training loss. Iterative methods, such as those used for the elastic net in Chapter 7, must be used. Sections 8.2 and 8.3 derive the results needed to efficiently estimate the optimal weights in a neural network.

8.2 Stochastic gradient descent

Gradient descent is an iterative algorithm for finding the minimum value of a function f. It updates to a new value by the formula

$$w_{new} = w_{old} - \eta \cdot \nabla_w f(w_{old}), \tag{8.10}$$

for a fixed learning rate η. At each step it moves in the direction the function locally appears to be decreasing the fastest, at a rate proportional to how fast it seems to be decreasing. Gradient descent is a good algorithm choice for neural networks. Faster second-order methods involve the Hessian matrix, which requires the computation of a square matrix with dimensions equal to the number of unknown parameters. Even relatively small neural networks have tens of thousands of parameters making storage and computation of the Hessian matrix infeasible. Conversely, in Section 8.3 we will see that the gradient can be computed relatively quickly.

Neural networks generally need many thousands of iterations to converge to a reasonable minimizer of the loss function. With large datasets and models, while still feasible, gradient descent can become quite slow. Stochastic gradient descent (SGD) is a way of incrementally updating the weights in the model without needing to work with the entire dataset at each step. To understand the derivation of SGD, first consider updates in ordinary gradient descent:

$$\left(w^{(0)} - \eta \cdot \nabla_w f \right) \rightarrow w^{(1)} \tag{8.11}$$

Notice that for squared error loss (it is also true for most other loss functions), the loss can be written as a sum of component losses for each observation. The gradient, therefore, can also be written as a sum of terms over all of the data points.

$$f(w) = \sum_i (\widehat{y}_i(w) - y_i)^2 \tag{8.12}$$

$$= \sum_i f_i(w) \tag{8.13}$$

$$\nabla_w f = \sum_i \nabla_w f_i \tag{8.14}$$

This means that we could write gradient descent as a series of n steps over each of the training observations.

$$\left(w^{(0)} - (\eta/n) \cdot \nabla_{w^{(0)}} f_1 \right) \rightarrow w^{(1)} \tag{8.15}$$

$$\left(w^{(1)} - (\eta/n) \cdot \nabla_{w^{(0)}} f_2 \right) \rightarrow w^{(2)} \tag{8.16}$$

$$\vdots \tag{8.17}$$

$$\left(w^{(n-1)} - (\eta/n) \cdot \nabla_{w^{(0)}} f_n \right) \rightarrow w^{(n)} \tag{8.18}$$

$$\tag{8.19}$$

The output $w^{(n)}$ here is exactly equivalent to the $w^{(1)}$ from before.

The SGD algorithm actually does the updates in an iterative fashion, but makes one small modification. In each step it updates the gradient with respect to the new set of weights. Writing η' as η divided by the sample size, we can write this as:

$$\left(w^{(0)} - \eta' \cdot \nabla_{w^{(0)}} f_1 \right) \rightarrow w^{(1)} \tag{8.20}$$

$$\left(w^{(1)} - \eta' \cdot \nabla_{w^{(1)}} f_2 \right) \rightarrow w^{(2)} \tag{8.21}$$

$$\vdots \tag{8.22}$$

$$\left(w^{(n-1)} - \eta' \cdot \nabla_{w^{(n)}} f_n \right) \rightarrow w^{(n)} \tag{8.23}$$

In comparison to the standard gradient descent algorithm, the approach of SGD should seem reasonable. Why work with old weights in each step when we already know what direction the vector w is moving? Notice that SGD does not involve any stochastic features other than being sensitive to the ordering of the dataset. The name is an anachronism stemming from the original paper of Robbins and Monro which suggested randomly selecting the data point in each step instead of cycling through all of the training data in one go [138].

In the language of neural networks, one pass through the entire dataset is called an *epoch*. It results in as many iterations as there are observations in the training set. A common variant of SGD, and the most frequently used in the training of neural networks, modifies the procedure to something between pure gradient descent and pure SGD. Training data are grouped into mini-batches, typically of about 32–64 points, with gradients computed and parameters updated over the entire mini-batch. The benefits of this tweak are two-fold. First, it allows for faster computations as we can vectorize the gradient calculations of the entire mini-batch. Secondly, there is also empirical research suggesting that the mini-batch approach stops the SGD algorithm from getting stuck in saddle points [28, 63]. This latter feature is particularly important because the loss function in a neural network is known to exhibit a dense collection of saddle points [41].

8.3 Backward propagation of errors

In order to apply SGD to neural networks, we need to be able to compute the gradient of the loss function with respect to all of the trainable parameters in the model. For dense neural networks, the relationship between any parameter and the loss is given by the composition of linear functions, the activation function σ, and the chosen loss function. Given that activation and loss functions are generally well-behaved, in theory computing the gradient function should be straightforward for a given network. However, recall that we need to have thousands of iterations in the SGD algorithm and that even small neural networks typically have thousands of parameters. An algorithm for computing gradients as efficiently as possible is essential. We also want an algorithm that can be coded in a generic way that can then be used for models with an arbitrary number of layers and neurons in each layer.

The backwards propagation of errors, or just *backpropagation*, is the standard algorithm for computing gradients in a neural network. It is conceptually based on applying the chain rule to each layer of the network in reverse order. The first step consists in inserting an input x into the first layer and then propagating the outputs of each hidden layer through to the final output. All of the intermediate outputs are stored. Derivatives with respect to parameters in the last layer are calculated directly. Then, derivatives are calculated showing how changing the parameters in any internal layer affect the output of just that layer. The chain rule is then used to compute the full gradient in terms of these intermediate quantities with one pass backwards through the network. The conceptual idea behind backpropagation can be applied to any hierarchical model described by a directed acyclic graph (DAG). Our discussion here will focus strictly on dense neural networks, with approaches to the more general problem delayed until Section 8.9.

We now proceed to derive the details of the backpropagation algorithm. One of the most important aspects in describing the algorithm is using a good notational system. Here, we will borrow from the language of neural networks in place of our terminology from linear models. Intercept terms become *biases*, slopes are described as *weights*, and we will focus on computing the gradient with respect to a single row of the data matrix X_i. Because we cannot spare any extra indices, X_i will be denoted by the lower case x. To use backpropagation with mini-batch SGD, simply compute the gradients for each sample in the mini-batch and add them together. Throughout, superscripts describe which layer in the neural network a variable refers to. In total, we assume that there are L layers, not including the input layer, which we denote as layer 0 (the input layer has no trainable weights so it is largely ignored in the backpropagation algorithm).

We will use the variable a to denote the outputs of a given layer in a neural network, known as the activations. As the name suggests, we will assume that the activate function as already been applied where applicable. The activations for the 0'th layer is just the input x itself,

$$a^0 = x. \tag{8.24}$$

For each value of l between 1 and L, the matrix $w_{j,k}^l$ gives the weights on the kth neuron in the $(l-1)$-st layer within the jth neuron in the lst layer. Likewise, the values b_j^l give the bias term for the jth neuron in the lst layer. The variable z is used to describe the output of each layer before applying the activation function. Put together, we then have the following relationship describing the a's, w's, b's, and z's:

$$z^l = w^l \circ a^{l-1} + b^l \tag{8.25}$$

$$a^l = \sigma(z^l) \tag{8.26}$$

These equations hold for all l from 1 up to (and including) L. Layer L is the output layer and therefore the activations a^L correspond to the predicted response:

$$a^L = \widehat{y}. \tag{8.27}$$

Finally, we also need a loss function f. As we are dealing with just a single observation at a time, we will write f as a function of only one input. For example, with mean squared error we would have:

$$f(y, a^L) = (y - a^L)^2 \tag{8.28}$$

The global loss function is assumed to be the sum over the losses of each individual observation. There is a great deal of terminology and notation here, but it is all needed in order to derive the backpropagation algorithm. You should make sure that all of the quantities make sense before proceeding.

The equations defining backpropagation center around derivatives with respect to the quantities z^l. As a starting point, notice that these derivatives give the gradient terms of the biases, b_j^l,

$$\frac{\partial f}{\partial b_j^l} = \frac{\partial f}{\partial z_j^l} \cdot \frac{\partial z_j^l}{\partial b_j^l} \tag{8.29}$$

$$= \frac{\partial f}{\partial z_j^l} \cdot (1) \tag{8.30}$$

$$= \frac{\partial f}{\partial z_j^l}. \tag{8.31}$$

To get the gradient with respect to the weights, we need to weight the derivatives with respect to z by the activations in the prior layer,

$$\frac{\partial f}{\partial w_{jk}^l} = \frac{\partial f}{\partial z_j^l} \cdot \frac{\partial z_j^l}{\partial w_{jk}^l} \tag{8.32}$$

$$= \frac{\partial f}{\partial z_j^l} \cdot a_k^{l-1}. \tag{8.33}$$

We see already why a forward pass through the data is required as a starting point; this forward pass gives us all the current activations a_k^{l-1}. At this point, we now have reduced the problem of computing the gradient to computing the derivatives of the responses z^l.

The derivatives in the Lth layer are straightforward to compute directly with the chain rule, taking advantage of the fact that a_j^L is a function of only z_j^L and no other activations in the Lth layer:

$$\frac{\partial f}{\partial z_j^L} = \sum_k \frac{\partial f}{\partial a_k^L} \cdot \frac{\partial a_k^L}{\partial z_j^L} \tag{8.34}$$

$$= \frac{\partial f}{\partial a_j^L} \cdot \frac{\partial a_j^L}{\partial z_j^L} \tag{8.35}$$

$$= \frac{\partial f}{\partial a_j^L} \cdot \sigma'(z_j^L) \tag{8.36}$$

The derivative of the loss function with respect to the activations in the last layer should be easy to compute since, as shown in Equation 8.28, the loss function is generally written directly in terms of the activations in the last layer.

The last and most involved step is to figure out how the derivatives with respect to z^l can be written as a function of derivatives with respect to z^{l+1}. Notice that if $l \neq L$, then we can write z^{l+1} in terms of the responses in the prior layer,

$$z_k^{l+1} = \sum_m w_{km}^{l+1} a_m^l + b_k^{l+1} \tag{8.37}$$

$$= \sum_m w_{km}^{l+1} \sigma(z_m^l) + b_k^{l+1}. \tag{8.38}$$

From this, we can take the derivative of z_k^{l+1} with respect to z_j^l,

$$\frac{\partial z_k^{l+1}}{\partial z_j^l} = w_{kj}^{l+1} \sigma'(z_j^l). \tag{8.39}$$

Finally, putting this together in the chain rule gives

$$\frac{\partial f}{\partial z_j^l} = \sum_k \frac{\partial f}{\partial z_k^{l+1}} \cdot \frac{\partial z_k^{l+1}}{\partial z_j^l} \tag{8.40}$$

$$= \sum_k \frac{\partial f}{\partial z_k^{l+1}} \cdot w_{kj}^{l+1} \sigma'(z_j^l). \tag{8.41}$$

We can simplify this using the Hadamard product $\odot$ (it computes element-wise products between matrices of the same size):

$$\frac{\partial f}{\partial z^l} = \left[(w^{l+1})^T \left(\frac{\partial f}{\partial z_j^{l+1}} \right) \right] \odot \sigma'(z^l) \tag{8.42}$$

And similarly for the last layer of the network,

$$\frac{\partial f}{\partial z^L} = \nabla_a f \odot \sigma'(z^L). \tag{8.43}$$

Putting these all together, backpropagation can be used to compute the gradient of f with respect to the b's and w's using the same order of magnitude number of steps as it takes to conduct forward propagation.

8.4   Implementing backpropagation

We now have all of the pieces required to train the weights in a neural network using SGD. For simplicity we will implement SGD without mini-batches (adding this feature is included as an exercise). We first make one minor change to the algorithm in Section 8.3. Notice that the current setup applies an activation function to the final layer of the network. If this activation is a ReLU unit, this makes it impossible to predict negative values. Typically, when fitting a neural network to a continuous response we do not use an activation in the final layer. In other words, $z^L = a^L$. Equivalently, we can define the final σ to be the identity function. This just makes the second term on the right-hand side of Equation 8.43, $\sigma'(z^L)$, equal to 1. We will need this activation again when doing classification in Section 8.7.

The code for training and predicting with neural networks is best split into individual functions. Our first step is to define a function that returns the weights and biases of a network in a usable format. We will store these parameters in a list, with one element per layer. Each element is itself a list containing one element **w** (a matrix of weights, w^l) and one element **b** (a vector of biases, b^l). The function `casl_nn_make_weights` creates such a list, filled with randomly generated values from a normal distribution.

```
# Create list of weights to describe a dense neural network.
#
# Args:
#     sizes: A vector giving the size of each layer, including
#            the input and output layers.
#
# Returns:
#     A list containing initialized weights and biases.
casl_nn_make_weights <-
function(sizes)
{
  L <- length(sizes) - 1L
  weights <- vector("list", L)
  for (j in seq_len(L))
  {
    w <- matrix(rnorm(sizes[j] * sizes[j + 1L]),
                ncol = sizes[j],
                nrow = sizes[j + 1L])
    weights[[j]] <- list(w=w,
                         b=rnorm(sizes[j + 1L]))
  }
  weights
}
```

Next, we need to define the ReLU function for the forward pass:

```
# Apply a rectified linear unit (ReLU) to a vector/matrix.
#
# Args:
#     v: A numeric vector or matrix.
#
# Returns:
```

```
#      The original input with negative values truncated to zero.
casl_util_ReLU <-
function(v)
{
  v[v < 0] <- 0
  v
}
```

And the derivative of the ReLU function for the backwards pass:

```
# Apply derivative of the rectified linear unit (ReLU).
#
# Args:
#     v: A numeric vector or matrix.
#
# Returns:
#     Sets positive values to 1 and negative values to zero.
casl_util_ReLU_p <-
function(v)
{
  p <- v * 0
  p[v > 0] <- 1
  p
}
```

We also need to differentiate the loss function for backpropagation. Here we use mean squared error (multiplied by 0.5).

```
# Derivative of the mean squared error (MSE) function.
#
# Args:
#     y: A numeric vector of responses.
#     a: A numeric vector of predicted responses.
#
# Returns:
#     Returned current derivative the MSE function.
casl_util_mse_p <-
function(y, a)
{
  (a - y)
}
```

We will write the code to accept a generic loss function derivative.

With the basic elements in place, we now need to describe how to take an input x and compute all of the responses z and activations a. These will also be stored as lists with one element per layer. Our function here accepts a generic activation function `sigma`.

```
# Apply forward propagation to a set of NN weights and biases.
#
# Args:
#     x: A numeric vector representing one row of the input.
#     weights: A list created by casl_nn_make_weights.
#     sigma: The activation function.
#
# Returns:
#     A list containing the new weighted responses (z) and
#     activations (a).
casl_nn_forward_prop <-
function(x, weights, sigma)
{
  L <- length(weights)
  z <- vector("list", L)
  a <- vector("list", L)
  for (j in seq_len(L))
  {
    a_j1 <- if(j == 1) x else a[[j - 1L]]
    z[[j]] <- weights[[j]]$w %*% a_j1 + weights[[j]]$b
    a[[j]] <- if (j != L) sigma(z[[j]]) else z[[j]]
  }

  list(z=z, a=a)
}
```

With the forward propagation function written, next we need to code a back-propagation function using the results from Equations 8.42 and 8.43. We will have this function accept the output of the forward pass and functions giving the derivatives of the loss and activation functions.

```
# Apply backward propagation algorithm.
#
# Args:
#     x: A numeric vector representing one row of the input.
#     y: A numeric vector representing one row of the response.
#     weights: A list created by casl_nn_make_weights.
#     f_obj: Output of the function casl_nn_forward_prop.
#     sigma_p: Derivative of the activation function.
#     f_p: Derivative of the loss function.
#
# Returns:
#     A list containing the new weighted responses (z) and
#     activations (a).
casl_nn_backward_prop <-
```

```
function(x, y, weights, f_obj, sigma_p, f_p)
{
  z <- f_obj$z; a <- f_obj$a
  L <- length(weights)
  grad_z <- vector("list", L)
  grad_w <- vector("list", L)
  for (j in rev(seq_len(L)))
  {
    if (j == L)
    {
      grad_z[[j]] <- f_p(y, a[[j]])
    } else {
      grad_z[[j]] <- (t(weights[[j + 1]]$w) %*%
                      grad_z[[j + 1]]) * sigma_p(z[[j]])
    }
    a_j1 <- if(j == 1) x else a[[j - 1L]]
    grad_w[[j]] <- grad_z[[j]] %*% t(a_j1)
  }

  list(grad_z=grad_z, grad_w=grad_w)
}
```

The output of the backpropagation function gives a list of gradients with respect to z and w. Recall that the derivatives with respect to z are equivalent to those with respect to the bias terms (Equation 8.31).

Using these building blocks, we can write a function `casl_nn_sgd` that takes input data, runs SGD, and returns the learned weights from the model. As inputs we also include the number of epochs (iterations through the data) and the learning rate `eta`.

```
# Apply stochastic gradient descent (SGD) to estimate NN.
#
# Args:
#     X: A numeric data matrix.
#     y: A numeric vector of responses.
#     sizes: A numeric vector giving the sizes of layers in
#            the neural network.
#     epochs: Integer number of epochs to computer.
#     eta: Positive numeric learning rate.
#     weights: Optional list of starting weights.
#
# Returns:
#     A list containing the trained weights for the network.
casl_nn_sgd <-
function(X, y, sizes, epochs, eta, weights=NULL)
{
```

```
  if (is.null(weights))
  {
    weights <- casl_nn_make_weights(sizes)
  }

  for (epoch in seq_len(epochs))
  {
    for (i in seq_len(nrow(X)))
    {
      f_obj <- casl_nn_forward_prop(X[i,], weights,
                             casl_util_ReLU)
      b_obj <- casl_nn_backward_prop(X[i,], y[i,], weights,
                             f_obj, casl_util_ReLU_p,
                             casl_util_mse_p)

      for (j in seq_along(b_obj))
      {
        weights[[j]]$b <- weights[[j]]$b -
                       eta * b_obj$grad_z[[j]]
        weights[[j]]$w <- weights[[j]]$w -
                       eta * b_obj$grad_w[[j]]
      }
    }
  }

  weights
}
```

We have also written in the ability to include an optional set of starting weights. This allows users to further train a fit model, possibly with a new learning rate or new dataset.

Finally, we need a function that takes the learned weights and a new dataset X and returns the fitted values from the neural network.

```
# Predict values from a training neural network.
#
# Args:
#     weights: List of weights describing the neural network.
#     X_test: A numeric data matrix for the predictions.
#
# Returns:
#     A matrix of predicted values.
casl_nn_predict <-
function(weights, X_test)
{
```

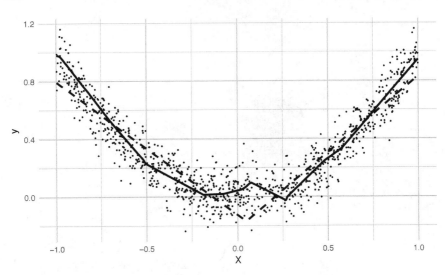

FIGURE 8.4: This scatter plot shows simulated data with a scalar input X and scalar output y (a noisy version of X^2). The dashed line shows the estimated functional relationship from a neural network with one hidden layer containing 5 hidden nodes. The solid line shows the estimated relationship using a model with 25 hidden nodes. Notice that the latter does a better job of fitting the quadratic relationship but slightly overfits near X equal to 0.1. Rectified linear units (ReLUs) were used as activation functions.

```
p <- length(weights[[length(weights)]]$b)
y_hat <- matrix(0, ncol = p, nrow = nrow(X_test))
for (i in seq_len(nrow(X_test)))
{
  a <- casl_nn_forward_prop(X_test[i,], weights,
                            casl_util_ReLU)$a
  y_hat[i, ] <- a[[length(a)]]
}

y_hat
}
```

The implementation here applies the forward propagation function and returns only the last layer of activations.

We can test the code on a small dataset using just a one column input X and one hidden layer with 25 nodes.

```
X <- matrix(runif(1000, min=-1, max=1), ncol=1)
y <- X[,1,drop = FALSE]^2 + rnorm(1000, sd = 0.1)
weights <- casl_nn_sgd(X, y, sizes=c(1, 25, 1),
```

```
                        epochs=25, eta=0.01)
y_pred <- casl_nn_predict(weights, X)
```

The output in Figure 8.4 shows that the neural network does a good job of capturing the non-linear relationship between X and y.

Coding the backpropagation algorithm is notoriously error prone. It is particularly difficult because many errors lead to algorithms that find reasonable, though non-optimal, solutions. A common technique is to numerically estimate the gradient of each parameter for a small model and then check the computed gradients compared to the numerical estimates. Specifically, we can compute

$$\nabla_w f(w_0) \approx \frac{f(w_0 + h) - f(w_0 - h)}{2h} \tag{8.44}$$

For a small vector h with only one non-zero parameter and compare this with the perturbation applied to all trainable parameters in the model.

The code to do this is relatively straightforward. Note that we need to compute the true gradient over all of the data points, not just for a single input x. The reason for this is that given the ReLU activations, many of the derivatives will be trivially equal to zero for many weights with a given input. Added up over all training points, however, this will not be the case for most parameters.

```
# Perform a gradient check for the dense NN code.
#
# Args:
#     X: A numeric data matrix.
#     y: A numeric vector of responses.
#     weights: List of weights describing the neural network.
#     h: Positive numeric bandwidth to test.
#
# Returns:
#     The largest difference between the empirical and analytic
#     gradients of the weights and biases.
casl_nn_grad_check <-
function(X, y, weights, h=0.0001)
{
  max_diff <- 0
  for (level in seq_along(weights))
  {
    for (id in seq_along(weights[[level]]$w))
    {
      grad <- rep(0, nrow(X))
      for (i in seq_len(nrow(X)))
      {
        f_obj <- casl_nn_forward_prop(X[i, ], weights,
```

```
                                          casl_util_ReLU)
        b_obj <- casl_nn_backward_prop(X[i, ], y[i, ], weights,
                                       f_obj, casl_util_ReLU_p,
                                       casl_util_mse_p)
        grad[i] <- b_obj$grad_w[[level]][id]
      }

      w2 <- weights
      w2[[level]]$w[id] <- w2[[level]]$w[id] + h
      f_h_plus <- 0.5 * (casl_nn_predict(w2, X) - y)^2
      w2[[level]]$w[id] <- w2[[level]]$w[id] - 2 * h
      f_h_minus <- 0.5 * (casl_nn_predict(w2, X) - y)^2

      grad_emp <- sum((f_h_plus - f_h_minus) / (2 * h))

      max_diff <- max(max_diff,
                      abs(sum(grad) - grad_emp))
    }
  }
  max_diff
}
```

We can now check our backpropagation code. Here, we will set the weights with just a single epoch so that the resulting gradients are not too small.

```
weights <- casl_nn_sgd(X, y, sizes=c(1, 5, 5, 1), epochs=1,
                       eta=0.01)
casl_nn_grad_check(X, y, weights)
```

```
[1] 5.911716e-12
```

Our numeric gradients match all of the simulated gradients up to a very small number, indicating that our computation of the gradient is accurate.

8.5   Recognizing handwritten digits

In most chapters we have saved applications with real datasets until the final sections. It is, however, hard to show the real usage of neural networks on simulated data. Here we will apply our neural network implementation to a small set of images in order to illustrate how neural networks behave on real datasets.

The MNIST dataset consists of a collection of 60,000 black and white images of handwritten digits. The images are scaled and rotated so that each

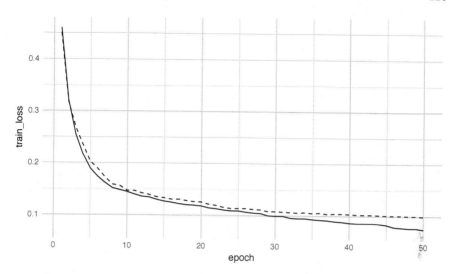

FIGURE 8.5: This plot shows the loss (root mean squared error) during training on both the training and validation sets at the end of each epoch. The solid line is the training error and the dashed line is the validation error.

digit is relatively centered. All images are converted to be 28 pixels by 28 pixels. In Section 8.10 we will see an extension of this dataset that includes handwritten letters. Here we work with a smaller version of MNIST that has been down-sampled to only 7 pixels by 7 pixels and includes only the numbers 0 and 1. This is a binary prediction problem. We can use the neural network from Section 8.4 for a continuous response by predicting the probability that an input is equal to 1 and treating the response as a continuous variable.

The dataset consists of 6000 images for each digits. We split the data in half to form testing and training sets; the response vector is also coded as a matrix as required by our neural network code. The input data, X_mnist is given as a three-dimensional array, which we flatten into a matrix by applying the cbind function over the rows.

```
X_mnist <- readRDS("data/mnist_x7.rds")
mnist <- read.csv("data/mnist.csv")

X_train <- t(apply(X_mnist[mnist$train_id == "train", , ], 1L,
                   cbind))
X_valid <- t(apply(X_mnist[mnist$train_id == "valid", , ], 1L,
                   cbind))
y_train <- matrix(mnist[mnist$train_id == "train", ]$class,
                  ncol=1L)
y_valid <- matrix(mnist[mnist$train_id == "valid", ]$class,
                  ncol=1L)
```

To learn the output, we will fit a linear model with two hidden layers, each containing 64 neurons. In order to plot its progress through the learning algorithm, the learning algorithm will be wrapped in a loop with the training and validation loss stored after every epoch.

```
results <- matrix(NA_real_, ncol = 2, nrow = 25)
val <- NULL
for (i in seq_len(nrow(results)))
{
  val <- casl_nn_sgd(X_train, y_train,
                     sizes=c(7^2, 64, 64, 1),
                     epochs=1L,
                     eta=0.001,
                     weights=val)

  y_train_pred <- (casl_nn_predict(val, X_train) > 0.5)
  y_train_pred <- as.numeric(y_train_pred)
  y_valid_pred <- (casl_nn_predict(val, X_valid) > 0.5)
  y_valid_pred <- as.numeric(y_valid_pred)
  results[i,1] <- sqrt(mean((y_train - y_train_pred)^2))
  results[i,2] <- sqrt(mean((y_valid - y_valid_pred)^2))
}
```

As we are treating this as a continuous response, the results are stored in terms of mean squared error. The values in `results` are plotted in Figure 8.5. We see that towards the end of the training algorithm the training set error improves but the validation error asymptotes to a value 0.1. A selection of images in the validation set that are misclassified by the algorithm are shown in Figure 8.6.

8.6 (⋆) Improving SGD and regularization

The straightforward SGD and backpropagation algorithms implemented in Section 8.4 performs reasonably well. While our R implementation will run quite slowly compared to algorithms written in a compiled language, the algorithms themselves could easily scale to problems with several hidden layers and millions of observations. There are, however, several improvements we can make to the model and the training algorithm that reduce overfitting, have a lower tendency to get stuck in local saddle points, or converge in a smaller number of epochs.

One minor change is to update the weight initialization algorithm. We want the starting activations a^l to all be of roughly the same order of magnitude with the initialized weights. Assuming that the activations in layer a^{l-1} are

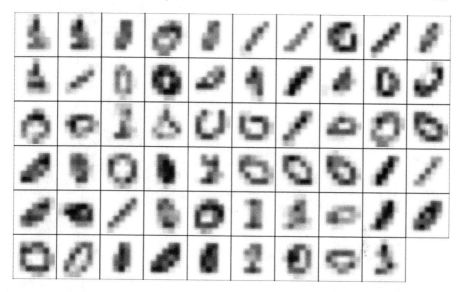

FIGURE 8.6: Misclassified samples from the down-sampled MNIST datasets with only 7-by-7 images.

fixed, we can compute the variance of z^l as a function of the weights in the lth layer,

$$\text{Var}(z_j^l) = \text{Var}\left(\sum_{k=1}^{M^{l-1}} \left[w_{ji}^l a_k^{l-1} + b_k^l\right]\right) \tag{8.45}$$

$$= \sum_{k=1}^{M^{l-1}} Var(w_{ji}^l)\left[a_k^{l-1}\right]^2 \tag{8.46}$$

Where M^l gives the number of neurons in the lth layer of the model. If we sample w_{ji}^{l-1} from the same distribution and assume that the activations in the $(l-1)$st layer are already of approximately the same magnitude, then we see that

$$\text{Var}(z_j^l) \propto M^{l-1} \cdot Var(w_{j1}^l) \tag{8.47}$$

This suggests that we sample the weights such that

$$w_{j1}^l \sim \mathcal{N}(0, \frac{1}{M^{l-1}}). \tag{8.48}$$

This is the recommendation proposed by Yann LeCun and Léon Bottou in a technical report on the subject of neural network initializers [103]. More complex procedures have been suggested by Kaiming He [73] and Xavier Glorot [67]. In our small example with just one hidden layer, the initializations did not have a large effect. For models with many layers, particularly if they are of vastly different sizes, the choice of initialization scheme can drastically improve the performance of the model learned by SGD.

We now integrate the new initialization scheme into a new version of the `casl_nn_make_weights` function. We keep the bias terms sampled from a standard normal distribution; it is also reasonable to set these equal to zero at the start.

```
# Create list of weights and momentum to describe a NN.
#
# Args:
#     sizes: A vector giving the size of each layer, including
#            the input and output layers.
#
# Returns:
#     A list containing initialized weights, biases, and
#     momentum.
casl_nn_make_weights_mu <-
function(sizes)
{
  L <- length(sizes) - 1L
  weights <- vector("list", L)
  for (j in seq_len(L))
  {
    w <- matrix(rnorm(sizes[j] * sizes[j + 1L],
                  sd = 1/sqrt(sizes[j])),
              ncol = sizes[j],
              nrow = sizes[j + 1L])
    v <- matrix(0,
              ncol = sizes[j],
              nrow = sizes[j + 1L])
    weights[[j]] <- list(w=w,
                    v=v,
                    b=rnorm(sizes[j + 1L]))
  }
  weights
}
```

This is a minor change that can be used as is in the `casl_nn_sgd` with no further modifications.

The large number of trainable parameters in a neural network can obviously lead to overfitting on the training data. Many of the tweaks to the basic neural network structure and SGD are directly related to addressing this concern. One common technique is *early stopping*. In this approach the validation error rate is computed after each epoch of the SGD algorithm. When the validation rate fails to improve after a certain number of epochs, or begins to degrade, the SGD algorithm is terminated. Early stopping is almost always used when training neural networks. While neural networks are often framed as predictive models described by an optimization problem, this early stopping criterion means that in practice this is not the case. Neural networks are trained with an algorithm motivated by an optimization problem, but are not directly attempting to actually solve the optimization task.

Another approach to address overfitting is to include a penalty term directly into the loss function. We can add an ℓ_2-norm, for example, as was done with ridge regression in Section 3.2,

$$f_\lambda(w, b) = f(w, b) + \frac{\lambda}{2} \cdot ||w||_2^2 \tag{8.49}$$

Notice here that we have penalized just the weights (slopes) but not the biases (intercepts). The gradient of f_λ can be written in terms of the gradient of f:

$$\nabla_w f_\lambda = \nabla_w f(w, b) + \lambda \cdot w \tag{8.50}$$

With this new form, the SGD algorithm is easy to modify. It is helpful to simplify the gradient updates in terms of a weighted version of the old values and a weighted version of the gradient.

$$w_{new} \leftarrow w_{old} - \eta \cdot \nabla_w f_\lambda(w_{old}) \tag{8.51}$$
$$\leftarrow w_{old} - \eta \cdot [\nabla_w(w_{old}) + \lambda \cdot w] \tag{8.52}$$
$$\leftarrow [1 - \eta\lambda] \cdot w_{old} - \eta \cdot \nabla_w f(w_{old}) \tag{8.53}$$

Other penalties, such as the ℓ_1-norm or penalties on the bias terms, can also be added with minimal difficulty.

Dropout is another clever technique for reducing over fitting during the training for neural networks. During the forward propagation phase, activations are randomly set to zero with some fixed probability p. During backpropagation, the derivatives of the corresponding activations are also set to zero. The activations are chosen separately for each mini-batch. Together, these modifications "prevents units from co-adapting too much," having much the same effect as an ℓ_2-penalty without the need to choose or adapt the parameter λ [148]. The dropout technique is only used during training; all nodes are turned on during prediction on new datasets (in order for the magnitude to work out, weights need to be scaled by a factor of $\frac{1}{1-p}$). Dropout is a popular technique used in most well-known neural networks and is relatively easy to incorporate into the forward and backpropagation algorithms.

Finally, we can also modify the SGD algorithm itself. It is not feasible to store second derivative information directly due to the large number of parameters in a neural network. However, relying only on the gradient leads to models getting stuck at saddle points and being very sensitive to the learning rate. To alleviate this problem, we can incorporate a term known as the *momentum* into the algorithm. The gradient computed on each mini-batch is used to update the momentum term, and the momentum term is then used to update the current weights. This setup gives the SGD algorithm three useful properties: if the gradient remains relatively unchanged over several steps it will 'pick-up speed' (momentum) and effectively use a larger learning rate; if the gradient is changing rapidly, the step-sizes will shrink accordingly; when passing through a saddle point, the built up momentum from prior steps helps propel the algorithm past the point. Expressing the momentum as the vector v, we then have update rules,

$$v_{new} \leftarrow v_{old} \cdot \mu - \eta \cdot \nabla_w f(w_{old}) \tag{8.54}$$

$$w_{new} \leftarrow [1 - \eta\lambda] \cdot w_{old} + v_{new}, \tag{8.55}$$

with μ some quantity between 0 and 1 (typically set between 0.7 and 0.9). Notice that this scheme requires only storing twice as much information as required for the gradient. Here, and in our implementation below, we apply momentum only to the weights. It is also possible, and generally advisable, to apply momentum terms to the biases as well.

Implementing an ℓ_2-penalty term and momentum only requires changes to the `casl_nn_sgd` function. Early stopping requires no direct changes and can be applied as is by running the SGD function for a single epoch, checking the validation rate, and then successively re-running the SGD for another epoch starting with the current weights. Dropout requires minor modifications to both `casl_nn_forward_prop` and `casl_nn_backward_prop`, which we leave as an exercise.

```
# Apply stochastic gradient descent (SGD) to estimate NN.
#
# Args:
#     X: A numeric data matrix.
#     y: A numeric vector of responses.
#     sizes: A numeric vector giving the sizes of layers in
#            the neural network.
#     epochs: Integer number of epochs to computer.
#     eta: Positive numeric learning rate.
#     mu: Non-negative momentum term.
#     12: Non-negative penalty term for 12-norm.
#     weights: Optional list of starting weights.
#
# Returns:
#     A list containing the trained weights for the network.
casl_nn_sgd_mu <-
function(X, y, sizes, epochs, eta, mu=0, 12=0, weights=NULL) {

  if (is.null(weights))
  {
    weights <- casl_nn_make_weights_mu(sizes)
```

```
    }

    for (epoch in seq_len(epochs))
    {
      for (i in seq_len(nrow(X)))
      {
        f_obj <- casl_nn_forward_prop(X[i, ], weights,
                                      casl_util_ReLU)
        b_obj <- casl_nn_backward_prop(X[i, ], y[i, ], weights,
                                       f_obj, casl_util_ReLU_p,
                                       casl_util_mse_p)

        for (j in seq_along(b_obj))
        {
          weights[[j]]$b <- weights[[j]]$b -
                            eta * b_obj$grad_z[[j]]
          weights[[j]]$v <- mu * weights[[j]]$v -
                            eta * b_obj$grad_w[[j]]
          weights[[j]]$w <- (1 - eta * l2) *
                            weights[[j]]$w +
                            weights[[j]]$v
        }
      }
    }

    weights
}
```

The function accepts two new parameters, mu and l2, to define the momentum and penalty terms, respectively.

Figure 8.7 shows the loss function during training both with and without momentum using the data from our simulation in Section 8.4. We see that the momentum-based algorithm trains significantly faster and seems to avoid a saddle point that the non-momentum based SGD gets trapped in. To illustrate the use of the ℓ_2-norm, we will apply the SGD algorithm using successively larger values of λ (l2).

```
l2_norm <- rep(NA_real_, 3)
l2_vals <- c(0, 0.01, 0.04, 0.05)
weights_start <- casl_nn_make_weights(sizes=c(1, 10, 1))
for (i in seq_along(l2_vals))
{
  weights <- casl_nn_sgd_mu(X, y, weights=weights_start,
                            epochs=10, eta=0.1,
                            l2=l2_vals[i])
  l2_norm[i] <- sum((weights[[1]]$w)^2)
```

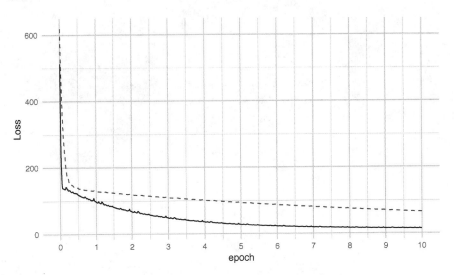

FIGURE 8.7: This plot shows the loss (mean squared error) during training on the training set as a function of the epoch number. The dashed line uses stochastic gradient descent whereas the solid line uses stochastic gradient descent plus momentum.

```
}
12_norm
```

```
[1] 6.942948e+00 1.674410e+00 3.672386e-02 7.117368e-07
```

We see that, as expected, the size of the weights decreases as the penalty term increases.

8.7 $(\star)$ Classification with neural networks

The structure of neural networks is well-suited to classification tasks with many possible classes. In fact, we mentioned in Section 5.6 that the original multinomial regression function in R (`multinom`) is contained in the **nnet** package. Here, we will show how to best use neural networks for classification and integrate these changes to our implementation.

With neural networks there is nothing special about the structure of the output layer. As with hidden layers, it is easy to have a multi-valued output layer. For classification tasks, we can convert a vector y of integer coded classes into a matrix Y containing indicator variables for each class,

$$
\begin{pmatrix} 2 \\ 4 \\ \vdots \\ 1 \end{pmatrix} \rightarrow \begin{pmatrix} 0 & 1 & 0 & 0 & 0 \\ 0 & 0 & 0 & 1 & 0 \\ \vdots & \vdots & \vdots & \vdots & \vdots \\ 1 & 0 & 0 & 0 & 0 \end{pmatrix}.
\tag{8.56}
$$

In neural network literature this is known as a *one-hot encoding*. It is equivalent to the indicator variables used in Section 2.1 applied to categorical variables in the data matrix X. Now, values from the neural network in the output layer can be regarded as probabilities over the possible categories.

Treating the output layer as probabilities raises the concern that these values may not sum to 1 and could produce negative values or values greater than one depending on the inputs. As we did with the logistic link function in Section 5.1, we need to apply a function to the output layer to make it behave as a proper set of probabilities. This will become the activation function σ in the final layer that we have in Equation 8.43. The activation we use is called the *softmax function*, defined as:

$$
a_j^L = \mathrm{softmax}(z_j^L)
\tag{8.57}
$$

$$
= \frac{e^{z_j}}{\sum_k e^{z_k}}.
\tag{8.58}
$$

It should be obvious from the definition that the returned values are all non-negative and sum to 1 (and therefore can never be greater than 1). While we could use squared error loss to train categorization models, it is not an ideal choice. We instead use a quantity known as categorical cross-entropy:

$$
f(a^L, y) = -\sum_k y_k \cdot log(a_k^L)
\tag{8.59}
$$

If the form of this seems surprising, notice that in two-class prediction this reduces to the log-likelihood for logistic regression. In fact, the multinomial distribution can be written as a multidimensional exponential family with the softmax function as an inverse link function and categorical cross-entropy as the log-likelihood function (see Section 5.2 for a description of exponential families and [125] for more details).

The derivative of the softmax function can be written compactly as a function of the Dirac delta operator δ_{ij}. The Dirac delta function is equal to 1 if the indices i and j match, and is equal to 0 otherwise. The softmax derivatives then become

$$
\frac{\partial a_j^L}{\partial z_i^L} = \frac{\delta_{ij} e^{z_j} \cdot (\sum_k e^{z_k}) - e^{z_j} \cdot e^{z_i}}{(\sum_k e^{z_k})^2}
\tag{8.60}
$$

$$
= a_j^L (\delta_{ij} - a_i^L).
\tag{8.61}
$$

With categorical cross-entropy, Equation 8.43 becomes a simple linear function of the activations a^L and the values of Y. Denoting y_i as the ith column of a particular row of Y, we have

$$\frac{\partial f}{\partial z_i^L} = -\sum_k y_k \cdot \frac{\partial}{\partial z_i^L}\left(log(a_k^L)\right) \tag{8.62}$$

$$= -\sum_k y_k \cdot \frac{1}{a_k^L} \cdot \frac{\partial a_k^L}{\partial z_i^L} \tag{8.63}$$

$$= -\sum_k y_k \cdot \frac{1}{a_k^L} \cdot a_k^L(\delta_{ik} - a_i^L) \tag{8.64}$$

$$= -\sum_k y_k \cdot (\delta_{ik} - a_i^L) \tag{8.65}$$

$$= a_i^L \cdot \left(\sum_k y_k\right) - \sum_k y_k \cdot \delta_{ik} \tag{8.66}$$

$$= a_i^L - y_i. \tag{8.67}$$

A change in the weighted response z^L has a linear effect on the loss function, a feature that stops the model from becoming too saturated with very small or very large predicted probabilities.

In order to implement a categorical neural network, we first need a softmax function to apply during forward propagation.

```
# Apply the softmax function to a vector.
#
# Args:
#     z: A numeric vector of inputs.
#
# Returns:
#     Output after applying the softmax function.
casl_util_softmax <-
function(z)
{
  exp(z) / sum(exp(z))
}
```

We then create a modified forward propagation function that takes advantage of the softmax function in the final layer.

```
# Apply forward propagation to for a multinomial NN.
#
# Args:
#     x: A numeric vector representing one row of the input.
#     weights: A list created by casl_nn_make_weights.
#     sigma: The activation function.
```

```
#
# Returns:
#     A list containing the new weighted responses (z) and
#     activations (a).
casl_nnmulti_forward_prop <-
function(x, weights, sigma)
{
  L <- length(weights)
  z <- vector("list", L)
  a <- vector("list", L)
  for (j in seq_len(L))
  {
    a_j1 <- if(j == 1) x else a[[j - 1L]]
    z[[j]] <- weights[[j]]$w %*% a_j1 + weights[[j]]$b
    if (j != L) {
      a[[j]] <- sigma(z[[j]])
    } else {
      a[[j]] <- casl_util_softmax(z[[j]])
    }
  }

  list(z=z, a=a)
}
```

Similarly, we need a new backpropagation function that applies the correct updates to the gradient of terms z^L from the final layer of the model.

```
# Apply backward propagation algorithm for a multinomial NN.
#
# Args:
#     x: A numeric vector representing one row of the input.
#     y: A numeric vector representing one row of the response.
#     weights: A list created by casl_nn_make_weights.
#     f_obj: Output of the function casl_nn_forward_prop.
#     sigma_p: Derivative of the activation function.
#
# Returns:
#     A list containing the new weighted responses (z) and
#     activations (a).
casl_nnmulti_backward_prop <-
function(x, y, weights, f_obj, sigma_p)
{
  z <- f_obj$z; a <- f_obj$a
  L <- length(weights)
  grad_z <- vector("list", L)
  grad_w <- vector("list", L)
```

```
for (j in rev(seq_len(L)))
{
  if (j == L)
  {
    grad_z[[j]] <- a[[j]] - y
  } else {
    grad_z[[j]] <- (t(weights[[j + 1L]]$w) %*%
                       grad_z[[j + 1L]]) * sigma_p(z[[j]])
  }
  a_j1 <- if(j == 1) x else a[[j - 1L]]
  grad_w[[j]] <- grad_z[[j]] %*% t(a_j1)
}

list(grad_z=grad_z, grad_w=grad_w)
}
```

Next, we construct an updated SGD function that calls these new forward
and backward steps. The momentum and ℓ_2-norm terms remain unchanged.

```
# Apply stochastic gradient descent (SGD) for multinomial NN.
#
# Args:
#     X: A numeric data matrix.
#     y: A numeric vector of responses.
#     sizes: A numeric vector giving the sizes of layers in
#            the neural network.
#     epochs: Integer number of epochs to computer.
#     eta: Positive numeric learning rate.
#     mu: Non-negative momentum term.
#     l2: Non-negative penalty term for l2-norm.
#     weights: Optional list of starting weights.
#
# Returns:
#     A list containing the trained weights for the network.
casl_nnmulti_sgd <-
function(X, y, sizes, epochs, eta, mu=0, l2=0, weights=NULL) {

  if (is.null(weights))
  {
    weights <- casl_nn_make_weights_mu(sizes)
  }

  for (epoch in seq_len(epochs))
  {
    for (i in seq_len(nrow(X)))
    {
```

```
      f_obj <- casl_nnmulti_forward_prop(X[i, ], weights,
                                  casl_util_ReLU)
      b_obj <- casl_nnmulti_backward_prop(X[i, ], y[i, ],
                                  weights, f_obj,
                                  casl_util_ReLU_p)

      for (j in seq_along(b_obj))
      {
        weights[[j]]$b <- weights[[j]]$b -
                          eta * b_obj$grad_z[[j]]
        weights[[j]]$v <- mu * weights[[j]]$v -
                          eta * b_obj$grad_w[[j]]
        weights[[j]]$w <- (1 - eta * 12) *
                          weights[[j]]$w +
                          weights[[j]]$v
      }
    }
  }

  weights
}
```

Finally, we also produce a new prediction function that uses the correct forward propagation function for classification.

```
# Predict values from training a multinomial neural network.
#
# Args:
#     weights: List of weights describing the neural network.
#     X_test: A numeric data matrix for the predictions.
#
# Returns:
#     A matrix of predicted values.
casl_nnmulti_predict <-
function(weights, X_test)
{

  p <- length(weights[[length(weights)]]$b)
  y_hat <- matrix(0, ncol=p, nrow=nrow(X_test))
  for (i in seq_len(nrow(X_test)))
  {
    a <- casl_nnmulti_forward_prop(X_test[i, ], weights,
                                  casl_util_ReLU)$a
    y_hat[i,] <- a[[length(a)]]
  }
```

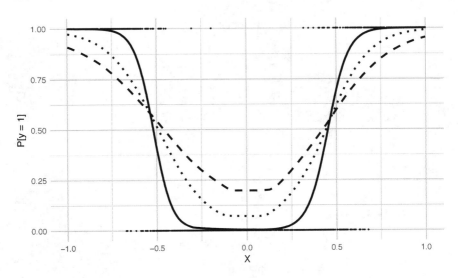

FIGURE 8.8: A scatter plot showing simulated data from a scalar quantity X and a binary variable y. The lines show fitted probabilities from a neural network with one hidden layer containing 25 neurons. Rectified linear units (ReLUs) were used on the hidden layer and softmax activations were used on the output layer to produce valid probabilities. The dashed line shows the fit at the end of the first epoch, the dotted line at the end of 2 epochs, and the solid line the solid line after 25 epochs. Stochastic gradient descent using momentum and categorical cross-entropy was used to the train the model.

```
   y_hat
}
```

The results of the prediction are a matrix with one row for each row in `X_text` and one column for each class in the classification task. To get the predicted class, can simply find whichever column contains the largest probability.

To illustrate this approach, we will simulate a small dataset with just two classes. We again restrict ourselves to a scalar input `X` in order to be able to plot the output in x-y space.

```
X <- matrix(runif(1000, min=-1, max=1), ncol=1)
y <- X[, 1, drop=FALSE]^2 + rnorm(1000, sd=0.1)
y <- cbind(as.numeric(y > 0.3), as.numeric(y <= 0.3))
weights <- casl_nnmulti_sgd(X, y, sizes=c(1, 25, 2),
                            epochs=25L, eta=0.01)
y_pred <- casl_nnmulti_predict(weights, X)
```

The predicted probabilities for various number of epochs are shown in Figure 8.8. During training, the predicted probabilities become more extreme as the algorithm gains confidence that certain inputs always lead to a particular category. On the boundary regions, the neural network correctly predicts a smooth continuum of probabilities. It also has no difficulty detecting the non-linear relationship between the probabilities and the input X (class 1 is most common when X is both relatively small or relatively large).

8.8 (⋆) Convolutional neural networks

As we have mentioned, prediction tasks with images as inputs constitute some of most popular applications of neural networks. When images are relatively small, it is possible to learn a neural network with individual weights placed on each input pixel. For larger images this becomes impractical. A model learned in this way does not take into account the spatial structure of the image, thus throwing away useful information. The solution to this problem is to include convolutional layers into the neural network. Convolutions apply a small set of weights to subsections of the image; the same weights are used across the image.

In the context of neural networks, a convolution can be described by a kernel matrix. Assume that we have the following kernel matrix, to be applied over a black and white image

$$K = \begin{pmatrix} 1 & 0 \\ 0 & -1 \end{pmatrix}. \tag{8.68}$$

The convolution described by this kernel takes every pixel value and subtracts is from the pixel value to its immediate lower right. Notice that this cannot be applied directly to pixels in the last row or column as there is no corresponding pixel to subtract from. The result of the kernel, then, is a new image with one fewer row and column. As shown in Figure 8.9, the convolution created by the kernel is able to capture edges in the original image.

A convolutional layer in a neural network applies several kernels to the input image; the weights, rather than being fixed, are themselves learned as part of the training algorithm. Subsequent layers of the network *flatten* out the multidimensional array and fit dense hidden layers as we have done in previous sections. The idea is that different convolutions will pick up different features. We have seen that one choice of a kernel matrix detects edges; learned convolutions can identify features such as oriented edges, descriptions of texture, and basic object types. If the input image is in color, the kernel matrix K must be a three-dimensional array with weights applied to each color channel. Similarly, we can apply multiple layers of convolutions to the image. The third dimension of kernels in the second layer has to match the size of the number of kernels in the second layer. Stacking convolutions allows neural networks to extract increasingly complex features from the input data.

For simplicity, we will implement a convolutional neural network with a

FIGURE 8.9: The left image is a black and white image of the Eiffel Tower (photo from WikiMedia by Arnaud Ligny). The right image shows the filter obtained by applying the kernel defined in Equation 8.68 to the original image.

single convolutional layer applied to a black and white image. We will also hard code the kernels to be of size 3-by-3 (the most common choice in image processing). As most image prediction tasks involve classification, we will build off of the implementation for multiclass classification from Section 8.7.

A mathematical definition of a convolution is relatively straightforward. Using the notation of Section 8.3, we will describe a neural network that replaces the first layer with a convolution. First, we will need to describe the input using two indices to represent the two spatial dimensions of the data,

$$a_{i,j}^0 = x_{i,j}. \tag{8.69}$$

Then, assuming we want to use kernels of size k_1-by-k_2, the output of the first hidden layer is given by

$$z_{i,j,k}^1 = \sum_{m=0}^{k_1} \sum_{n=0}^{k_2} w_{m,n}^1 \cdot a_{i,j}^0 + b_k^1. \tag{8.70}$$

The indices i and j represent the height and width of an input pixel and k indicates the kernel number. To define the rest of the neural network, we re-parametrize z^1 as:

$$z_q^1 = z_{i,j,k}^1, \quad q = (i-1) \cdot W \cdot K + (j-1) \cdot K + j \tag{8.71}$$

with K the total number of kernels and W the width of the input image. Using z_q^1, all of our previous equations for dense neural networks hold, including those for backpropagation.

By applying the definition in Equation 8.70, it is relatively easy to compute the forward pass step in training neural networks with a single convolutional layer. How does the backpropagation step work? Notice that equations for $\frac{\partial f}{\partial z_q^1}$ still hold with no changes. The only steps required are computing the gradient of the parameters in layer 1 in terms of the derivatives with respect to z_q^1. Once again, this is achieved by applying the chain rule, first for the intercepts

$$\frac{\partial f}{\partial b_k^1} = \sum_i \sum_j \frac{\partial f}{\partial z_{i,j}^1} \cdot \frac{\partial z_{i,j}^1}{\partial b_k^1} \tag{8.72}$$

$$= \sum_i \sum_j \frac{\partial f}{\partial z_{i,j}^1} \tag{8.73}$$

and then for the weights

$$\frac{\partial f}{\partial w_{m,n,k}^1} = \sum_i \sum_j \frac{\partial f}{\partial z_{i,j}^1} \cdot \frac{\partial z_{i,j}^1}{\partial w_{m,n,k}^1} \tag{8.74}$$

$$= \sum_i \sum_j \frac{\partial f}{\partial z_{i,j}^1} \cdot a_{i+m,j+n}^0. \tag{8.75}$$

Given that a weight or bias term in the convolutional layers affects all of the outputs in a given filter, it should seem reasonable that our derivatives now involve a sum over many terms. The only slight complication involves making sure that we compute the derivatives of z indexed as vector, but then apply them to the next layer as an array with three dimensions. This is conceptually simple but some care must be taken in the implementation.

The weights for the first layer of our convolutional neural network need an array of dimension 3-by-3-by-K, where K is the number of kernels. The second layer of weights accepts $(W-2) \cdot (H-2) \cdot K$ inputs, where W and H are the width and height of the input image. We will implement `casl_nn_make_weights` such that the first size gives the size of the outputs in the first convolution and the second size gives the number of filters in the convolution.

```
# Create list of weights and momentum to describe a CNN.
#
# Args:
#     sizes: A vector giving the size of each layer, including
```

```
#                  the input and output layers.
#
# Returns:
#     A list with initialized weights, biases, and momentum.
casl_cnn_make_weights <-
function(sizes)
{
  L <- length(sizes) - 1L
  weights <- vector("list", L)
  for (j in seq_len(L))
  {
    if (j == 1)
    {
      w <- array(rnorm(3 * 3 * sizes[j + 1]),
                 dim=c(3, 3, sizes[j + 1]))
      v <- array(0,
                 dim=c(3, 3, sizes[j + 1]))
    } else {
      if (j == 2) sizes[j] <- sizes[2] * sizes[1]
      w <- matrix(rnorm(sizes[j] * sizes[j + 1],
                        sd=1/sqrt(sizes[j])),
                  ncol=sizes[j],
                  nrow=sizes[j + 1])
      v <- matrix(0,
                  ncol=sizes[j],
                  nrow=sizes[j + 1])
    }

    weights[[j]] <- list(w=w,
                         v=v,
                         b=rnorm(sizes[j + 1]))
  }
  weights
}
```

The output is, as before, a list containing the weights, velocities, and bias terms.

The forward propagation step now needs to convolve the weights in the first layer with the input image.

```
# Apply forward propagation to a set of CNN weights and biases.
#
# Args:
#     x: A numeric vector representing one row of the input.
#     weights: A list created by casl_nn_make_weights.
#     sigma: The activation function.
```

```
#
# Returns:
#     A list containing the new weighted responses (z) and
#     activations (a).
casl_cnn_forward_prop <-
function(x, weights, sigma)
{
  L <- length(weights)
  z <- vector("list", L)
  a <- vector("list", L)
  for (j in seq_len(L))
  {
    if (j == 1)
    {
      a_j1 <- x
      z[[j]] <- casl_util_conv(x, weights[[j]])
    } else {
      a_j1 <- a[[j - 1L]]
      z[[j]] <- weights[[j]]$w %*% a_j1 + weights[[j]]$b
    }
    if (j != L)
    {
      a[[j]] <- sigma(z[[j]])
    } else {
      a[[j]] <- casl_util_softmax(z[[j]])
    }
  }

  list(z=z, a=a)
}
```

The `casl_util_conv` function called by the implementation of forward propagation function is defined as follows.

```
# Apply the convolution operator.
#
# Args:
#     x: The input image as a matrix.
#     w: Matrix of the kernel weight.
#
# Returns:
#     Vector of the output convolution.
casl_util_conv <-
function(x, w) {
  d1 <- nrow(x) - 2L
  d2 <- ncol(x) - 2L
```

```
d3 <- dim(w$w)[3]
z <- rep(0, d1 * d2 * d3)
for (i in seq_len(d1))
{
  for (j in seq_len(d2))
  {
    for (k in seq_len(d3))
    {
      val <- x[i + (0:2), j + (0:2)] * w$w[,,k]
      q <- (i - 1) * d2 * d3 + (j - 1) * d3 + k
      z[q] <- sum(val) + w$b[k]
    }
  }
}

z
}
```

Notice that this code involves a triple loop, so it will be relatively slow in native R code. Custom libraries, which we discuss in Section 8.9, provide fast implementations of the convolution operations.

The backpropagation code is where the real work of the convolutional neural network happens. The top layers proceed as before, but on the first layer we need to add up the contributions from each output to the weights $w_{m,n}^1$. We also need to store the gradient of the bias terms as this is no longer trivially equal to the gradient with respect to the terms z.

```
# Apply backward propagation algorithm for a CNN.
#
# Args:
#     x: A numeric vector representing one row of the input.
#     y: A numeric vector representing one row of the response.
#     weights: A list created by casl_nn_make_weights.
#     f_obj: Output of the function casl_nn_forward_prop.
#     sigma_p: Derivative of the activation function.
#
# Returns:
#     A list containing the new weighted responses (z) and
#     activations (a).
casl_cnn_backward_prop <-
function(x, y, weights, f_obj, sigma_p)
{
  z <- f_obj$z; a <- f_obj$a
  L <- length(weights)
  grad_z <- vector("list", L)
  grad_w <- vector("list", L)
```

```
for (j in rev(seq_len(L)))
{
  if (j == L)
  {
    grad_z[[j]] <- a[[j]] - y
  } else {
    grad_z[[j]] <- (t(weights[[j + 1]]$w) %*%
                    grad_z[[j + 1]]) * sigma_p(z[[j]])
  }
  if (j == 1)
  {
    a_j1 <- x

    d1 <- nrow(a_j1) - 2L
    d2 <- ncol(a_j1) - 2L
    d3 <- dim(weights[[j]]$w)[3]
    grad_z_arr <- array(grad_z[[j]],
                        dim=c(d1, d2, d3))
    grad_b <- apply(grad_z_arr, 3, sum)
    grad_w[[j]] <- array(0, dim=c(3, 3, d3))

    for (n in 0:2)
    {
      for (m in 0:2)
      {
        for (k in seq_len(d3))
        {
          val <- grad_z_arr[, , k] * x[seq_len(d1) + n,
                                       seq_len(d2) + m]
          grad_w[[j]][n + 1L, m + 1L, k] <- sum(val)
        }
      }
    }

  } else {
    a_j1 <- a[[j - 1L]]
    grad_w[[j]] <- grad_z[[j]] %*% t(a_j1)
  }

}

list(grad_z=grad_z, grad_w=grad_w, grad_b=grad_b)
}
```

Notice that the code is simplified by constructing an array version of the gradient, `grad_z_arr`, while working on the first layer parameters.

Finally, we put these parts together in the stochastic gradient descent function. Care needs to be taken to correctly update the bias terms in the first layer. Thankfully, R's vector notation allows us to write one block of code can be used to update the weights regardless of whether they are stored as an array (the convolutional layer) or a matrix (the dense layers).

```r
# Apply stochastic gradient descent (SGD) to a CNN model.
#
# Args:
#     X: A numeric data matrix.
#     y: A numeric vector of responses.
#     sizes: A numeric vector giving the sizes of layers in
#            the neural network.
#     epochs: Integer number of epochs to computer.
#     eta: Positive numeric learning rate.
#     mu: Non-negative momentum term.
#     l2: Non-negative penalty term for l2-norm.
#     weights: Optional list of starting weights.
#
# Returns:
#     A list containing the trained weights for the network.
casl_cnn_sgd <-
function(X, y, sizes, epochs, rho, mu=0, l2=0, weights=NULL) {

  if (is.null(weights))
  {
    weights <- casl_cnn_make_weights(sizes)
  }

  for (epoch in seq_len(epochs))
  {
    for (i in seq_len(nrow(X)))
    {
      f_obj <- casl_cnn_forward_prop(X[i,,], weights,
                                     casl_util_ReLU)
      b_obj <- casl_cnn_backward_prop(X[i,,], y[i,], weights,
                                      f_obj, casl_util_ReLU_p)

      for (j in seq_along(b_obj))
      {
        grad_b <- if(j == 1) b_obj$grad_b else b_obj$grad_z[[j]]
        weights[[j]]$b <- weights[[j]]$b -
                          rho * grad_b
        weights[[j]]$v <- mu * weights[[j]]$v -
```

```
                              rho * b_obj$grad_w[[j]]
          weights[[j]]$w <- (1 - rho * 12) *
                            weights[[j]]$w +
                              weights[[j]]$v

      }
    }
  }

  weights
}
```

Note too that we have to be careful to now index the input X as a three-dimensional array (one dimension for the samples and two for the spatial dimensions). We will also write a prediction function for the convolutional network.

```
# Predict values from training a CNN.
#
# Args:
#     weights: List of weights describing the neural network.
#     X_test: A numeric data matrix for the predictions.
#
# Returns:
#     A matrix of predicted values.
casl_cnn_predict <-
function(weights, X_test)
{

  p <- length(weights[[length(weights)]]$b)
  y_hat <- matrix(0, ncol=p, nrow=nrow(X_test))
  for (i in seq_len(nrow(X_test)))
  {
    a <- casl_cnn_forward_prop(X_test[i, , ], weights,
                               casl_util_ReLU)$a
    y_hat[i, ] <- a[[length(a)]]
  }

  y_hat
}
```

The only differences here are the indices on X and the particular variant of the forward propagation code applied.

To verify that our convolutional neural network implementation is reasonable, we will apply it to our small MNIST classification task. We can now do proper multiclass classification, so the first step is to construct a response matrix y_mnist with two columns.

```
y_mnist <- mnist$class[mnist$class %in% c(0, 1)]
y_mnist <- cbind(1 - y_mnist, y_mnist)
y_train <- y_mnist[1:6000,]
y_valid <- y_mnist[6001:12000,]
```

In the convolutional neural network, we now need to keep the dataset `X_train` as a three-dimensional array.

```
X_train <- X_mnist[seq_len(6000), , ]
X_valid <- X_mnist[seq(6001, 12000), , ]
```

We then pass the training data directly to the neural network training function `casl_cnn_sgd`. Our network includes 5 kernels and one hidden dense layer with 10 neurons. The output layer has two neurons to match the number of columns in `y_train`.

```
out <- casl_cnn_sgd(X_train, y_train,
                    c(5 * 5, 5, 10, 2),
                    epochs=5L, rho=0.003)
pred <- casl_cnn_predict(out, X_valid)
table(pred[, 2] > 0.5, y_valid[, 2])
```

```
              0    1
FALSE      2962    3
TRUE         13 3022
```

The results of the model on the validation set show only 16 misclassified points out of a total 6000, an impressive result that greatly improves on the dense neural network from Section 8.5. The predictive power is particularly impressive given that we are working only with images containing 49 total pixels.

The output of the first convolutional layer will be significantly larger, by a factor of K, than the input image. In our small test case this is a not a concern. As we consider larger input images and a corresponding increase in the number of kernels, this can quickly become an issue. The solution is to include *pooling* layers into the neural network. These pooling operators reduce the resolution of the image after applying a convolution. Most typically, they result in a halving of the width and height of the data. Pooling can be performed by grouping the pixels into 2-by-2 squares and either taking the average of the activations (*mean-pooling*) or the maximum of the activations (*max-pooling*). Backpropagation can be applied to these pooling layers by summing the derivatives over the pool for mean-pooling or assigning the derivative to the pixel with the largest intensity for max-pooling. A common pattern in convolutional neural networks involves applying a convolutional layer following by a pooling layer several times, producing many filters that capture high-level features with a relatively low resolution. We will see an example of this in our application in Section 8.10.

8.9 Implementation and notes

In this chapter we have seen that it is relatively easy to implement neural networks using just basic matrix operations in R. If we want to work with larger datasets, this approach will only carry us so far. The code will run relatively slowly as it executes many nested loops. It also cannot as written take advantage of faster GPU implementations, which are highly optimized for computing tensor products and utilized in almost all research papers published with neural networks. Perhaps most critically, the code as written needs to be completely re-factored whenever we make any minor change to the architecture other than the number of hidden layers and neurons. Fortunately, there are several libraries purpose-built for building neural networks.

The R package **tensorflow** [7] provides access to the TensorFlow library [1]. This is achieved by calling the corresponding Python library by the same name. TensorFlow provides the low-level functionality for working efficiently with multidimensional arrays and a generic form of backpropagation. Models written in TensorFlow are compiled directly to machine code, and can be optimized with CPU or GPU processors. Keras is a higher-level library built on top of TensorFlow. It is available in R through the **keras** package. This library, which we will make use of in the following applications, allows for building neural networks out of layer objects. The corresponding backpropagation algorithm is computing automatically and compiled into fast machine code through TensorFlow. It provides support for many common tweaks to the basic neural network framework, including convolutional neural networks (CNNs) and recurrent neural networks (RNNs).

8.10 Application: Image classification with EMNIST

8.10.1 Data and setup

Here we will work with the EMNIST dataset. This is a recent addition to the MNIST classification data. In place of hand-written digits the EMNIST data consists of handwritten examples of the 26 letters in the Latin alphabet. Both upper and lower case letters are included, but these are combined together into 26 classes. The goal is to use the pixel intensities (a number between 0 and 1) over a 28-by-28 grid to classify the letter.

The data comes in two different parts. This first simply indicates the identity of the letter and whether it is in the training set or the testing set.

```
emnist <- read.csv("data/emnist.csv", as.is=TRUE)
head(emnist)
```

FIGURE 8.10: Example images from the EMNIST dataset, in alphabetical order (y and z not shown). The images are usually coded as white on a black background; we have flipped them to look best in print.

```
      obs_id train_id class class_name
1 id_000001     test     6          g
2 id_000002    train     9          j
3 id_000003    valid    14          o
4 id_000004    train    23          x
5 id_000005    train     5          f
6 id_000006    valid    23          x
```

The actual pixel data is contained in a four-dimensional array.

```
x28 <- readRDS("data/emnist_x28.rds")
dim(x28)
```

```
[1] 124800    28    28     1
```

The first dimension indicates the samples; there are over 120 thousand observations in the dataset. The next two dimensions indicate the height and width of the image. In the final dimension, we simply indicate that this image is black and white rather than color (a color image would have three channels in the fourth position). A plot showing some example letters is given in Figure 8.10.

The first step in setting up the data is to convert the categorical variable class into a matrix with 26 columns. We will make use of the **keras** function to_categorical; notice that it expects the first category to be zero.

```
library(keras)
Y <- to_categorical(emnist$class, num_classes=26L)
emnist$class[seq_len(12)]
```

```
[1]  6  9 14 23  5 23 24 17 16  1  5  4
```

The first few rows and columns of resulting matrix should be as expected given the first 12 categories.

```
Y[seq_len(12), seq_len(10)]
```

```
      [,1] [,2] [,3] [,4] [,5] [,6] [,7] [,8] [,9] [,10]
 [1,]    0    0    0    0    0    0    1    0    0     0
 [2,]    0    0    0    0    0    0    0    0    0     1
 [3,]    0    0    0    0    0    0    0    0    0     0
 [4,]    0    0    0    0    0    0    0    0    0     0
 [5,]    0    0    0    0    0    1    0    0    0     0
 [6,]    0    0    0    0    0    0    0    0    0     0
 [7,]    0    0    0    0    0    0    0    0    0     0
 [8,]    0    0    0    0    0    0    0    0    0     0
 [9,]    0    0    0    0    0    0    0    0    0     0
```

[10,]	0	1	0	0	0	0	0	0	0	0
[11,]	0	0	0	0	0	1	0	0	0	0
[12,]	0	0	0	0	1	0	0	0	0	0

Next, we need to flatten the pixel data x28 into a matrix. This is achieved by applying the cbind function of the rows to the array. We then split the data into training and testing sets.

```
X <- t(apply(x28, 1, cbind))

X_train <- X[emnist$train_id == "train",]
X_valid <- X[emnist$train_id != "train",]
Y_train <- Y[emnist$train_id == "train",]
Y_valid <- Y[emnist$train_id != "train",]
```

The **keras** library will allow us to give a validation set in addition to the training set. This allows users to observe how well the model is doing during training, which can take a while. This aids in the process of early stopping as we can kill the training if the model appears to start overfitting.

8.10.2 A shallow network

Our first task will be to fit a shallow neural network without any hidden nodes. This will help to explain the basic functionality of the **keras** package and allow us to visualize the learned weights. The first step in building a neural network is to call the function keras_model_sequential. This constructs an empty model that we can then add layers to.

```
model <- keras_model_sequential()
```

Next, we add layers to the model using the %>% function. Here we just add one dense layer with one neuron per column in the output Y. We also need to specify that the input matrix has 28^2 columns and that we want to apply the softmax activation to the top layer.

```
model %>%
  layer_dense(units=26, input_shape=c(28^2)) %>%
  layer_activation(activation="softmax")
```

Notice that **keras** has an un-R like object-oriented calling structure. We do not need to save the model with the <- sign; the object model is mutable and changes directly when adding layers. The same calling mechanism works for training. If we manually terminate the SGD algorithm during training, the weights up to that point are not lost.

Next, we need to compile the model using the compile function. This is where we specify the loss function (categorical_crossentropy), the optimization function (SGD, with an η of 0.01 and momentum term of 0.8), and what metrics we want printed with the result.

```
model %>% compile(loss='categorical_crossentropy',
              optimizer=optimizer_sgd(lr=0.01,
                                     momentum=0.80),
              metrics=c('accuracy'))
```

The model has now been compiled to machine code. Options exist within the R package to compile for GPU architectures if these are available.

Finally, we run the function `fit` on the model to train the weights on the training data. We specify the number of epochs and also pass the the validation data.

```
history <- model %>%
  fit(X_train, Y_train, epochs=10,
      validation_data=list(X_valid, Y_valid))
```

After fitting the model, we can then run the `predict_classes` to get the predicted classes on the entire dataset X.

```
emnist$predict <- predict_classes(model, X)
tapply(emnist$predict == emnist$class, emnist$train_id,
       mean)
```

```
    train      valid
0.7259135 0.7122115
```

The accuracy rate here is over 70%, which is actually quite good given the lack of any hidden layers in the network. We can visualize the learned weights for each letter, as shown in Figure 8.11. Notice that the positive weights often seem to trace features of each letter.

8.10.3   A deeper network

To achieve better results we need to use a larger and deeper neural network. The flexibility of the **keras** library makes this easy to code, though of course the algorithm takes significantly longer to run. Here we build a neural network with 4 hidden layers all having 128 neurons. Each layer is followed by a rectified linear unit and dropout with a probability of 25%.

```
model <- keras_model_sequential()
model %>%
  layer_dense(128, input_shape=c(28^2)) %>%
  layer_activation(activation="relu") %>%
  layer_dropout(rate=0.25) %>%

  layer_dense(128, input_shape=c(28^2)) %>%
  layer_activation(activation="relu") %>%
```

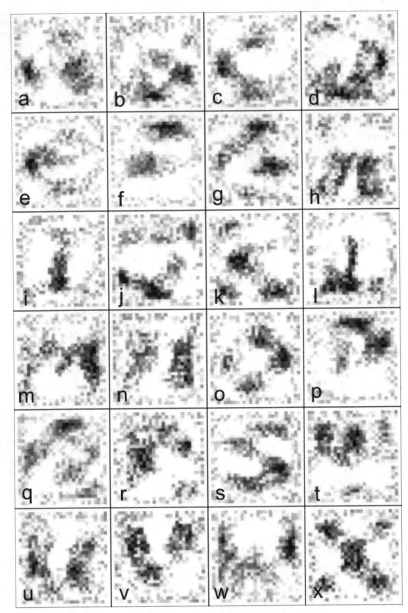

FIGURE 8.11: Visualization of positive weights in a neural network with no hidden layers. The dark pixels indicate strong positive weights and white pixels indicate zero or negative weights. Notice that the positive weights seem to sketch out critical features of the letters, with some (such as M, S, U, and X) sketching the entire letter shape.

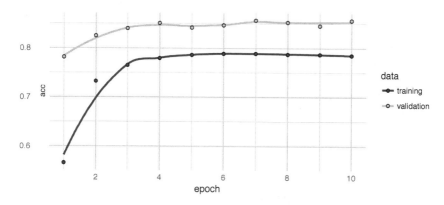

FIGURE 8.12: Accuracy of the dense neural network on the training and validation sets at the end of each epoch in the SGD algorithm. The training accuracy is lower than the validation accuracy due to dropout being turned on in the former but removed in the latter.

```
layer_dropout(rate=0.25) %>%

layer_dense(128, input_shape=c(28^2)) %>%
layer_activation(activation="relu") %>%
layer_dropout(rate=0.25) %>%

layer_dense(128, input_shape=c(28^2)) %>%
layer_activation(activation="relu") %>%
layer_dropout(rate=0.25) %>%

layer_dense(units=26) %>%
layer_activation(activation="softmax")

model %>% compile(loss='categorical_crossentropy',
                optimizer=optimizer_rmsprop(lr=0.001),
                metrics=c('accuracy'))

history <- model %>%
  fit(X_train, Y_train, epochs=10,
      validation_data=list(X_valid, Y_valid))
```

Figure 8.12 shows the training and validation accuracy during training. The validation accuracy is better due to the fact that dropout is turned off for the validation set, but not for the training set during training.

The final model improves on the shallow network, achieving an accuracy of around 85% on the testing set.

```
emnist$predict <- predict_classes(model, X)
tapply(emnist$predict == emnist$class, emnist$train_id,
       mean)
```

```
    train    valid
0.8754968 0.8570192
```

Looking at the most confused classes, we see that just a few pairs of letters are causing most of the problems.

```
emnist$predict <- letters[emnist$predict + 1]
tab <- dplyr::count(emnist, train_id, predict, class_name,
                    sort = TRUE)
tab <- tab[tab$train_id == "valid" &
           tab$predict != tab$class_name,]
tab
```

```
   train_id predict class_name     n
1  valid    i       l            566
2  valid    q       g            276
3  valid    l       i             96
4  valid    e       c             69
5  valid    o       d             69
6  valid    u       v             67
7  valid    g       q             59
8  valid    o       a             59
9  valid    v       y             59
10 valid    o       q             57
```

Distinguishing between 'i' and 'l' (probably the upper case version of the first for the lower case version of the second) and 'q' and 'g' seem to be particularly difficult.

8.10.4 A convolutional neural network

In order to improve our predictive model further, we will need to employ convolutional neural networks. Thankfully, this is relatively easy to do in **keras**. To start, we will now need to have the data X un-flattened:

```
X <- array(x28, dim = c(dim(x28), 1L))
X_train <- X[emnist$train_id == "train", , , , drop=FALSE]
X_valid <- X[emnist$train_id != "train", , , , drop=FALSE]
```

Then, we build a keras model as usual, but use the convolutional layers `layer_conv_2d` and `layer_max_pooling_2d`. Options for these determine the number of filters, the kernel size, and options for padding the input.

FIGURE 8.13: Examples of mis-classified test EMNIST observations by the convolutional neural network.

```
model <- keras_model_sequential()
model %>%
  layer_conv_2d(filters = 32, kernel_size = c(2,2),
                  input_shape = c(28, 28, 1),
                  padding = "same") %>%
  layer_activation(activation = "relu") %>%
  layer_conv_2d(filters = 32, kernel_size = c(2,2),
                  padding = "same") %>%
  layer_activation(activation = "relu") %>%
  layer_max_pooling_2d(pool_size = c(2, 2)) %>%
  layer_dropout(rate = 0.5) %>%

  layer_conv_2d(filters = 32, kernel_size = c(2,2),
                  padding = "same") %>%
  layer_activation(activation = "relu") %>%
  layer_conv_2d(filters = 32, kernel_size = c(2,2),
                  padding = "same") %>%
  layer_activation(activation = "relu") %>%
  layer_max_pooling_2d(pool_size = c(2, 2)) %>%
  layer_dropout(rate = 0.5) %>%

  layer_flatten() %>%
  layer_dense(units = 128) %>%
  layer_activation(activation = "relu") %>%
  layer_dense(units = 128) %>%
  layer_activation(activation = "relu") %>%
  layer_dropout(rate = 0.5) %>%
  layer_dense(units = 26) %>%
  layer_activation(activation = "softmax")
```

Before passing the convolutional input into the dense layers at the top of the network, the layer `layer_flatten` is used to convert the multidimensional input into a two-dimensional output.

Compiling and fitting a convolutional neural network is done exactly the same way that dense neural networks are used in **keras**.

```
model %>% compile(loss = 'categorical_crossentropy',
                  optimizer = optimizer_rmsprop(),
                  metrics = c('accuracy'))

history <- model %>%
  fit(X_train, Y_train, epochs = 10,
      validation_data = list(X_valid, Y_valid))
```

The prediction accuracy is now significantly improved, with a testing accuracy of over 90%.

```
emnist$predict <- predict_classes(model, X)
tapply(emnist$predict == emnist$class, emnist$train_id,
       mean)
```

```
   train      valid
0.9241667 0.9207372
```

Figure 8.13 shows a selection of those images which are still incorrectly classified (known as *negative examples*). While many of these are recognizable at first glance by human readers, many of them are quite unclear with at least two feasible options for the letter.

8.11 Exercises

1. Using the **keras** package functions, use a neural network to predict the tip percentage from the NYC Taxicab dataset in Section 3.6.1. How does this compare to the ridge regression approach?

2. Write a function to check the derivatives of the CNN backpropagation routine from Section 8.8. Does it match the analytic derivatives?

3. The **keras** package contains the function `application_vgg16` that loads the VGG16 model for image classification. Load this model into R and print out the model. In each layer, the number of trainable weights is listed. What proportion of trainable weights is in the convolutional layers? Why is this such a small portion of the entire model? In other words, why do dense layers have many more weights than convolutional layers?

4. Adjust the kernel size, and any other parameters you think are useful, in the convolutional neural network for EMNIST in Section 8.10.4. Can you improve on the classification rate?

5. Implement dropout in the dense neural network implementation from Section 8.4.

6. Change the implementation of backpropagation from Section 8.4 to include a mini-batch of size 16. You can assume that the data size is a multiple of 16.

7. Add an ℓ_1-penalty term in addition to the ℓ_2-penalty term in the code from Section 8.6. You will need to first work out analytically how the updates should be performed.

8. Write a function that uses mean absolute deviation as a loss function, instead of mean squared error. Test the use of this function with a simulation containing several outliers. How well do neural networks and SGD perform when using robust techniques?

9. Rewrite the functions from Section 8.8 to allow for a user supplied kernel size (you may assume that it is square).

10. Implement zero padding in the convolutional neural network implementation from Section 8.8.

9

Dimensionality Reduction

9.1 Unsupervised learning

The dimensionality of data has played a key role in our study of statistical learning. Techniques such as ridge regression, the lasso, elastic net, additive models and the entire machinery of neural networks can all be viewed as attempts to deal with the impending "curse of dimensionality." Yet, high-dimensional spaces have also provided some of the most powerful tools for predictive modeling. In Chapter 4, we expanded our treatment of linear regression to capture non-linear relationships. The resulting splines, and their penalized variants, were eventually derived as a particular form of basis expansion. Rather than develop new techniques for non-linear regression we instead projected our data into a new larger space. Chapter 6 expanded this approach to multivariate data.

In this chapter, we undertake a brief study of several methods in which the dimension of our training data matrix X is explicitly modified. Specifically, we will study functions ϕ such that

$$\phi : \mathbb{R}^p \to \mathbb{R}^d. \tag{9.1}$$

Our ultimate end goal will be to perform dimensionality reduction, where the dimension d is much less than p. While our focus will be on reducing the dimension of the data matrix, several of the techniques presented here will require us to first project into a larger space before reducing into a smaller one. This seemingly contradictory approach is used because, as was the case with linear smoothers, projection into a larger space often linearizes effects that are difficult to discern in lower dimensions.

Unlike most of the other techniques in this text, dimensionality reduction comes from the field of *unsupervised learning*. That is, the techniques do not depend on the presence of a known response vector y. Our only object of study is the data matrix X. The reason for including this topic in a text about supervised learning comes from the central importance that the issue of dimensionality holds over the study of predictive modeling. In many applications, even when one does have a response variable, it can be useful to first apply unsupervised dimensionality reduction to the data matrix. Then, the representation of the data into a smaller space makes it possible to use predictive techniques that are subject to the curse of dimensionality. In fact, techniques such as factorization machines apply dimensionality reduction and predictive modeling into a single unified algorithm. Dimensionality reduction is, of course, also useful in data visualization. While not a central focus of our work, visualizing data is an important part of the larger task of data analysis. We will make use of visualization techniques as a useful way of describing the effects of various methods for projecting our data into a smaller subspace.

Our study of dimensionality reduction will draw from two different approaches. The first closely mirrors those throughout the text so far. Matrix operations are used to preserve metric properties of the higher-dimensional object in lower dimensions. We have already seen one example of this in our derivation of principal components from Section 3.4. The second will make use of stronger probabilistic assumptions in order to capture distributional properties of the original data in a smaller subspace.

9.2 Kernel functions

Our first dimensionality reduction technique will consider ways of identifying low-dimensional structures within higher-dimensional spaces. As mentioned, identifying these structures often requires first considering our data as projected into a larger space. For example, take the simulated data in the first panel of Figure 9.1. The data are roughly situated along a spiraling curve from the center of the plot. Recall that ordinary principal component analysis projects data into a hyperplane that captures the largest possible variation present in the original data. In this simulation, almost any line through the origin would capture about half of the variation; the rest of the variation comes through a line perpendicular to the first. A better way to capture the variation in this data would be to map each observation to the nearest point on the dashed spiral. The variation along this one-dimensional space dominates the variation perpendicular to it.

The general problem with projecting to the dashed line is that, typically, we do have access to such a line. With higher dimensions, more noise, and larger datasets, it is not feasible to identify such structures by hand. How might we automate the identification of a non-linear, lower-dimensional structure to which the data should be projected into? One solution is to use the same technique employed with fitting non-linear relationships using regression splines. We can project the data into a higher-dimensional space that linearizes the relationship. In this example, we could use a second-order polynomial expansion:

$$\phi(x) = \left(x_1^2, x_2^2, x_1 \cdot x_2\right). \tag{9.2}$$

The second panel of Figure 9.1 shows the first two coordinates of the data

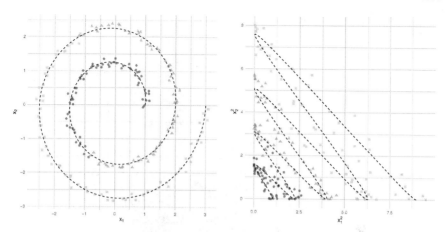

FIGURE 9.1: The first figure shows 200 sample points generated as uniform noise deviating from the dashed spiral. The second figure shows the same data as projected into two of the three components of the feature space induced by the polynomial kernel (with d equal to 2 and c equal to 0). The lightness of the grey points and the three sizes are used to assist in understanding how points in the first image match up to points in the second.

points in this projected space. We see that now the dashed line in the first figure corresponds to a periodic curve moving monotonically away from the origin in the second figure. In the new space, we can draw a line that separates, for example, the dark circles from the other shapes. This implies that logistic regression could be successfully used in the kernel space to classify the shape of the points. While this particular expansion seems specifically constructed for just this example, including high-order terms can linearize many forms of non-linearity.

We will return to the task of dimensionality reduction in the projected space in Section 9.3. Here, we will derive properties of various functions ϕ that are useful in bringing forward non-linear relationships in the original data. In order to capture a large set of possible relationships, we need the *feature space*—the space that ϕ maps into—to be much larger than the original dataset. As we have seen, many applications in statistical learning start with data that already consist of a large number of dimensions. Our primary goal here will be to develop techniques for learning about the feature space without having to explicitly compute and store the projections.

To start with, consider a slightly modified projection from that given in Equation 9.2:

$$\phi(x) = \left(x_1^2, x_2^2, \sqrt{2} \cdot x_1 \cdot x_2\right). \tag{9.3}$$

The $\sqrt{2}$ term will help to simplify our calculations without substantively changing the key geometric properties of the feature space. Now, consider taking an inner product of two vectors in the feature space, namely

$$\phi(x)^t \cdot \phi(z) = x_1^2 z_1^2 + x_2^2 z_2^2 + 2x_1 z_1 x_2 z_2 \qquad (9.4)$$

$$= (x_1 z_1 + x_2 z_2)^2 \qquad (9.5)$$

$$= (x^t z)^2. \qquad (9.6)$$

Conveniently, an inner product in the three-dimensional function space can be written as a function of an inner product in the two-dimensional input space. We call the function that directly maps two input vectors into the inner product in the feature space a *kernel function* (not to be confused with the kernels from linear smoothers in Section 4.3 or convolutional neural networks in Section 8.8). Here, the kernel function K is given by

$$K(x, z) = (x^t z)^2, \quad K : \mathbb{R}^2 \times \mathbb{R}^2 \to \mathbb{R}. \qquad (9.7)$$

If we can construct dimensionality reduction techniques that depend only on inner products, the kernel function allows us to make use of a larger space without needing to explicitly compute or store projections into the feature space. With some carefully constructed logic, this approach, called the *kernel trick*, can be applied to common dimensionality reduction techniques such as principal component analysis as well as predictive models such as linear and logistic regression.

The power and pervasiveness of the kernel trick has often led to academic texts glossing over the connection between the kernel function and the feature space. In many cases, a specific kernel function is given directly without deriving it from an inner product in a larger space. This is an understandable approach. Partially, this is because computation of the final estimator does not require any reference to the function ϕ if we already have access to the kernel function. Also, it can often be difficult or cumbersome to describe a general form for ϕ in cases where K is easily given in a closed form. For example, a generalization of the kernel from Equation 9.3 yields the *polynomial kernel*:

$$K(x, z) = (x^t z + c)^d \qquad (9.8)$$

which can be defined for any input dimension of x and z, positive order d, and non-negative free parameter c. The free parameter c helps to control the influence of higher-order terms over lower-order terms. However, we should not lose sight of the fact that the kernel is just a computational artifact of working in a larger dimensional space.

As long as we can compute inner products, there is no restriction on the dimension of the feature space. In fact, the most commonly used kernel function corresponds to projection into an infinite dimensional space. Consider for the moment a simplified example where our input dataset contains a univariate input. Take the function ϕ defined component-wise as

$$\phi_k(x) = x^k \times \left[\frac{2^k}{k!}\right]^{1/2} \times \left[e^{-x^2}\right], \quad k = 0, 1, 2, \ldots \qquad (9.9)$$

The first term shows that this projection is a scaled version of a power expansion. The second term dampens the first by ensuring that for a sufficiently large k the magnitude of the terms decays to zero. The final term, which is fixed for all values of k, normalizes the projection in a specific way. The reason for the particular dampening and scaling terms given here can be motivated by computing the ℓ_2-norm of a feature vector:

$$||\phi_k(x)||_2^2 = \sum_{k=0}^{\infty} x^{2k} \cdot \frac{2^k}{k!} \cdot \left[e^{-x^2}\right]^2 \qquad (9.10)$$

$$= \left[e^{-x^2}\right]^2 \cdot \sum_{k=0}^{\infty} x^{2k} \cdot \frac{2^k}{k!} \qquad (9.11)$$

$$= \left[e^{-x^2}\right]^2 \cdot \left[e^{x^2}\right]^2 = 1. \qquad (9.12)$$

The last step comes from applying the Taylor series expansion of e^{x^2} around zero. So, the form of this ϕ function has been chosen to map any input into a vector with norm 1; in other words, the map ϕ maps into the unit ball of the space ℓ_∞.

How can we define the kernel function of the map defined in Equation 9.9? A closed form solution comes from a derivation similar to the one that showed that all vectors have a unit norm. Specifically, we have

$$\phi(x)^t \cdot \phi(z) = e^{-x^2} \cdot e^{-z^2} \cdot \sum_{k=0}^{\infty} \frac{2^k x^k z^k}{k!} \qquad (9.13)$$

$$= e^{-x^2} \cdot e^{-z^2} \cdot e^{2xz} \qquad (9.14)$$

$$= \exp\left\{-(x-z)^2\right\}, \qquad (9.15)$$

by once again making use of the Taylor series. We can generalize this to inputs of other dimensions by defining the kernel function for a free parameter $\gamma > 0$ as:

$$K(x, z) = \exp\left\{-\gamma \cdot ||x - z||_2^2\right\}. \qquad (9.16)$$

Setting γ to one and using a univariate input results in the inner product defined in Equation 9.15. As with the polynomial kernel, we should remember that this kernel defines an inner product in a higher-dimensional space. However, it is difficult to write out this space for a general input dimension and value of the free parameter. The kernel defined in Equation 9.16 is called either the *radial basis function* (RBF) or the *Gaussian kernel*. The first name signifies that the inner product depends only on the overall norm of the distance between the inputs (in other words, the radius of the difference as a vector rooted at the origin). The second term derives from setting $\gamma = 1/(2\sigma^2)$ and recognizing the kernel as a scaled version of the normal distribution.

9.3   Kernel principal component analysis

In the prior section we saw how the kernel trick can be used to compute any algorithm projected into a high-dimensional feature space if that algorithm can be described in terms of inner products between observations. Previously, we initially derived the principal components of a matrix in Chapter 3 by describing how they capture variance in the original space. The definition was not written in terms of inner products. Is there a way of re-writing the algorithm that would allow for use of the kernel trick?

A clue comes from the second derivation of principal components from Equation 2.4 using the singular values of the matrix X. Recall that if we write the data matrix X in its SVD decomposition as

$$X = U\Sigma V^t, \tag{9.17}$$

then the kth principal component is given as the kth column of the matrix

$$U\Sigma. \tag{9.18}$$

The outer product matrix XX^t written in terms of the singular value decomposition is

$$XX^t = U\Sigma^2 U^t. \tag{9.19}$$

The eigenvalues of XX^t are the squared singular values of X, and the eigenvectors are the left-singular vectors. Importantly, this outer product matrix can be written component-wise in terms of inner products between observations:

$$(XX^t)_{i,j} = x_i^t x_j. \tag{9.20}$$

Therefore, the principal components are defined entirely in terms of inner products between samples of the data and thus are perfectly suited to make use of the kernel trick.

Our derivation above is written to compute principal components in the original input space. To write out an algorithm for principal components in a larger feature space we define a feature matrix W (though we never need to explicitly construct it due to the kernel trick) and its singular value decomposition

$$w_j = \phi(x_j), \quad W = U_W \Sigma_W V_W^t. \tag{9.21}$$

Then, consider the matrix M of inner products

$$M_{i,j} = w_j^t w_i = K(x_i, x_j), \tag{9.22}$$

which we will compute using the kernel function K. Finally, take the eigenvalue decomposition of M

$$M = Q_M \Lambda_M Q_M^t. \tag{9.23}$$

Then, the kth principal component in the feature space is given by the kth column of the matrix

$$Q_M \cdot \Lambda_M^{1/2}, \tag{9.24}$$

defined for any k no greater than the sample size n.

The only technical detail to fill in concerns the standardization of the data projected into the feature space. Usually variables are centered around zero prior to performing principal component analysis. We instead want to work with the variables

$$\widetilde{\phi}(x_i) = \phi(x_i) - \frac{1}{n} \sum_{k=1}^{n} \phi(x_k) \tag{9.25}$$

and the kernel matrix

$$\widetilde{M}_{i,j} = \widetilde{\phi}(x_i)^t \widetilde{\phi}(x_j). \tag{9.26}$$

We can compute this in matrix form directly (assuming a symmetric kernel function) as

$$\widetilde{M} = M - \frac{2}{n} \cdot \mathbb{1}_n \cdot M + \frac{1}{n^2} \cdot \mathbb{1}_n \cdot M \cdot \mathbb{1}_n \tag{9.27}$$

where $\mathbb{1}_n$ is a square n-by-n matrix of ones. In the linear case, columns are also standardized to have unit variance before applying principal component analysis. This, however, is not sensible to do in the feature space. In the radial basis kernel, for example, the components should decay in magnitude by construction; otherwise, the infinite number of components would lead to infinitely large inner products.

With the kernel principal component analysis algorithm now defined, we can proceed to implement the algorithm in R. The first step is to define a function that computes the kernel matrix M. While a kernel function is defined between two observations, we will be able to use vectorized R functions by constructing a single function that computes the entire matrix all in one shot. For example, the polynomial basis can use the function `tcrossprod` to quickly compute all of the required inner products in the input space.

Kernel Principal Components (k-PCA)

Let X be an $n \times p$ data matrix and K a kernel function. Define the $n \times n$ matrix M as $M_{i,j} = K(x_i, x_j)$ and the normalized matrix $\widetilde{M}$ by:

$$\widetilde{M} = M - \frac{2}{n} \cdot \mathbb{1}_n \cdot M + \frac{1}{n^2} \cdot \mathbb{1}_n \cdot M \cdot \mathbb{1}_n$$

where $\mathbb{1}_n$ is a square n-by-n matrix of ones. The first k kernel principal components are given by the first k columns of $Q\Lambda^{1/2}$, where $Q\Lambda Q^t$ is the eigenvalue decomposition of $\widetilde{M}$.

```
# Compute square kernel matrix for polynomial kernel.
#
# Args:
#     X: A numeric data matrix.
#     d: Integer degree of the polynomial.
#     c: Numeric constant to modify the kernel.
#
# Returns:
#     A square kernel matrix with the nrow(X) rows and columns.
casl_util_kernel_poly <-
function(X, d=2L, c=0)
{
  cross <- tcrossprod(X, X)
  M <- (cross + c)^d
  M
}
```

Similarly, the radial basis function can call the R function `stats::dist` to compute all pairwise Euclidean distances between the data points.

```
# Compute square kernel matrix for a radial kernel.
#
# Args:
#     X: A numeric data matrix.
#     gamma: Positive tuning parameter to set kernel shape.
#
# Returns:
#     A square kernel matrix with the nrow(X) rows and columns.
casl_util_kernel_radial <-
function(X, gamma=1)
{
  d <- as.matrix(dist(X))^2
  M <- exp(-1 * gamma * d)
```

```
      M
}
```

Next, we need a function to take a raw kernel matrix M and return the matrix $\bar{M}$ corresponding to the centered coordinates.

```
# Calculate normalized kernel matrix.
#
# Args:
#     M: A raw kernel matrix.
#
# Returns:
#     A normalized version of the kernel matrix.
casl_util_kernel_norm <-
function(M)
{
  n <- ncol(M)
  ones <- matrix(1 / n, ncol = n, nrow = n)
  Ms <- M - 2 * ones %*% M  + ones %*% M %*% ones
  Ms
}
```

Finally, we provide a function to compute the actual principal components.

```
# Compute kernel version of PCA matrix.
#
# Args:
#     X: A numeric data matrix.
#     k: Integer number of components to return.
#     kfun: The kernel function.
#     ...: Other options passed to the kernel function.
#
# Returns:
#     The nrow(X)-by-k kernel PCA matrix.
casl_kernel_pca <-
function(X, k=2L, kfun=casl_util_kernel_poly, ...)
{
  M <- casl_util_kernel_norm(kfun(X, ...))
  e <- irlba::partial_eigen(M, n = k)
  pc <- e$vectors %*% diag(sqrt(e$values))
  pc[]
}
```

The code here allows users, using the ... notation in R, to provide tuning parameters that effect the kernel function. We have used the **irlba** package to efficiently compute only those principal components required by the computation.

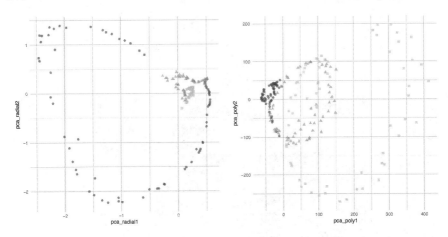

FIGURE 9.2: Plots of the first two kernel principal components of the simulated data from Figure 9.1. The first plot uses a radial basis kernel with γ equal to 1. The second uses a polynomial kernel of degree 2 with free parameter c equal to 1.

Now, we can apply the kernel principal component function to some simulated data. Specifically, we generate data from:

```
set.seed(1)
n <- 200

theta <- seq(0, 4 * pi, length.out = n)
r <- seq(1, 3, length.out = n)
Q <- cbind(cos(theta) * r, sin(theta) * r)
X <- Q + matrix(runif(n * 2, min = -0.15, max = 0.15), ncol = 2)
```

We will apply both the radial and polynomial kernels, grabbing two components each.

```
pca_radial <- casl_kernel_pca(X, k=2L, gamma=25,
                              kfun=casl_util_kernel_radial)
pca_poly <- casl_kernel_pca(X, k=2L, c=1,
                            kfun=casl_util_kernel_poly)
```

Figure 9.2 shows the results of these principal component expansions. Both do a better job of separating the data points than principal components would in the input space. The polynomial kernel more accurately maps the dashed line from the simulated data to a linear subspace in the original plot.

Our implementation of kernel principal components has glossed over one important detail. Once we have used a training set to define a set of principal components, how can we apply the same mapping to a new set of data? In the linear case, this is just a matter of multiplying the new dataset by the *loadings matrix* V; this is equivalent to the right-singular vectors of X. Unfortunately in kernel PCA we do not have direct access to the matrix V because we are only working with the outer product XX^t. Also, the matrix V is either very large (polynomial kernel) or infinitely large (the radial kernel). Fortunately some matrix manipulations show that it is possible to apply the kernel principal components to a new dataset without having to explicitly project into the feature space. Let W_{new} be the projection of the new data into the feature space and $U\Lambda V^t$ be the singular value decomposition of the original W matrix. Also, let $\widetilde{M}_{new}$ be the (normalized) matrix of inner products between the new samples (rows) and the training samples (columns). We see that the principal components $W_{new}V$ can be computed as

$$W_{new}V = W_{new}V\lambda\lambda^{-1} \tag{9.28}$$

$$= W_{new}V\lambda U^t U \lambda^{-1} \tag{9.29}$$

$$= W_{new}W\lambda^{-1} \tag{9.30}$$

$$= \widetilde{M}_{new}U\lambda^{-1}. \tag{9.31}$$

Being able to compute values from new data is an important task when using kernel principal components. Notice that the current algorithm requires us to compute, store, and decompose an $n \times n$ matrix. For large datasets this can become infeasible. A common solution is to take a smaller random or representative sample of the data containing $m \ll n$ points. The algorithm then requires only storing an $m \times m$ matrix. Applying the loadings, as above, to the larger dataset requires only an $n \times m$ matrix $\widetilde{M}$. Assuming the sample is reasonably representative of the whole, the principal components from this method will be nearly as useful at a significantly smaller computational cost.

While we have focused here on principal component analysis, many other algorithms in statistical learning can be written in terms of inner products and therefore used along with the kernel trick. However, now that we know how to produce kernel principal components we can make use of another shortcut. Start by picking a kernel of interest and produce a reasonably large set of kernel principal components. Then, apply any supervised learning algorithm to the projected data. This is similar to the kernel regression algorithm developed in Chapter 3, which we showed was closely connected to regularization with an ℓ_2-penalty. Working with the principal components has several benefits over directly kernelizing a predictive algorithm. First of all, for large datasets we can use the computational trick mentioned in the prior paragraph to estimate the components with a random sample and then project the remainder of the data using significantly less computational resources. Secondly, the non-kernelized methods are often faster, particularly given that in this case they will be working with a relatively small number of input variables. Finally, this approach does not require us to reimplement any other algorithms or limit ourselves to only those methods that can be written in terms of inner products. Implementing this scheme is left as an exercise.

9.4   Spectral clustering

When using the kernel trick our primary object of study is the $n \times n$ matrix M of inner products between all observations. If the feature space is scaled so that each observation has unit length (such as in the radial kernel basis), the matrix M can be viewed as a specific example of a similarity matrix. An element $M_{i,j}$ is close to 1 if two observations are very close together, zero if they are perpendicular to one another, and -1 if they are as far apart as possible. Here, we develop a dimensionality reduction method that only requires a matrix of similarity measures. This is a less restrictive class than a distance matrix or set of inner products; in fact, in general any $n \times n$ matrix is a valid set of similarity scores. We will show an example of such a measurement that is particularly useful for identifying low-dimensional structures in high-dimensional spaces.

For our purposes, it will be useful to put two mild conditions on the allowed similarity matrices. We will refer to an $n \times n$ matrix S as a *similarity matrix* if it is symmetric and it only contains non-negative elements. Neither of these properties should pose any major difficulties. If we have a desired similarity score (such as inner products) that can contain negative elements, this can be dealt with by adding the smallest possible value to every element of the matrix. When working with a similarity score that may be non-symmetric, we can symmetrize the matrix by adding it to its own transpose. An adjacency matrix is a particular type of similarity matrix corresponding to a graph structure with nodes defined by the observations. Specifically, the elements are defined as

$$S_{i,j} = \begin{cases} 1, & \text{if } (i,j) \text{ is an edge} \\ 0, & \text{otherwise.} \end{cases} \tag{9.32}$$

If we are working with an undirected graph, this will correspond to a symmetric similarity matrix. Viewing the adjacency matrix as a set of similarity scores, a node is similar to its neighbors and dissimilar to all other nodes. If we have a graph with edge weights, the entries 1 in the matrix can be replaced with the corresponding weights. Notice that any similarity matrix can be viewed as an adjacency matrix for a graph with edge weights defined by the similarity scores.

We now consider the $n \times n$ matrix, denoted by L, known as the graph Laplacian. Before jumping to its full definition, we will start by considering its action as a quadratic form for any n-dimensional vector v

$$v^t L v = \frac{1}{2} \sum_{i=1}^{n} \sum_{j=1}^{n} S_{i,j} (v_i - v_j)^2. \tag{9.33}$$

This value will be small if v does not differ very much between observations that are similar according to the matrix S. For our analysis, the most important feature of L will be understanding the set of smallest eigenvalues and corresponding eigenvectors. Recall that the smallest eigenvalue will correspond to a unit eigenvector v that minimizes the quadratic form of a matrix. What does this look like for a graph Laplacian?

For simplicity, assume that we have a similarity matrix S defined by an unweighted adjacency matrix. A connected component of a graph is any set of nodes that contains all of its neighbors; in other words, it is the set of nodes that can all be reached by following along edges of the graph. If we take an n-dimensional vector v and set it to a constant on a connected component and zero otherwise, notice that the quadratic form in Equation 9.33 will be equal to zero. There is no pair i and j for which both $S_{i,j}$ and $(v_i - v_j)$ are both non-zero. By construction, the quadratic form defined by L can never be less than zero, and therefore v is a minimal eigenvector. The corresponding eigenvalue is 0. So, a graph with m connected components has a graph Laplacian whose m smallest eigenvalues are equal to zero.

Now, consider the similarity matrix corresponding to a connected graph (one that has only one connected component). The eigenvector associated with the smallest eigenvalue, v_1, will put equal weight on every node and corresponds to an eigenvalue λ_1 of 0. What does the eigenvector v_2 of the second smallest eigenvalue λ_2 look like? By definition, it will minimize the quadratic form in Equation 9.33 subject to being perpendicular to v_1 (most importantly, it cannot be constant). To achieve this, v_2 will put similar weights on nodes that are highly connected to one another and will attempt to differ only between areas of the graph that are poorly connected. We can see this as an attempt to approximate the behavior of the minimal eigenvalues over connected components. Here, finally, we see the utility of this approach: the eigenvector v_2 is a one-dimensional score in which similar observations are close together and dissimilar observations are far apart. It provides a form of dimensionality reduction based only on similarity scores. We can continue on to compute v_3, which will identify sub-clusters that are perpendicular to v_2, and so forth to yield a dimensionality reduction scheme to any desired dimension $k \leq n$.

The approach we have outlined here regarding the smallest eigenvalues of the graph Laplacian falls under a collection of related techniques known as *spectral clustering*. While developed for clustering points on a graph, as we have seen they are actually broadly useful for dimensionality reduction. Clustering of data can proceed in one of two ways:

1. Perform dimensionality reduction to a k-dimensional space and then apply any generic clustering algorithm (such as k-means) to the reduced data

2. Assuming the graph is connected, apply a cutoff value (often the median) to the elements in the second smallest eigenvector to split the data into two groups. Then, apply the same algorithm separately over the detected subcomponents. Iterate until the clusters are as small as desired.

Spectral Clustering Dimensionality Reduction

Let S be an $n \times n$ symmetric matrix of positive similarity scores between observations. Define the degree matrix D, a diagonal matrix given by

$$D_{i,j} = \begin{cases} \sum_j S_{i,j}, & i = j \\ 0, & \text{otherwise,} \end{cases}$$

and the graph Laplacian $L = D - S$. The kth spectral dimension is the eigenvector of L corresponding to the kth smallest positive eigenvalue.

Clustering is not our primary focus, so we leave the implementation of these two approaches as an exercise. While we have motivated the properties of graph Laplacian eigenvalues using unweighted adjacency matrices as similarity scores, the same logic holds for any arbitrary similarity matrix S.

In order to implement dimensionality reduction using spectral clustering, we will need an explicit formula to define the matrix L. It will be useful to define the degree matrix D, a diagonal matrix given by

$$D_{i,j} = \begin{cases} \sum_k S_{i,k}, & i = j \\ 0, & \text{otherwise.} \end{cases} \tag{9.34}$$

Notice then, that our quadratic form from Equation 9.33 can be rewritten in matrix form as

$$v^t L v = \frac{1}{2} \sum_{i=1}^n \sum_{j=1}^n S_{i,j}(v_i - v_j)^2 \tag{9.35}$$

$$= \frac{1}{2} \sum_{i=1}^n \sum_{j=1}^n S_{i,j} v_i^2 + \frac{1}{2} \sum_{i=1}^n \sum_{j=1}^n S_{i,j} v_i^2 + \sum_{i=1}^n \sum_{j=1}^n S_{i,j} v_i \cdot v_j \tag{9.36}$$

$$= \sum_{i=1}^n \sum_{j=1}^n S_{i,j} v_i^2 + \sum_{i=1}^n \sum_{j=1}^n S_{i,j} v_i \cdot v_j \tag{9.37}$$

$$= \sum_{i=1}^n v_i^2 \cdot D_{i,i} + \sum_{i=1}^n \sum_{j=1}^n S_{i,j} v_i \cdot v_j \tag{9.38}$$

$$= v^t D v - v^t S v. \tag{9.39}$$

Motivated by this, we will define the graph Laplacian as

$$L = D - S. \tag{9.40}$$

Put simply, the off-diagonal elements are the negative similarity scores and the diagonal elements are set to make all row and column sums zero.

The final element needed to implement spectral clustering is a similarity

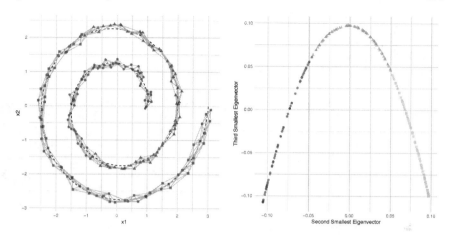

FIGURE 9.3: The first figure shows 200 sample points generated as uniform noise deviating from the dashed line with points connected by a line segment based on the symmetric 8-nearest neighbors algorithm. The second visualizes the same points projected into the first two spectral components.

metric. As mentioned, we could use a modified version of the kernel inner products or invert some distance metric. One of the powerful aspects of spectral clustering, however, is the ability to use other similarity scores that cannot be easily described as an inner product space. For example, we could use a score known as *symmetric k-nearest neighbors*. With a fixed k, define N_i as the set containing the indices of the closest observations to the input i; here we will assume close is defined by Euclidean distance, but this is not required. Then, the symmetric k-nearest neighbors similarity matrix is defined as

$$S_{i,j} = \begin{cases} 1, & \text{if } i \in N_j \text{ and } j \in N_i \\ 0, & \text{otherwise.} \end{cases} \tag{9.41}$$

In other words, two points are connected if they are mutually in each other's set of k-nearest neighbors. The corresponding graph structure often does a good job of approximating manifold structures in higher dimensions.

We start an implementation of spectral clustering by defining the similarity metric. This is very easy to do as a double loop over all of the data points, but this would be very inefficient in R. Instead, we will use the `stats::dist` function and then apply `sort` over each row.

```
# Compute symmetric k-nearest neighbors similarity matrix.
#
# Args:
#     X: A numeric data matrix.
#     k: Number of nearest neighbors to consider.
#
```

```
# Returns:
#     A square similarity matrix with nrow(X) rows & columns.
casl_util_knn_sim <-
function(X, k=10L)
{
  d <- as.matrix(dist(X, upper = TRUE))^2
  N <- apply(d, 1L, function(v) v <= sort(v)[k + 1L])
  S <- (N + t(N) >= 2)
  diag(S) <- 0
  S
}
```

Now we implement spectral clustering. The algorithm is in fact deceptively simple to implement as most of the computational time occurs in the eigenvalue decomposition. The only subtle point is to use a numeric cutoff (here, 1e-8) to identify the zero eigenvalues. These will not be exactly equal to zero due to the nature of working with floating point arithmetic.

```
# Perform spectral clustering using a similarity matrix.
#
# Args:
#     S: A symmetric similarity matrix.
#     k: Number of components to return.
#
# Returns:
#     A matrix with nrow(S) rows and k columns.
casl_spectral_clust <-
function(S, k=1)
{
  D <- diag(apply(S, 1, sum))
  L <- D - S
  e <- eigen(L)
  Z <- e$vector[,rev(which(e$value > 1e-8))[seq_len(k)]]
  Z
}
```

With these functions implemented, we can run spectral clustering over our simulated data.

```
S <- casl_util_knn_sim(X, k=8L)
Z <- casl_spectral_clust(S, k=2L)
```

A visualization of both S and Z are given in Figure 9.3. The left panel shows which data points are connected via the symmetric k-nearest neighbors algorithm. On the right, the first two spectral clustering dimensions are visualized. Notice that the first component unravels the data along the spiral almost perfectly.

9.5 t-Distributed stochastic neighbor embedding (t-SNE)

The first two examples of dimensionality reduction presented in this chapter have started by producing an $n \times n$ matrix describing the relationship between observations in the observed data. Kernel-based methods used a matrix of inner products and spectral clustering made use of a generic similarity function. A general strategy for reducing the input dimension of a dataset after having selected such a matrix proceeds as follows: construct a projection of the original data such that the $n \times n$ matrix computed in the lower-dimensional space (such as inner products or similarity scores) are preserved as closely as possible. Choices for the particular metric to use and the set of allowed projections lead to many dimensionality reduction algorithms, including multidimensional scaling, Isomap, Sammon mapping, and Maximum Variance Unfolding.

The method we turn to now works by defining an $n \times n$ matrix corresponding to a probability distribution over transitions between input data points. In turn, the goal is to find a low-dimensional projection of the input points that yields a similar matrix of transition probabilities. This method, *t-Distributed Stochastic Neighbor Embedding* (t-SNE), has recently become popular in the statistical learning literature. It is particularly good at taking very-high-dimensional input data, such as a term frequency matrix or matrix of image pixels, and spreading out input data into two dimensions in a way that is useful for visualizing data before, during, and after applying predictive models.

As a first step in describing the t-SNE algorithm, we need to describe a set of conditional probabilities. We will define an $n \times n$ matrix of transition probabilities, which we can envision as a random walk on the observations in our training data. The probability of transitioning into observation j, given that we are currently at observation i will be modeled proportional to a Gaussian distribution:

$$P_{j|i} = \frac{\exp\left\{-||x_i - x_j||_2^2/2\sigma_i^2\right\}}{\sum_{i \neq k} \exp\left\{-||x_i - x_k||_2^2/2\sigma_i^2\right\}}, \quad j \neq i. \tag{9.42}$$

We will set $P_{i|i}$ to zero by definition. The choice of σ_i will be discussed shortly. From these conditional probabilities, we can then produce a full matrix of probabilities by

$$P_{i,j} = \frac{P_{j|i} + P_{i|j}}{2n}. \tag{9.43}$$

This definition is used, which approximates rather than being the actual random walk probabilities, because in high dimensions the actual transitions can become very low for outlying points.

The goal of t-SNE is to estimate a set of points $\{y_j\}_{j=1}^n$ in a low-dimensional subspace of size d with a similar set of transition probabilities.

The transitions in the lower-dimensional space are defined slightly differently as

$$Q_{i,j} = \frac{\left(1 + ||y_i - y_j||_2^2\right)^{-1}}{\sum_{l \neq k} \left(1 + ||y_l - y_k||_2^2\right)^{-1}}, \quad j \neq i. \tag{9.44}$$

As in the lower-dimensional space, $Q_{i,i}$ is always set to zero. This definition differs in two ways: (1) the normalization defines the actual set of transition probabilities over the data and (2) a Cauchy distribution (equivalently, and hence the name, a T-distribution with one degree of freedom) is used in place of a Gaussian density function. The first modification only serves to simplify the calculation and can be used because we will construct the lower-dimensional space to not include any outliers. The T-distribution helps to address several problems observed in the original SNE implementation, namely crowding of points in the input space and vanishing gradients in the optimization steps.

With the two probability distributions constructed, the next step is to describe what it means for two distributions to be close to one another. Here, we use the Kullback–Leibler divergence, a common measurement of the discrepancy between probability distributions motivated by information theoretic concepts [39]. The divergence between our two discrete distributions, which we use as our cost function C, is given by

$$C = D_{KL}(P||Q) = \sum_{i \neq j} P_{i,j} \cdot \log_2 \left(\frac{P_{i,j}}{Q_{i,j}}\right). \tag{9.45}$$

We will minimize this cost function using gradient descent. The gradient of the cost function has a relatively simple closed form given by

$$\nabla_{y_i} C = 4 \cdot \sum_{j \neq i} (P_{i,j} - Q_{i,j}) \cdot \left(1 + ||y_i - y_j||_2^2\right)^{-1} \cdot [y_i - y_j] \tag{9.46}$$

where each $\nabla_{y_i} C$ is itself a d-dimensional gradient. The derivation is relatively extensive, however, and we will not derive it here. See the appendix to the original t-SNE paper for further details [110]. The gradient can then be used in conjunction with gradient descent, optionally using tricks developed in Sections 8.2 and 8.6.

The final detail that needs to be addressed is how to compute the factors σ_i^2 in Equation 9.42. The goal is to scale each term so that the effective number of neighbors is relatively constant for each term. An easy way to measure this is by computing the *perplexity* of the conditional probability distribution defined by $P_{j|i}$. The complexity of the probability distribution is defined as

$$\text{Perplexity}(P_{\cdot|i}) = 2^{-1 \cdot \sum_j P_{j|i}}. \tag{9.47}$$

The goal is to determine values σ_i^2 such that these perplexities are approximately the same for each i. The final output of t-SNE is not sensitive to small

---

T-Distributed Stochastic Neighbors (t-SNE)

Let X be an $n \times p$ dimensional data matrix. For a set of positive values σ_i^2, compute

$$P_{j|i} = \frac{\exp\left\{-||x_i - x_j||_2^2/2\sigma_i^2\right\}}{\sum_{i \neq k} \exp\left\{-||x_i - x_k||_2^2/2\sigma_i^2\right\}}, \quad j \neq i$$

setting $P_{i|i}$ equal to zero, and

$$P_{i,j} = \frac{P_{j|i} + P_{i|j}}{2n}$$

for a fixed dimension d, let $\{y_i\}_{i=1}^n$ be a collection of vectors randomly initialized with a multivariate normal distribution centered around the origin. Then, iteratively modify the values y_i using gradient descent according to the gradient given by

$$\nabla_{y_i} C = 4 \cdot \sum_{j \neq i} (P_{i,j} - Q_{i,j}) \cdot \left(1 + ||y_i - y_j||_2^2\right)^{-1} \cdot [y_i - y_j]$$

where

$$Q_{i,j} = \frac{\left(1 + ||y_i - y_j||_2^2\right)^{-1}}{\sum_{l \neq k} \left(1 + ||y_l - y_k||_2^2\right)^{-1}}, \quad j \neq i$$

defining $Q_{i,i}$ to be zero. At convergence, the vectors $\{y_i\}_{i=1}^n$ represent a t-SNE embedding of the original dataset X.

---

differences in the perplexity value, and therefore we can set σ_2^2 using a simple binary search. In the original paper, the authors suggest that generally the perplexity should be set between 5 and 50. Most implementations of t-SNE allow the user to pre-specify the target perplexity values.

It will be advantageous to split our implementation of t-SNE into two functions. The first takes a matrix X together with a set of perplexity scores and returns the matrix $P_{i,j}$ from Equation 9.43.

```
# Compute t-SNE probability score matrix.
#
# Args:
#     X: A numeric data matrix.
#     perplexity: Desired perplexity score for all variables.
#
# Returns:
#     A matrix of probability densities.
casl_tsne_p <-
```

```
function(X, perplexity=15)
{

  D <- as.matrix(dist(X))^2
  P <- matrix(0, nrow(X), nrow(X))
  svals <- rep(1, nrow(X))

  for (i in seq_along(svals))
  {
    srange <- c(0, 100)
    tries <- 0

    for(j in seq_len(50))
    {
      Pji <- exp(-D[i, -i] / (2 * svals[i]))
      Pji <- Pji / sum(Pji)
      H <- -1 * Pji %*% log(Pji, 2)

      if (H < log(perplexity, 2))
      {
        srange[1] <- svals[i]
        svals[i] <- (svals[i] + srange[2]) / 2
      } else {
        srange[2] <- svals[i]
        svals[i] <- (svals[i] + srange[1]) / 2
      }
    }
    P[i, -i] <- Pji
  }

  return(0.5 * (P + t(P)) / sum(P))
}
```

Next, we write the user-level function that runs the full t-SNE algorithm.

```
# Compute t-SNE embeddings.
#
# Args:
#     X: A numeric data matrix.
#     perplexity: Desired perplexity score for all variables.
#     k: Dimensionality of the output.
#     iter: Number of iterations to perform.
#     rho: A positive numeric learning rate.
#
# Returns:
#     An nrow(X) by k matrix of t-SNE embeddings.
```

```
casl_tsne <-
function(X, perplexity=30, k=2L, iter=1000L, rho=100) {

  Y <- matrix(rnorm(nrow(X) * k), ncol = k)
  P <- casl_tsne_p(X, perplexity)
  del <- matrix(0, nrow(Y), ncol(Y))

  for (inum in seq_len(iter))
  {
    num <- matrix(0, nrow(X), nrow(X))
    for (j in seq_len(nrow(X))) {
      for (k in seq_len(nrow(X))) {
        num[j, k] = 1 / (1 + sum((Y[j,] - Y[k, ])^2))
      }
    }
    diag(num) <- 0
    Q <- num / sum(num)

    stiffnesses <- 4 * (P - Q) * num
    for (i in seq_len(nrow(X)))
    {
      del[i, ] <- stiffnesses[i, ] %*% t(Y[i, ] - t(Y))
    }

    Y <- Y - rho * del
    Y <- t(t(Y) - apply(Y, 2, mean))
  }

  Y
}
```

Our implementation is fairly minimal. Additional tweaks to the basic algorithm such as a momentum term, early exaggeration, and simulated annealing can help to make the algorithm perform better.

The t-SNE algorithm is usually most useful when applied to high-dimensional data, but we can apply it here to the simulated spiral dataset in order to illustrate its usage and the effect of the perplexity term.

```
Y10 <- casl_tsne(X, perplexity = 10)
Y25 <- casl_tsne(X, perplexity = 25)
```

Plots of the results are shown in Figure 9.4. The lower perplexity model does split apart the data along the spiral, but breaks up the data to lose the relationship between the clusters. The higher perplexity example is fairly similar to the original data, replacing the spiral with two concentric semi-circles. In

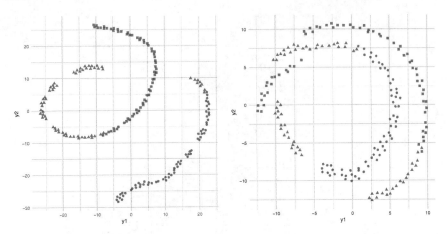

FIGURE 9.4: t-SNE projections of the simulated data shown in Figure 9.1. The left panel shows t-SNE with perplexity equal to 10 and the right panel shows perplexity equal to 25.

Section 9.8 we will see a more powerful example of t-SNE applied to a high-dimensional dataset.

9.6 Autoencoders

We have motivated the study of dimensionality reduction through applications to predictive modeling. By using unsupervised learning techniques to reduce the complexity of the training data, we can then (hopefully) build fast and predictive models in a smaller feature space. Our final dimensionality reduction techniques flips this relationship around; here, we will use a predictive model to assist in the construction of dimensionality reduction.

An *autoencoder* is any model, though almost always some type of neural network, trained with the same data used as both an input and an output. That is, the model attempts to learn a function f such that

$$f(x_i) \approx x_i, \quad i = 1, \ldots, n. \tag{9.48}$$

Of course, this approach is most interesting when the class of possible models is sufficiently restricted to disallow our estimate of f to be equal to the identity function. In the context of neural networks, an autoencoder typically has two distinguishing features:

1. A loss function describing how well an element x_i is approximated by $f(x_i)$. Typically squared error will suffice, but some applications may use an ℓ_1-penalty or probability divergence as appropriate.

2. A hidden layer, known as the *bottleneck layer*, which has only $k < p$ neurons.

The bottleneck layer is what stops the algorithm from being able to make every layer equal to an identity function. We can view all of the layers prior to the bottleneck as the *encoding* of the data and all of the layers after it as a *decoding* of the data. Often the encoding layers expand the dimension of the data before contracting it, in much the same way that kernel principal components unravel the data in the feature space before projecting down into the principal components. In fact, a neural network autoencoder with only a single hidden layer and sigmoid activation functions can be shown to be asymptotically equivalent to principal component analysis [29].

The connection between autoencoders and dimensionality reduction can be made explicit by considering the activations in the bottleneck layer. Here, we have a k-dimensional representation of the original dataset X. Further, this representation can be decoded in order to approximate the original dataset. Therefore, the bottleneck representation in a well-trained neural network serves as a good low-dimensional approximation of the original data. While it also possible to construct and investigate bottleneck layers in other neural networks (as we did in Section 8.10), these dimensionality reductions will be specific to a particular learning task. In contrast, the bottleneck representations from an autoencoder should be applicable broadly to any new classification task on the data.

A final benefit of autoencoders is that they require no changes to the generic neural network implementations we have from Chapter 8. While many variations exist on the basic approach here, such as variational autoencoders, a basic autoencoder can be derived from our own functions or the various methods provided in the **keras** package. We employ an autoencoder to the application in Section 9.8.

9.7   Implementation and notes

Implementations in R exist for all of the approaches presented in this chapter. Several packages provide access to kernel principal components. For example, advanced options, including many other kernalized algorithms, are included in the package **kernlab** [91]. While fast and feature rich, this package does require some close reading of the documentation to use even in simple applications. Several packages implement variants of t-SNE, though not all of these are very efficient. We will make use of the package **Rtsne** [95] in Section 9.8, which implements the much faster Barnes–Hut algorithm for working with large datasets [161]. As already mentioned in Section 9.6, neural network autoencoders can be easily implemented using the **keras** R library [6].

While spectral clustering is implemented in several packages, including

kernlab, none that we are familiar with provide the actual spectral dimensions. Only the resulting cluster indices are returned. Fortunately, our `casl_spectral_clust` is already quite efficient for reasonably sized datasets. We will see in Chapter 10 a final modification that will assist in making this function run over even larger datasets.

9.8   Application: Classifying and visualizing fashion MNIST

9.8.1   Data

In Section 8.5 we introduced the MNIST dataset, a collection of small black and white images of handwritten digits used to illustrate concepts in computer vision. The application section of Chapter 8 then used a variation of this, the EMNIST dataset, of handwritten letters. Here, we present our final variation of this data: Fashion MNIST [181]. Like the original, it is contains 60 thousand 28x28 greyscale images from 10 categories. The images depict 10 classes of clothing types. We use it here because images are a common example of data that require non-linear dimensionality reduction. The format of the data is exactly the same as the MNIST dataset and can be read in with similar code.

```
X <- readRDS("data/emnist_x28.rds")
X <- array(X, dim = c(nrow(X), 28^2))
emnist <- read.csv("data/emnist.csv")
y <- emnist$class
```

The small size and relative uniformity of the collection makes it possible to learn this dimensionality reduction with relatively modest hardware and dataset sizes.

One major goal of dimensionality reduction is to visualize the projected results. We found through experimentation that it is hard to represent more than 1000 points or 5 categories given the constraints (greyscale and limited resolution) dictated by the printing of this text. Therefore, in this section we have taken just a small sample of the Fashion MNIST dataset limited to only the five categories (trouser; pullover; dress; coat; sandal) and 1000 samples.

```
set.seed(1)
id <- sample(which(y %in% c(1,2,3,4,5)), 1000)
X <- X[id,]
y <- y[id]
```

Examples of the categories are included in Figure 9.10.

The overall theme of this text is to build predictive models. One way that we motivated the study of dimensionality reduction in Section 9.1 was through

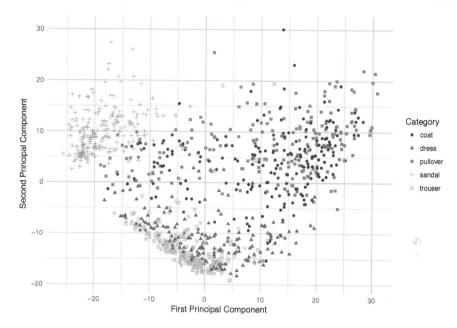

FIGURE 9.5: First two principal components from a sample of the Fashion MNIST dataset, with classes distinguished by shape and shade of grey. Sandals are well separated from the other classes, but there is a reasonable degree of overlap with the other classes. These become separable in higher dimensions.

the benefit of dimensionality reduction to use as part of a predictive modeling pipeline. Because of these, we will include for each dimensionality reduction technique in this section the results of using the projected dimensions to learn predictive models. To simplify matters, we will only look at a training set; this is reasonable because our dimensionality reduction algorithms are only using structural data about the data matrix X. That is, they cannot *cheat* by over-fitting to the training response in the same way that supervised learning can. For each technique we will look at three simple models: (1) multinomial regression, (2) k-nearest neighbors with k equal to 3, and (3) k-nearest neighbors with k equal to 10. The custom function `accuracy` (see online supplemental materials for the full code) takes the reduced dataset and computes the accuracy of these three models for several different dimensions. In our discussion, the appropriateness of the various models will be discussed in the context of each dimensionality reduction algorithm.

9.8.2 Principal component analysis

In order to have some baseline with which to compare our non-linear models, here we run a straightforward principal component analysis. The principal components of the flattened data matrix X passed to `stats::prcomp` and the results are passed to our custom function `accuracy`.

```
out <- prcomp(X[,apply(X, 2, sd) != 0], scale = TRUE)
Z <- out$x
accuracy(Z, y)
```

	d=2	d=3	d=5	d=10	d=20
multinomial	0.694	0.735	0.812	0.909	0.925
knn,k=3	0.660	0.722	0.800	0.866	0.877
knn,k=10	0.691	0.725	0.798	0.866	0.871

The accuracy of the models increases steadily with increased access to the dimensions of principal components. The nearest neighbors models are competitive but slightly outperformed by the multinomial model. The first two principal components are shown in Figure 9.5. We see that sandals are easily separated from the other points. Pullovers and coats are difficult to distinguish, at least without additional dimensions. Trousers and dresses are similarly projected, though trousers are pushed to one side of the cluster making it seemingly possible to produce a reasonably predictive model for separating these classes. Overall, at least with a high enough number of dimensions, this relatively small classification task does not seem too difficult. The goal of the non-linear methods will be to try to spread the dataset out more uniformly so that high classification rates can occur in a smaller number of dimensions.

9.8.3   Kernel principal component analysis

Next, we will apply kernel principal components. For this, we use the **kernlab** package with a radial basis kernel (the default). The package makes it easy to modify the tuning parameter of the kernel; here we reduced the sigma parameter by an order of magnitude (compared to the default) after evaluating the original output.

```
Z <- kernlab::kpca(X, kpar=list(sigma=0.005))@rotated
accuracy(Z, y)
```

	d=2	d=3	d=5	d=10	d=20
multinomial	0.659	0.761	0.811	0.842	0.903
knn,k=3	0.626	0.719	0.802	0.817	0.867
knn,k=10	0.647	0.751	0.824	0.826	0.854

The results are relatively disappointing compared to the linear principal component analysis in terms of the resulting predictive power of the projections. In three dimensions it performs slightly better, similarly for dimensions 2 and 5, and slightly worse in the two highest dimensions (10 and 20). Figure 9.6 shows a scatterplot of the data in the first two kernel principal components.

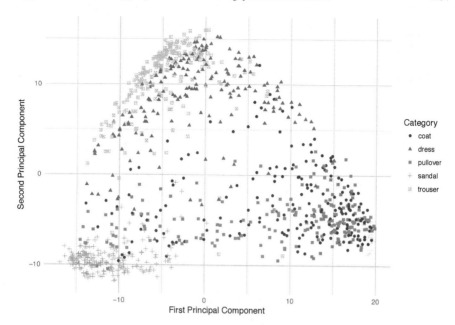

FIGURE 9.6: First two principal components from a sample of the Fashion MNIST dataset using kernel principal components with a radial basis kernel with σ set to 0.005 (equivalently, $\gamma = 100$ in the parametrization of Equation 9.16). Classes of clothing are distinguished by shape and the shade of grey.

At least visually, the results do seem to spread out the data points in a more systematic way compared to the linear principal component analysis. The data points seem drawn to three poles representing the sandal class, the class of tops (coats and pull-overs), and the class of bottoms (dresses and and trousers).

9.8.4 Spectral clustering

For spectral clustering, we will apply a standard nearest neighbors similarity matrix by attaching each point to its 6 closest neighbors. We then perform spectral clustering of similarity matrix, grabbing the first 20 spectral dimensions. Here, we use our own function `casl_spectral_clust` because, as mentioned, there are no existing packages that return the actual spectral dimensions required for our algorithm.

```
S <- casl_util_knn_sim(X, k=6L)
Z <- casl_spectral_clust(S, k=20L)
accuracy(Z, y)
```

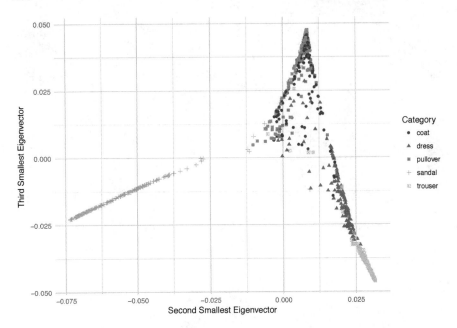

FIGURE 9.7: The first two spectral clustering dimensions, the two eigenvectors corresponding to the smallest non-zero eigenvalues, applied to a sample of the Fashion MNIST dataset. For similarity scores we link observations with their 6 nearest neighbors. Note that most of the interesting separation of the data occurs in the first spectral dimension, where the second smallest eigenvector as the smallest corresponds to an eigenvalue of zero.

	d=2	d=3	d=5	d=10	d=20
multinomial	0.782	0.795	0.802	0.813	0.823
knn,k=3	0.779	0.765	0.785	0.795	0.795
knn,k=10	0.777	0.781	0.780	0.788	0.804

Here we see that all three models are noticeably more predictive when only using 2 or 3 dimensions compared to either of the principal component algorithms. However, the higher dimensions perform worse. The plot in Figure 9.7 provides some motivation as to why this occurs. Almost all of the predictive power comes, in fact, from the first spectral dimension, which nearly perfectly separates the sandals and trousers from the remainder of the data. As mentioned in Section 9.4, spectral clustering is often used hierarchically: the data are split using just the first component, and then the algorithm is applied separately on each component. In this plot we see an example of why this might be useful. The spectral algorithm seems to be very good at picking the first few dimensions but less adept at picking the remaining ones.

FIGURE 9.8: Computed t-SNE dimensions applied to a sample of the Fashion MNIST dataset into two dimensions for four different values of the perplexity score. Note that higher perplexity more completely separates the output but begins to lose local topological relationships between the clusters. Visually, these seem the most appealing of all the dimensionality algorithms. In fact, t-SNE is most often employed for data visualization.

9.8.5 t-Distributed stochastic neighbor embedding

In order to apply the t-SNE algorithm, we make use of the function `Rtsne` from the package of the same name. For example, to fit the model with just two dimensions and a perplexity of 30, we use the code:

```
Z <- Rtsne(X, perplexity=30, dims=2L)[["Y"]]
```

Unlike principal component analysis and spectral clustering, t-SNE does not produce an ordered set of dimensions from which we can subsequently pick and choose. Instead, for each dimension of interest we need to run a completely new model. The algorithm is not designed to produce high-dimensional output, so here we use it only for the three smallest test dimensions.

```
                d=2    d=3    d=5
multinomial  0.807  0.814  0.833
knn,k=3      0.844  0.863  0.860
knn,k=10     0.854  0.841  0.833
```

In terms of predictive power, these results are the best of all of the algorithms we have tested so far across all three models. However, its primary usage is restricted to data visualization because dimensionality reduction for predictive purposes typically requires the ability to project into higher spaces. Compare for example the $d = 5$ model here and the $d = 20$ model using ordinary principal component analysis.

Figure 9.8 illustrates how the perplexity score effects the output of the algorithm. As the perplexity increases the data separates the data points clump together into groups that map surprisingly well onto the five classes. However, with lower perplexity scores the topological relationships between groups are more clearly shown. Finding a good perplexity score is a key step in making use of t-SNE. We choose a value of 30 for our classification analysis after looking at the plots as a balance between these two extremes. It is interesting to note that the nearest neighbors algorithm performs better than the multinomial, a reversal of the pattern seen in the other classifiers. This is a result of the fact that t-SNE primarily cares about local similarity and disregards points that are far apart. We can also see this in the plots.

9.8.6   Autoencoder

Our next task is to use the **keras** library to build an autoencoder. To do this, we construct a neural network where the input layer and output layer have the same dimensionality. For our purpose, we also want a hidden bottleneck layer that has only a very small number of neurons. The architecture we chose includes 5 hidden layers with the bottleneck layer right in the middle. Here we build the model that has a bottleneck of just 2 dimensions (other dimensions require reconstructing the model with a different number of units for the middle layer). The layers are connected with straightforward rectified linear units and compiled using the mean squared error loss.

```
library(keras)
model <- keras_model_sequential()
model %>%
  layer_dense(units = 64, input_shape = c(ncol(X))) %>%
  layer_activation(activation = "relu") %>%
  layer_dense(units = 16, input_shape = c(ncol(X))) %>%
  layer_activation(activation = "relu") %>%
```

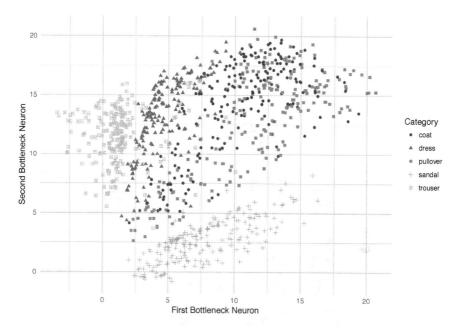

FIGURE 9.9: Dimensionality reduction of a sample of the Fashion MNIST dataset using a dense neural network autoencoder with a bottleneck layer of dimension two. The scatter plot visualizes the two neurons in the bottleneck. While not trained on the clothing category, the dimensionality reduction does a good job of separating the five classes.

```
layer_dense(units = 2) %>%
layer_activation(activation = "relu") %>%
layer_dense(units = 16, input_shape = c(ncol(X))) %>%
layer_activation(activation = "relu") %>%
layer_dense(units = 64) %>%
layer_activation(activation = "relu") %>%
layer_dense(units = ncol(X))

model %>% compile(loss = 'mse',
                  optimizer = optimizer_rmsprop())
model %>% fit(X, X, epochs = 100)
```

In order to actually see the output of the bottleneck layer, we need to construct a new model that terminates at the bottleneck. We have:

```
model2 <- keras_model_sequential()
model2 %>%
  layer_dense(units = 64, input_shape = c(ncol(X))) %>%
  layer_activation(activation = "relu") %>%
```

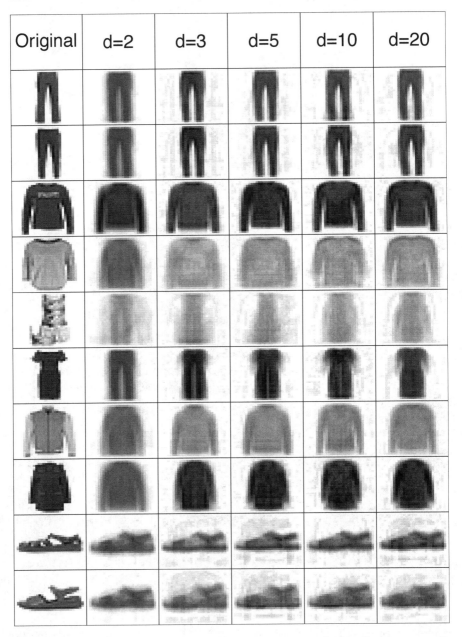

FIGURE 9.10: Reconstructed images from an autoencoder applied to a sample of images from the Fashion MNIST data. Ten examples are shown here, two from each of the five categories.

```
layer_dense(units = 16, input_shape = c(ncol(X))) %>%
layer_activation(activation = "relu") %>%
layer_dense(units = 2)
```

Next, we copy the weights from the first 5 layers (remember that the activations count as layers for **keras**) into this new model and predict the results using the data matrix X.

```
for (j in 1:5) {
  model2$layers[[j]]$set_weights(model$layers[[j]]$get_weights())
}
Z <- predict(model2, X)
```

Now, Z is a matrix with two columns and one row for each observation. Iterating this over the five dimensions of interest gives the following classification results:

	d=2	d=3	d=5	d=10	d=20
multinomial	0.772	0.815	0.820	0.847	0.886
knn,k=3	0.771	0.776	0.781	0.803	0.835
knn,k=10	0.779	0.795	0.784	0.801	0.849

It performs slightly worse than the t-SNE model for low dimensions and slightly worse than linear principal components in the higher dimensions. That is, it seems to be a good compromise in terms of predictive power across both the lower and higher dimensions of our case study. A plot of the components in Figure 9.9 shows a similar relationship between the classes found in the spectral clustering. The classes are arranged in a roughly monotonic ordering: trousers, dresses, coats, pullovers, and sandals. In part this makes sense because dresses are somewhere between the shape of trousers and coats and sandals look very distinctive compared to the other four classes.

In order to understand an autoencoder it is useful to see some of the reconstructed images. Several reconstructions across the five dimensions used here are shown in Figure 9.10. Across all dimensions, the overall shapes are well reconstructed. Some smaller details are missing, however, particularly in the lower dimensionality reconstructions. For example, the zipper on the first jacket and the straps on both sandals are completely missing. The $d = 2$ reconstructions are noticeably more blurry compared to the other examples. Looking closely, you can see a phantom jacket showing up in the $d = 5$ reconstruction of the two sandals. Likely some of the weights in the decoder part of the network work to reconstruct a jacket shape, and the sandals examples accidentally activate a portion of these.

9.8.7 Supervised neural network

In Section 9.6 we mentioned that a bottleneck layer can be included in any neural network in order to perform supervised dimensionality reduction. Here we

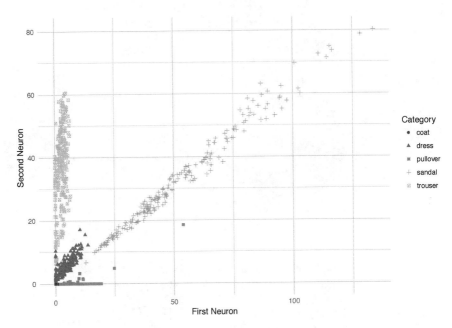

FIGURE 9.11: A neural network was fit on a sample of the Fashion MNIST dataset with the goal of predicting the output class from one of five categories. A bottleneck layer with just two neurons was included in the network architecture; this plot shows a scatter plot of the observations. While leading to a nearly perfect separation of the five categories, this does not provide as nice of a visualization as some of the unsupervised learning algorithms.

implement such a model in order to compare to the unsupervised approaches. The model is similar to the autoencoder, but now includes only three hidden layers.

```
Y <- keras::to_categorical(y - 1)
model <- keras_model_sequential()
model %>%
  layer_dense(units = 64, input_shape = c(ncol(X))) %>%
  layer_activation(activation = "relu") %>%
  layer_dense(units = 2) %>%
  layer_activation(activation = "relu") %>%
  layer_dense(units = 8) %>%
  layer_activation(activation = "relu") %>%
  layer_dense(units = ncol(Y)) %>%
  layer_activation(activation = "softmax")


model %>% compile(loss = 'categorical_crossentropy',
```

```
                     optimizer = optimizer_rmsprop(),
                     metrics = c('accuracy'))
model %>%
  fit(X, Y, epochs = 100)
```

As before, we need to construct a new model and copy over the weights to find the actual predictions of the bottleneck. Using these dimensions, we find the following prediction scores.

```
> mat
             d=2   d=3   d=5  d=10  d=20
multinomial 1.000 1.000 0.988 1.000 1.000
nn,k=3      0.997 0.998 0.988 0.997 0.990
nn,k=10     0.998 0.998 0.976 0.996 0.983
```

The supervised approach outperforms the others in terms of predictive power by a wide margin. In fact, across all dimension sizes all three models are able to predict the five classes almost perfectly. It would be reasonable to expect that the corresponding plots do a good job of visually splitting the points. As seen in Figure 9.11 this is not exactly true. While it is possible to split the points well algorithmically from this plot, it is stretched and laid out in a way that is very unhelpful for visualization. The coats data points, for example, are all projected to the origin. So, while the predictive power is useful, this example illustrates why supervised approaches are not always preferred for dimensionality reduction: the resulting reduction is often overfit to one classification task and fails to capture other interesting variations in the underlying input space.

9.9   Exercises

1. Taking the polynomial kernel definition from Equation 9.8, compute the implied mapping ϕ for d equal to 2 for an arbitrary value of c and dimension p of the input data.

2. Using the answer from question 1, write an R function that takes a value for c and returns the value of ϕ for each input.

3. Verify by simulation that inner products produced by projecting via the function ϕ from your solution to question 2 match the kernel function in Equation 9.8.

4. The text glossed over the details in the computation of Equation 9.27. Show how this is derived from the requirement in Equation 9.25.

5. Rewrite the function `casl_kernel_pca` to include a matrix `Xnew` such that `X` is used to construct the principal components but the components for all of `Xnew` are returned.

6. We described a scenario for using the output of the kernel principal components to run a regression analysis. Design a simulation that tests the efficacy of this approach for logistic regression. Compare the results of using the kernel principal components with running a logistic regression on the entire feature set. (Hint: use a polynomial kernel and manually project the data using the function you wrote in question 2.)

7. Rerun the spectral clustering algorithm on the simulated spiral data using only the 4 symmetric nearest neighbors. The resulting graph is disconnected. What happens now in the plot of the first two spectral dimensions?

8. As you should have seen in the prior question, our spectral clustering does not work very well when the underlying graph is disconnected. Modify the function `casl_spectral_clust` to include an option that applies spectral clustering independently to each connected component. Return these components, pasted back together, along with an extra column indicating which connected component an observation is in. Retest with the simulated spiral data using only the 4 symmetric nearest neighbors.

9. Implement a function that applies the hierarchical spectral clustering algorithm to cluster and input dataset `X`. As an input parameter, simply ask how many splits are required.

10. Include a momentum term into the function `casl_tsne`, similar to the neural network implementation given in Section 8.6.

11. Rerun the supervised learning neural network model given in Section 9.8.7 with a bottleneck containing just 2 dimensions several times. You should notice that the neural network often gets 'stuck'; sometimes it produces a model no better than random guessing (20% accuracy), other times it produces a model with 40% or 60% accuracy, and sometimes it produces a near perfect fit. What is causing this erratic behavior? Implement a fix and test that it produces a model that almost always produces a near-perfect fit of the response.

10

Computation in Practice

10.1 Reference implementations

One of the primary motivating goals of this text was to motivate and produce a set of reference implementations for several popular classes of algorithms in supervised learning. As discussed in Chapter 1, we feel that understanding the computational details behind statistical modeling algorithms is an important aspect of learning how to build predictive models.

The theme of this chapter is not to explain how to move from our reference implementations to code that fills in all of the details often seen in production level code. For this there are a plethora of general-purpose and language-specific guides for software design that can be consulted. Looking at the source code of published projects is also a useful strategy for learning good code style. Our goal here is to discuss algorithmic details that will allow our current implementations to run faster or to be applied to larger datasets. That is, we are still focused on reference implementations rather than production level code containing all of the details required for widespread, general purpose usage. However, we want to add important algorithmic modifications to the code we already have to make it more efficient by either running faster, consuming less memory, or allowing for parallel execution across threads, cores, or across a connected cluster of machines.

Throughout this text we have made efforts in our implementations to comment on or include efficient algorithms for solving specific problems. The distinction here is that we now focus on general purpose approaches to working with data efficiently that can be applied across many algorithms. These modifications take one of the following forms: (1) the use of new data structures, (2) splitting algorithms to run on data too large to fit in memory and in parallel computing settings, or (3) replacing exact quantities by approximations. Depending on the nature of the modification, we will either integrate it into a method we have already seen or implement it as a separate function that can be called via other functions. We do not aim to offer an exhaustive treatment of every known trick for modifying algorithms to be more efficient. A detailed study of matrix approximation techniques could easily expand to several volumes on its own. Our treatment of these modifications attempts to illustrate some of the most commonly used methods and design patterns on which other techniques often build.

10.2   Sparse matrices

A large proportion of the algorithms we have discussed can be described, at least in part, in terms of basic matrix manipulations. This point, in fact, has been one central theme of the text. As such, many of the sections in this chapter focus on ways to work more efficiently with matrices. We start here by describing structures known as *sparse matrices*. These have already been mentioned in several implementation sections and even used in the NYC Taxi dataset application in Section 3.6. The importance of sparse matrix representations however demands some further treatment, which we give here.

A matrix is called *sparse* if most of its entries are zero. Matrices that are not sparse are called *dense*. The *sparsity* of a matrix is a number indicating the proportion of elements that are equal to zero. There is no formal threshold for the minimal sparsity value before a matrix can be called sparse. Typically a 'sparse' matrix has at least 90% of its terms equal to zero, with most applications having a sparsity of 99% or more. Sparse matrices arise in many statistical learning applications. When constructing a numeric data matrix as indicator variables from a categorical input, as seen in Section 2.1, only one variable in each row of X is allowed to be non-zero. Datasets representing counts are often sampled from zero-inflated distributions; term frequency matrices, for instance, often have less than 0.01% of their entries equal to a non-zero value. The kernel and similarity matrices we saw in Chapter 9 can become sparse in two ways. Some metrics such as symmetric nearest neighbors lead to naturally sparse similarity scores. Others, such as the radial kernel, lead to many very small entries, which can be hard thresholded to zero with minimal loss of performance.

All of the algorithms that work for dense matrices can be applied just as well to matrices that contain mostly zero values. Why bother with worrying about sparse matrices? Building special cases for sparse matrices can yield large savings in both memory storage and computation. For instance, consider the simple example of a square n-by-n diagonal matrix X,

$$X = X_{ij} = \begin{cases} \text{non-zero} & \text{if } i = j, \\ 0 & \text{otherwise,} \end{cases} \qquad i, j = 1, 2, \ldots, n.$$

The matrix is clearly sparse since only n out of n^2 total entries are non-zero. Now consider computing the matrix vector product $y = X\beta$ for some vector $\beta = [\beta_i]$, $i = 1, 2, \ldots, n$. The usual matrix multiplication procedure uses *all* the values in X, most of which are zero and contribute nothing to the answer. That approach requires n^2 total scalar arithmetic operations to compute.

Instead, we can skip the zero values and simply compute each element of the product vector y using $y_i = x_i \beta_i$, needing only n scalar arithmetic operations.

Moreover, instead of storing all n^2 scalar values of the matrix X—most of which are zero—we can save on memory by storing X in a compressed form. That means store the non-zero values of X together with a recipe for recovering their positions in X along with the overall dimensions of the matrix. In our simple diagonal matrix case we could, for instance, store the n non-zero diagonal values plus a single number indicating which diagonal the values run along for a total of only $n+1$ scalars worth of storage. This is the gist of sparse matrix operations: save memory and computation by omitting zero values and thereby only computing what they need to.

A dense matrix is typically stored in memory as a vector; that is, a sequence of sequential bytes representing a stream of numbers. Two additional attributes give the number of rows and columns in order to understand how the vector should be treated when conducting matrix operations. Elements from the vector are implicitly thought of as filling in a matrix row-by-row (row major) or column-by-column (column major) depending on the conventions of a programming language or specific library. For a concrete example of this, note that we can get an element in the first column and second row of a matrix A in R, that is `A[2, 1]`, equivalently by treating the matrix as a vector: `A[2]`.

When working with a sparse matrix it is possible to store only the non-zero terms with all other entries can be assumed to be zero by default. The added complication of storing only non-zero terms is the need to indicate where in the matrix each term belongs. We cannot assume that the elements 'fill-in' the matrix because not all the terms are being stored. A common and simple way to do this is through the use of a coordinate list where three arrays of equal length are constructed:

- $i \rightarrow$ an array of row indices

- $j \rightarrow$ an array of column indices

- $x \rightarrow$ an array of the non-zero values

In the coordinate list format the matrix has an element x_k in the position (i_k, j_k) and zeros otherwise. Typically the overall dimensions of the matrix are also included in the data structure. The coordinate list makes it easy to cycle through all of the non-zero elements in a matrix. Depending on how the arrays are stored, it is possible to add or remove elements from the matrix without having to construct a complete copy of the data. A popular variation of the coordinate list format for storing non-zero elements is the compressed sparse row (CSR) format. It uses slightly less storage space, but most importantly gives random access to any row of the matrix. This random access greatly improves the ease of implementing many algorithms. However, it is not easy to iteratively add or remove elements in CSR format. Similarly, we can take advantage of the structure of diagonal, banded, and triangular matrices to efficiently store their data in custom formats. Typically sparse matrix libraries allow users to construct matrices in a coordinate list format and handle conversion to CSR or other formats internally. We will limit ourselves to a discussion of the coordinate list format here for simplicity.

If a matrix has n rows, p columns, and d non-zero entries, the dense matrix representation will take $n \cdot p$ elements to store whereas the coordinate list format requires only $3 \cdot d$ elements.[1] Depending on the matrix sparsity, this could result in a significant reduction in the memory or disk space required to store the matrix. Also, some common matrix operations can be easily adapted to work with data stored in a coordinate list. Multiplication by a scalar value, for example, is a trivial operation on the x component of the list. It is also relatively straightforward to add together two sparse matrices component-wise. For example, storing sparse matrices as data frames in R leads to the following implementation of addition using the `merge` function:

```
# Add two sparse matrices stored as triplet lists.
#
# Args:
#     a: A list describing a sparse matrix.
#     b: A list describing a sparse matrix.
#
# Returns:
#     A list describing the sum of a and b as a sparse matrix.
casl_sparse_add <-
function(a, b)
{
  c <- merge(a, b, by = c("i", "j"), all = TRUE,
            suffixes = c("", "2"))
  c$x[is.na(c$x)] <- 0
  c$x2[is.na(c$x2)] <- 0
  c$x <- c$x + c$x2
  c[, c("i", "j", "x")]
}
```

This function can be used as follows, noting that it takes care of entries that are only in one matrix as well as entries that are non-zero in both.

```
df <- data.frame(i = c(1, 2), j = c(1, 1), x = c(4.4, 1.2))
casl_sparse_add(df, df)
```

```
  i j   x
1 1 1 8.8
2 2 1 1.2
3 2 2 1.2
```

---

[1]We will assume that the indices require the same space to store as the entries to simplify the discussion. This may not be the case, however, if the matrix requires double storage or can store values in a single byte (such as pixel intensities). The major theme of our discussion, however, is unchanged by these details.

If we assume that entries are ordered first by the row index and then by the column index (this is a more formal part of the CSR format), a variation of the merge sort algorithm can be used to perform matrix addition even faster. Notice that nowhere in our implementation does the actual size of the matrix come into play. We work only with the non-zero components, and therefore the computationally complex depends only on the number of non-zero elements and not the overall matrix size.

An algorithm for sparse matrix multiplication can be derived similarly. As with matrix addition, an easy way to implement the algorithm is to do a merge over the two coordinate lists. Here we want to match the column index of the first matrix with the row index of the second.

```
# Multiply two sparse matrices stored as triplet lists.
#
# Args:
#     a: A list describing a sparse matrix.
#     b: A list describing a sparse matrix.
#
# Returns:
#     A list describing the product of a and b as a sparse matrix.
casl_sparse_multiply <-
function(a, b)
{
  colnames(b) <- c("i2", "j2", "x2")
  c <- merge(a, b, by.x = "j", by.y = "i2",
             all = FALSE, suffixes = c("1", "2"))
  c$x <- c$x * c$x2
  c$key <- paste(c$i, c$j, sep = "-")
  x <- tapply(c$x, c$key, sum)
  key <- strsplit(names(x), "-")
  d <- data.frame(i = sapply(key, getElement, 1),
                  j = sapply(key, getElement, 2),
                  x = as.numeric(x))
  d
}
```

The function can be used as:

```
df1 <- data.frame(i = c(1, 2), j = c(1, 1), x = c(4.4, 1.2))
df2 <- data.frame(i = c(1, 2), j = c(1, 1), x = c(2.1, 3.7))
casl_sparse_multiply(df1, df2)
```

```
  i j    x
1 1 1 9.24
2 2 1 2.52
```

As with addition, the CSR format allows for a similar but faster implementation of the same idea where the `merge` command can be replaced by a hashed join function.

In practice one should at all costs avoid writing custom functions such as `casl_sparse_multiply` to implement sparse matrix operations. It is almost always better to call well-tested code such as Tim Davis' SuiteSparse Library [42], or wrappers to this code such as provided by the R **Matrix** package [19]. The reason to have a rough understanding of the underlying algorithms is so that we know how to rewrite our own algorithms to avoid constructing dense matrices whenever possible.

From these simple examples it should be clear that in many applications it is important to implement algorithms that work with sparse matrix formats as well as dense ones. It would seem that this is a simple drop in replacement where operations on matrices are replaced with functions that do one thing if the matrix is dense and another if it is sparse. Sometimes this is the case, but often more effort is required. When adding two matrices together, typically the sparsity of the output will be reduced because the sparse entries in the first matrix may not line up entirely with those in the second. Similarly, the product of two sparse matrices might be less sparse if the elements do line up well. However, in both cases we would expect the output to still be considered a sparse matrix. Other operations are more destructive. A common matrix operation in predictive modeling is to scale and center the columns of the data matrix X. The scaling factor does not break sparsity, but centering the columns does. In general, centering a sparse matrix would create an output where almost every value is non-zero. Suddenly, a sparse matrix that takes up only 1MB can expand to 1GB in size by something as innocuous as centering its columns. Clearly another approach is needed to make sparse matrices work in the context of algorithms that require centered columns of data.

As a simple example of avoiding the construction of a dense matrix from a sparse one, take the task of computing the starting value of λ in linear regression with an ℓ_1-penalty as in Chapter 7. With standardized variables Z_j this is given by the simple formula $\max_j |Y^t Z_j|$. As we have seen, however, constructing the centered columns X_j would turn a sparse matrix into a dense one. As an alternative, we could instead work with the raw values X_j and simply build the standardization into the algorithm. Assume that we have the mean and standard deviations of the columns defined as

$$Z_j = \frac{X_j - \mu_j}{\sigma_j} \tag{10.1}$$

Then, we see that we can compute λ_1 from:

$$\lambda_1 = \max_j |Y^t Z_j| \tag{10.2}$$

$$= \max_j |Y^t X_j \cdot \sigma_j^{-1} - Y^t \mathbb{1} \cdot \mu_j \cdot \sigma_j^{-1}| \tag{10.3}$$

$$= ||Y^t X \cdot \operatorname{diag}(\sigma_1^{-1}, \dots, \sigma_p^{-1}) - Y^t \cdot \operatorname{diag}(\mu_1 \sigma_1^{-1}, \dots, \mu_p \sigma_p^{-1})||_\infty. \tag{10.4}$$

The dot product $Y^t X$ is just a vector of length p and so are all of the other terms in the equation. Therefore, this approach computes the λ_1 without filling-in all of the non-zero terms. We can implement this easily in R for both dense and sparse matrices using the **Matrix** package's format for sparse matrices.

```
# Compute starting lambda for lasso regression.
#
# Args:
#     X: A sparse or dense numeric matrix.
#     y: The numeric response vector.
#
# Returns:
#     The numeric value of lambda max.
casl_sparse_lmax <-
function(X, y)
{
  svals <- apply(X, 2, sd)
  mvals <- apply(X, 2, mean)
  v <- (t(y) %*% X) / svals - sum(y) * mvals / svals
  max(abs(v))
}
```

We can test this with both a dense and sparse version of the of the matrix X.

```
set.seed(1)
Xd <- matrix(as.numeric(runif(2000) > 0.9), ncol = 20)
Xs <- Matrix::Matrix(Xd, sparse = TRUE)
y <- Xd[,1] * 0.1 + rnorm(100, sd = 0.1)
c(casl_sparse_lmax(Xd, y), casl_sparse_lmax(Xs, y))
```

```
[1] 1.881498 1.881498
```

Lasso regression is often used with datasets that have a very large number of columns. These data matrices are often very sparse, so ensuring that the implementation retains the sparse structure of the data matrix is important. It is also sensitive to the scale and mean of the columns. Therefore, the strategy employed in the function `casl_sparse_lmax` is used throughout the implementation of the elastic net algorithm found in the **glmnet** package. Notice that the end result of this modified computation is the same as the naïve approach for scaled dense matrices, but the formal algorithm is slightly modified from the original.

10.3   Sparse generalized linear models

As a more involved example using sparse matrices, consider implementing
generalized linear regression with a sparse data matrix X. The SVD-based
method employed in our implementation in Section 2.4 destroys sparsity in
the model matrix. That is, the singular value matrices U and V are generally
dense even if X is sparse. If we want to avoid constructing sparse matrices
when solving generalized linear models some modifications to our original
algorithms are needed.

There is an available algorithm for computing a sparse form of the QR-
decomposition. While the matrix Q itself is dense, it is possible to write Q
as a product of sparse transformations. With these transformations we can
compute products $Q^t z$ and Qz while never needing to form the full matrix Q.
In R, this algorithm is available through the overloaded function qr from the
Matrix package. For example, here we compute the dense and sparse variants
of the QR decomposition for a very sparse matrix:

```
library(Matrix)
Xd <- matrix(0, ncol=100, nrow=4000)
Xd[sample(seq_along(Xd), 10000)] <- 1
Xs <- Matrix(Xd, sparse=TRUE)
qrd <- base::qr(Xd)
qrs <- Matrix::qr(Xs)
```

Looking at the size of the objects reveals that the sparse QR decomposition
is retaining some sparse structure in the output.

```
       Dense         Sparse
X      "3125.2 Kb"   "13.5 Kb"
qr(X)  "3127.1 Kb"   "122.8 Kb"
qrR    "78.3 Kb"     "5.7 Kb"
qrQ    "3125.2 Kb"   "3126.1 Kb"
```

For example, the QR decomposition object itself is about 30 times smaller in
the sparse case. The only object that is the same size is if we construct the
actual matrix Q, which as mentioned is not necessary.

The sparse QR decomposition is an excellent tool for working with very
sparse matrices. It is possible to use it to implement an iteratively reweighted
least squares algorithm to solve generalized linear models. There are two short-
comings that make this approach non-optimal for general use. The size of the
sparse QR object grows quickly as we decrease in the sparsity of the data
matrix. In the simulation above, if we increase the number of non-zero terms
by a factor of 10, the size of the sparse decomposition increases by about a
factor of 24 and is virtually the same size as the dense variant. Secondly, the
computational cost of computing Qz or $Q^t z$ is equivalent to computing the

dense products. While avoiding the memory requirements of creating dense matrices, the sparse QR offers minimal computational benefits.

We introduced the QR and SVD approaches to solve least squares problems by illustrating how they help to avoid computational issues that arise when directly solving the normal equations. There is no computational penalty to these two approaches and so they are preferred in the general case. When working with sparse matrices, there is a penalty for these decompositions because we need to work with dense matrices Q, U, and V (or a virtual counterpart of these, as in the sparse QR decomposition). This suggests that, at least in cases where numerical issues are a minimal concern, it would be best to return to the normal equations to solve sparse least squares problems. If the data matrix X is sparse and the weight matrix W is diagonal, the quantities

$$A = X^t W X, \quad z = X^t W y, \tag{10.5}$$

can be computed quickly and with minimal memory requirements using sparse matrix multiplication algorithms. Assuming the input space p is reasonable in size (small enough that we can store a p-by-p matrix in memory) we can proceed to solve the least squares problem by creating a dense version of A and solving

$$Ab = z \tag{10.6}$$

for b using any general solver for a linear system such as the Cholesky or LU-decompositions. The computation of this algorithm avoids constructing dense $n \times p$ matrices or performing the equivalent of dense matrix computations.

We will now make use of this approach to build a general purpose sparse iteratively reweighted least squares algorithm for a sparse model matrix X. We use the Cholesky decomposition to solve the weighted least squares problem in the inner loop of the algorithm.

```
# Compute GLM regression with a sparse data matrix.
#
# Args:
#     X: A sparse or dense numeric matrix.
#     y: The numeric response vector.
#     family: Instance of an R 'family' object.
#     maxit: Integer maximum number of iterations.
#     tol: Numeric tolerance parameter.
#
# Returns:
#     The estimated regression vector.
casl_sparse_irwls <-
function(X, y, family=binomial(), maxit=25L, tol=1e-08)
{
  p <- ncol(X)
```

```
  b <- rep(0, p)
  for(j in seq_len(maxit))
  {
    eta <- as.vector(X %*% b)
    g <- family$linkinv(eta)
    gprime <- family$mu.eta(eta)
    z <- eta + (y - g) / gprime
    W <- as.vector(gprime^2 / family$variance(g))
    bold <- b
    XTWX <- crossprod(X,X * W)
    wz <- W * z
    XTWz <- as.vector(crossprod(X, wz))

    C <- chol(XTWX, pivot=TRUE)
    if(attr(C,"rank") < ncol(XTWX))
    {
      stop("Rank-deficiency detected.")
    }
    piv <- attr(C, "pivot")
    s <- forwardsolve(t(C), XTWz[piv])
    b <- backsolve(C,s)[piv]

    if(sqrt(crossprod(b-bold)) < tol) break
  }
  list(coefficients=b, iter=j)
}
```

Because it is possible that this implementation will be less numerically stable, we added a check to ensure that the rank of the Cholesky decomposition matches p, the number of columns in X.

To test the function `casl_sparse_irwls`, we will construct a sparse data matrix and corresponding response vector.

```
set.seed(1)
n <- 1000; p <- 30
Xd <- matrix(rnorm(n*p), ncol = p)
Xd[sample(n*p, n*p*0.9)] <- 0
Xd[,1] <- 1
Xs <- Matrix(Xd, sparse = TRUE)
b <- c(1,1,1,rep(0,p-3))
mu <- 1 / (1 + exp(-Xd %*% b))
y <- as.numeric(runif(n) <= mu)
```

We then compare this implementation to our sparse function with the `glm` function in R.

```
beta_d <- glm.fit(Xd, y, family = binomial())
```

```
beta_s <- casl_sparse_irwls(Xs, y)
max(abs(beta_d$coefficients - beta_s$coefficients))
```

```
[1] 2.384683e-09
```

So the sparse solution using the Cholesky decomposition closely matches
the result from iteratively reweighted least squares using the QR decomposition. Note that if an input X stored in a dense format is given to
`casl_sparse_irwls`, the function produces an error because the output of
`chol` will not contain attributes `rank` or `pivot`. A minimal amount of additional logic in the function would allow the same code to process sparse and
dense matrices together, though with a dense matrix the QR-decomposition
may be preferred for numerical stability.

Another approach to computing regression functions with sparse matrices
is to replace an exact computation with an approximation. For example, it is
possible to compute a truncated singular value decomposition from a sparse
matrix without constructing a large dense matrix. This can be used, with a
reasonable number of principal components, to apply principal component regression. We will show an example of the sparse PCA using the **irlba** package
in our application in Section 10.8.

10.4   Computation on row chunks

Our reference implementations all take for granted the fact that the data
matrix can be stored and accessed in memory. In the case of large datasets,
this may not be possible. Very large datasets may not even fit on a single
machine. Applying statistical learning algorithms to large datasets requires
modified algorithms that can be written to work with only a subset of the data
at any given time. As was the case for sparse matrix algorithms, sometimes the
modification is relatively minor and other times a completely new approach
is required.

A common paradigm for working with large datasets is to split the dataset
into M sets of row-wise chunks,

$$X = \begin{bmatrix} X_1 \\ X_2 \\ \vdots \\ X_M \end{bmatrix}, \tag{10.7}$$

such that any X_k can fit into memory.[2] We will assume that the response

---

[2]This is possible as long as a single row is small enough to fit into memory. We are not
aware of any practical example where a single row is too large to allow for such a partition.

variable y is also broken up into corresponding chunks $[y_1, \ldots, y_M]$. Algorithms can then iterate over these chunks of the rows by loading and processing one X_k at a time. If no one machine can hold all of the chunks, the algorithmic iteration instead offloads the computation to whatever machine happens to holds the data. When the computation applied to each chunk is independent of the other chunks, the iteration can be done in parallel—this approach is called *split-apply-combine* [176].

Stochastic gradient descent, as developed in Section 8.2, can be applied almost trivially to a sequential application over row chunks by setting the mini-batches to be equal to each X_k. Weighted least squares, with a diagonal weight matrix, can be modified to work over the chunks X_k with only a modest amount of work. As with the sparse case in Section 10.3, we will not be able to compute a QR-decomposition of X and instead work on a decomposition of $X^t X$. By writing the weights matrix as

$$
W = \begin{bmatrix}
W_1 & 0 & \cdots & 0 \\
0 & W_2 & & 0 \\
\vdots & & \ddots & \vdots \\
0 & 0 & \cdots & W_M
\end{bmatrix},
\tag{10.8}
$$

notice that the inner products needed for the weighted normal equations become a simple sum over their chunked version

$$
X^t W X = \sum_{k=1}^{M} X_k^t W_k X_k
\tag{10.9}
$$

$$
X^t y = \sum_{k=1}^{M} X_k^t W_k y_k.
\tag{10.10}
$$

We can then compute $X^t W X$ by filling in a $p \times p$ matrix of zeros and adding the contribution of each chunk in sequence. Once we have these values solving for the regression vector proceeds as usual. A similar procedure can be applied to iteratively re-weighted least squares because we can compute the new weights and effective response locally on each chunk.

To implement a generalized linear model that operates over partitions of X, we need to provide iterative access to a chunk of data. Here we implement this approach assuming that the data matrix is saved as a single large comma separated file with the response vector equal to the first column of the data. We open the file, read the first chunk to find the dimensionality of the data, and the cycle over the file until the end is reached.

```
# Compute GLM regression with data stored on disk.
#
# Args:
#     filename: Path to a file containing the dataset.
#     nmax: Maximum number of rows in the chunk.
```

```
#       family: Instance of an R 'family' object.
#       maxit: Integer maximum number of iterations.
#       tol: Numeric tolerance parameter.
#
# Returns:
#       A list containing the regression vector beta and
#       number of iterations.
casl_blockwise_irwls <-
function(filename, nmax, family=gaussian(), maxit=25L,
         tol=1e-08)
{
  b <- NULL

  for(j in seq_len(maxit))
  {
    fin <- file(filename, "r")
    X <- read.table(fin, sep=",", nrows=nmax)
    p <- ncol(X) - 1L
    if (is.null(b)) b <- rep(0, p)

    XTWX <- matrix(0, ncol = p, nrow = p)
    XTWz <- rep(0, p)
    while(!is.null(X))
    {
      X <- as.matrix(X)
      y <- X[, 1]; X <- X[, -1, drop=FALSE]
      eta <- as.vector(X %*% b)
      g <- family$linkinv(eta)
      gprime <- family$mu.eta(eta)
      z <- eta + (y - g) / gprime
      W <- as.vector(gprime^2 / family$variance(g))
      XTWX <- XTWX + crossprod(X,X * W)
      XTWz <- XTWz + as.vector(crossprod(X, W * z))

      X <- tryCatch(read.table(fin, sep=",", nrows=nmax),
                    error=function(e) NULL)
    }
    close(fin)

    bold <- b
    C <- chol(XTWX, pivot=TRUE)
    if(attr(C,"rank") < ncol(XTWX))
    {
      stop("Rank-deficiency detected.")
    }
```

```
    piv <- attr(C, "pivot")
    s <- forwardsolve(t(C), XTWz[piv])
    b <- backsolve(C,s)[order(piv)]

    if(sqrt(crossprod(b-bold)) < tol) break
  }

  list(coefficients=b, iter=j)
}
```

An implementation written to operate over a cluster would look similar, but the code would need to be specific to the software used to communicate over a particular cluster.

In order to test the code, we first need to simulate a dataset and save the results as a comma separated file. An estimate of the regression vector will be saved to compare with the incremental algorithm's result.

```
set.seed(1)
n <- 1000; p <- 30
X <- matrix(rnorm(n*p), ncol = p)
b <- c(1,1,1,rep(0,p-3))
mu <- 1 / (1 + exp(-X %*% b))
y <- as.numeric(runif(n) <= mu)
Z <- cbind(y, X)
write.table(Z, "input.csv", sep=",",
            row.names=FALSE, col.names=FALSE)
beta_local <- glm.fit(X, y, family = binomial())
```

The dataset here is not actually very large, but still illustrates the behavior of the algorithm if we set the chunk size to only 100.

```
beta_chunk <- casl_blockwise_irwls("input.csv", nmax=100L,
                                   family = binomial())
max(abs(beta_chunk$coef - beta_local$coef))
```

```
[1] 1.063496e-10
```

And we see that the two implementations produce very similar resulting regression vectors.

One caveat to the algorithm `casl_blockwise_irwls` is that it requires the storage of the $p \times p$ matrix $X^t W X$. This is reasonable for p around $10,000$ on most current hardware, and a few times larger on high performance machines. For very large values of p it is unlikely that unpenalized regression will be a reasonable approach. Instead, the elastic net or a similar algorithm is required to reduce the input space. Fortunately the elastic net algorithm in Chapter 7 can also be modified to operate over row chunks of the data but does not require the full calculation of the matrix $X_k^t W_k X_k$. When working with data distributed over a cluster of machines, the overhead cost of transmitting many $p \times p$ matrices can become a problem for even reasonably sized p. Several schemes exist, with a variety of underlying theoretical guarantees, where $X_k^t W_k X_k$ is transmitted through a low-rank approximation [188].

10.5   Feature hashing

Many techniques that we have studied involve transforming the data matrix X into another matrix Z. We have given relatively little attention to how the matrix X can be formed from raw data. If the original dataset contains a non-numeric variable, we briefly described in Section 2.1 how this can be transformed into numeric indicator variables. The same technique was applied to a categorical response in Section 8.7 in order to build neural networks for classification tasks. We did not go into much detail about how this transformation is performed because it can generally be done automatically using functions such as R's `stats::model.matrix` or the `to_categorical` in **keras**. Many of the details in this process, such as picking factor contrasts, are important for statistical inference but do not affect the predictive power of the model.

The task of converting a dataset into a numeric data matrix becomes nontrivial when moving from a single in-memory dataset to one split into row chunks. Turning a categorical variable into indicator values requires knowing at the start all of the available categories. Unless we can specify ahead of time the closed set of categories a variable comes from, this will require an (expensive) initial pass through the dataset to collect all of the unique values. The problem is amplified with streaming data, where there is no possibility of a first pass across all observations. A similar issue arises if we need to build a model that can predict values from data unseen at training time.

The technique known as *feature hashing*—or the *hashing trick*—offers a solution to these potential problems, particularly when the number of possible categories is large and unbounded [173]. A hash function is a deterministic map from any string to an integer in a fixed range. As a simple example, consider mapping the letters a through z to the numbers 1–26. A hash function can be defined by mapping each letter in a string to its corresponding number, adding all of the letters together, and giving the result modulo the desired size of the hash. Better hash functions exist that map more uniformly into the target space and deal with capital letters and other punctuation marks. We can make use of a hash function ϕ to convert categorical data into a data matrix by picking a starting hash size H, and matching the category v to column $\phi(v)$ in the data matrix. This allows for the construction of a data matrix without prior knowledge of the categories.

One of the most common applications of feature hashing is for the construction of term frequency matrices, which we will use as an example through the remainder of this section. A term frequency matrix is a data matrix where each observation corresponds to some raw textual input and the columns capture information about the terms in the text. For example, we might have one column in the data matrix X corresponding to each unique word across all of the documents. The entry $X_{i,j}$ counts the number of times word j is used in document i. With feature hashing this matrix can be built incrementally without first knowing all of the words used in every document. Specifically, we can iterate over all of the words $w_{i,k}$ in document i in order to modify X by

$$X_{i,\phi(w_{i,k})} = X_{i,\phi(w_{i,k})} + 1. \qquad (10.11)$$

After filling this in for every document, the matrix X will give a term frequency matrix for the entire dataset. Term frequency matrices tend to be very sparse, and this updating scheme can be implemented in a sparse matrix format.

There is always the possibility that two words of interest will be mapped to the same index position. This is known as a hash collision. There is no complete fix to avoid this issue, but the possibility of collisions can be minimized by choosing a hash function with a sufficiently large hash size. Recent theoretical and empirical results have also shown that a moderate amount of collision does not significantly effect the prediction power of a model [36]. In our implementation we will also see a way of avoiding collisions for a set of most frequent or predetermined words.

In order to implement feature hashing, we first need a good hashing function. For this, we will use the **digest** package and the 32-bit MurmurHash algorithm. The code here converts any input text string into a string of eight characters.

```
library(digest)
h <- digest("statistic", algo = "murmur32")
h
```

```
[1] "152b8d04"
```

This string should be interpreted as a number in base-16,[3] which is equal to the integer 355175684 in normal base-10 notation. The hash size here of 16^8 will generally be too large for statistical applications, but we can modify the hash size by taking the output of MurmurHash modulo our desired hash size. This is made easiest if we pick a hash size that can be written as $16^k = 2^{4k}$, as the modulo function becomes a simple truncation of the number to its first k digits. Using the `strtoi` function to convert a hexadecimal (base-16) string to a number in R, we now get the hashed value of the word "statistic" using a hash size of 65536:

```
k <- 4
```

---

[3]In common base-16 notation, the letters a–f stand for the numbers 10–15.

```
strtoi(stringi::stri_sub(h, -k, -1), base = 16L)
```

```
[1] 36100
```

We will stick to hash sizes of the form 16^k for simplicity in our implementation. Now, we will wrap this up as a function that returns a word index given the string and desired hash size.

```
# Map a string object to a numeric hash value.
#
# Args:
#     word: A string object to hash.
#     k: Size of the hash, given as 2^(4*k).
#
# Returns:
#     A numeric hash value from 0 to 2^(4*k) - 1.
casl_hash_word2index <- function(word, k=4L)
{
  h <- digest::digest(word, algo="murmur32")
  v <- strtoi(stringi::stri_sub(h, -k, -1), base=16L)
  v
}
```

This can be tested by passing several words to the function.

```
sapply(c("stat", "stats", "star"), casl_hash_word2index)
```

```
 stat stats  star
49023 36229 13131
```

Notice that the values of the three terms are all significantly different despite the input words all being very similar; this is a desirable property of hash functions, particularly when used for cryptographic applications.

A modification of the iterative construction of X in Equation 10.11 is often used to ensure that the expected value of each inner product $X_j^t X_k$ is equal to zero. Define a new hash function ζ that maps words uniformly into the set $\{+1, -1\}$; we can then modify the definition of X to be

$$X_{i,\phi(w_{i,k})} = X_{i,\phi(w_{i,k})} + \zeta(w_{i,k}). \tag{10.12}$$

So, we add a counter to X if ζ is positive and subtract it if it is negative. To implement this, we will use a MurmurHash with a different seed and check just the first hexadecimal digit to determine the sign of the word.

```
# Map a string object to a sign function.
#
# Args:
```

```
#      word: A string object to hash.
#
# Returns:
#      A numeric value of either +1 or -1.
casl_hash_word2sign <- function(word)
{
  h <- digest::digest(word, algo="murmur32", seed=1)
  s <- ifelse(stringi::stri_sub(h, 1, 1) %in% seq(0, 7), 1, -1)
  s
}
```

The output can again be tested on our three words.

```
sapply(c("stat", "stats", "star"), casl_hash_word2sign)
```

```
stat stats  star
   1     1    -1
```

So, when observing the word "stat" in a document i, the value in $X_{i,49023}$ is increased by one. If "star" is observed, the value in $X_{i,13131}$ is decreased by one.

Our next step is to build a lookup table to take a vector of words and return a data frame mapping each word to an index and a sign value.

```
# Map multiple string objects to a numeric hash values.
#
# Args:
#      words: A vector of string objects to hash.
#      k: Size of the hash, given as 2^(4*k).
#
# Returns:
#      A data frame giving hash values and signs.
casl_hash_matrix <- function(words, k=4L)
{
  hash <- data.frame(tok=words, hash=0, sign=0,
                     stringsAsFactors = FALSE)
  for (i in seq_along(hash$tok))
  {
    hash$hash[i] <- casl_hash_word2index(hash$tok[i], k)
    hash$sign[i] <- casl_hash_word2sign(hash$tok[i])
  }
  hash
}
```

It is not necessary to build and store the entire lookup table because the map from a word to its index and sign can be computed using the deterministic hash function. The benefit of building the table is two-fold: it improves the run-time because looking up on a table is much faster than computing the hash function, and it allows us to compute the inverse map of the hash function. The latter is useful to detect bad collisions and to interpret the results of models that are computed based on the hashed features.

Using the function `casl_hash_matrix`, we now build a function that takes input text and returns a data frame of the hashed values for each word. To split the text into tokens, we use the function `stri_split_boundaries` from the **stringi** package. We also convert all of the terms to lower case.

```
# Map text to numeric hash values.
#
# Args:
#     text: A character vector of inputs.
#     k: Size of the hash, given as 2^(4*k).
#
# Returns:
#     A data frame giving hashed values.
casl_fhash <- function(text, k=4L)
{
  tok <- stringi::stri_trans_tolower(text)
  tok <- stringi::stri_split_boundaries(tok, type="word",
                                        skip_word_none=TRUE)
  id <- mapply(function(u,v) rep(u, v),
               seq_along(tok), sapply(tok, length))
  id <- as.vector(unlist(id))
  df <- data.frame(tok=unlist(tok), id=id,
              hash=0L, sign=0L,
              stringsAsFactors = FALSE)
  hash <- casl_hash_matrix(unique(df$tok), k=k)
  id <- match(df$tok, hash$tok)
  df$hash <- hash$hash[id]
  df$sign <- hash$sign[id]
  df
}
```

To test this function we will input four short strings, making one word ("fish") occur multiple times to verify that it is consistently mapped to the same hash values.

```
text <- c("One fish", "Two fish", "Red fish", "Blue fish fish")
casl_fhash(text)
```

```
    tok id  hash sign
1   one  1 51954    1
2  fish  1 22325   -1
3   two  2 34674   -1
```

```
4 fish  2 22325    -1
5  red  3 11866    -1
6 fish  3 22325    -1
7 blue  4  7027    -1
8 fish  4 22325    -1
9 fish  4 22325    -1
```

We see, for example, that "fish" is mapped to the hash value of 22325 and a sign of -1.

Eventually, we want to turn the output of `casl_fhash` into a sparse matrix. The function `Matrix::sparseMatrix` makes this very easy because when given a matrix as a coordinate list, by default it adds any terms together that have the same coordinates. Therefore if we assign the value of x in the coordinate list to be the sign of the word, we can move directly from our data frame to a sparse matrix.

```
# Construct a data matrix using hash function.
#
# Args:
#     df: The output of casl_fhash.
#     k: Size of the hash, given as 2^(4*k).
#
# Returns:
#     Sparse matrix representation of the data matrix.
casl_fhash_matrix <-
function(df, k=4L)
{
  sparseMatrix(i=df$id, j=(df$hash + 1L), x=df$sign,
               dims = c(max(df$id), 16^k))
}
```

Notice that we do need to know the value of k when constructing the matrix to ensure that its dimensionality will match others made on any other chunks using the same hash values. In order to view the matrix representation, we will decrease k to be equal to one (a maximum hash size of 16). We also convert the sparse matrix to a dense one in the print out.

```
as.matrix(casl_fhash_matrix(casl_fhash(text, k = 1), k = 1))
```

```
     [,1] [,2] [,3] [,4] [,5] [,6] [,7] [,8] [,9]
[1,]    0    0    1    0    0   -1    0    0    0
[2,]    0    0   -1    0    0   -1    0    0    0
[3,]    0    0    0    0    0   -1    0    0    0
[4,]    0    0    0   -1    0   -2    0    0    0
     [,10] [,11] [,12] [,13] [,14] [,15] [,16]
[1,]     0     0     0     0     0     0     0
```

[2,]	0	0	0	0	0	0	0
[3,]	0	-1	0	0	0	0	0
[4,]	0	0	0	0	0	0	0

We see that "fish" is mapped to the 6th column, and the algorithm correctly decremented the 6th column value for the fourth document twice because fish occurred twice that document. Also notice that "one" and "two" were both mapped to the third column, but their sign function differs, allowing the matrix to distinguish between the colliding keys.

As a final modification of the function `casl_fhash`, we want to offer an improvement to the problem of hash collisions. In order to do this, we allow for an input that allows the users to pass a starting hash table. The values in this table can, but do not need to, match those determined by the actual hash function ϕ. This is useful because we can map known important terms to integers outside of the hash size. It also useful when applying the function many times because we can construct a starting hash table and only need to look up terms that are new in the next chunk of data.

```
# Construct a data matrix using hash function.
#
# Args:
#     text: The input text vector.
#     k: Size of the hash, given as 2^(4*k).
#     hash: Either NULL or cached hash data frame.
#
# Returns:
#     A list containing the data frame of hashed values and the
#     updated hash table.
casl_fhash_cache <-
function(text, k=4L, hash=NULL)
{
  tok <- stringi::stri_trans_tolower(text)
  tok <- stringi::stri_split_boundaries(tok, type="word",
                                        skip_word_none=TRUE)
  id <- mapply(function(u,v) rep(u, v),
               seq_along(tok), sapply(tok, length))
  id <- as.vector(unlist(id))
  df <- data.frame(tok=unlist(tok), id=id,
                   hash=0L, sign=0L,
                   stringsAsFactors = FALSE)

  if (!is.null(hash))
  {
    words <- setdiff(df$tok, hash$tok)
  } else {
    words <- unique(df$tok)
```

```
}

if (length(words) > OL)
{
  hash <- rbind(hash, casl_hash_matrix(words, k = k))
}

id <- match(df$tok, hash$tok)
df$hash <- hash$hash[id]
df$sign <- hash$sign[id]
list(df=df, hash=hash)
}
```

Given its high frequency in our table, we will map "fish" to the number 16 (the hash only maps to values 0 to 15) and apply the new function casl_fhash_cache.

```
hash <- data.frame(tok = c("fish"),
                   hash = 16,
                   sign = 1,
                   stringsAsFactors = FALSE)
casl_fhash_cache(text, hash = hash, k = 1)[["df"]]
```

```
   tok id hash sign
1  one  1    2    1
2 fish  1   16    1
3  two  2    2   -1
4 fish  2   16    1
5  red  3   10   -1
6 fish  3   16    1
7 blue  4    3   -1
8 fish  4   16    1
9 fish  4   16    1
```

We see that now the term "fish" will never collide with any other terms in the data. Note that we would need to modify the code for casl_fhash_matrix to allow the output matrix to include an extra column. A more involved application of feature hashing will be given in Section 10.8.

10.6   Data quality issues

Models are built upon data observations and data observations might be imperfect. For instance some data may be missing, misread as weird values, or simply contain some amount of measurement error. And as the amount of data analyzed increases, it can sometimes be more likely to encounter one or all of these problems.

The simplest approach to dealing with missing data omits observations with missing values, recording which observations were omitted for posterity. The R language, for example, includes systematic approaches for this. Alternatively, missing data may be imputed in many applications; see, for instance the paper by King, Honaker, Joseph, and Scheve [94] for a good overview of available methods.

Sometimes data are mis-recorded in a way that is obviously detectable, for instance a negative value for a variable that should only be non-negative. Or some variables may contain numeric error flags like NaN or Inf. Such cases are sometimes referred to as *outliers* and can be detected and removed prior to analysis. Outlier removal is necessarily very domain-specific; there are few generic approaches. Methods often use robust statistical techniques mentioned in Chapters 2 and 6 to help identify potentially bad data.

We usually focus on modeling errors in data associated with a measured response variable. For example, given a data matrix X and observed response y, ordinary least squares can be written as the constrained optimization problem

$$\min_{\beta, \Delta y} \|\Delta y\|_F, \qquad \text{subjet to } X\beta = y + \Delta y,$$

explicitly including the residual error Δy in the optimization objective, where $\| \cdot \|_F$ indicates the Euclidean norm in the case that y and Δy are vectors or the Frobenius matrix norm if they are matrices. Less usually considered are cases with errors in the data themselves–that is, in the model matrix X. When errors in the response *and* the data are considered, the ordinary least squares formulation becomes

$$\min_{\beta, \Delta y, \Delta X} \|[\Delta X \ \Delta y]\|_F, \qquad \text{subject to } (X + \Delta X)\beta = y + \Delta y, \qquad (10.13)$$

where $[\Delta X \ \Delta y]$ means column-wise concatenation of ΔX and Δy. The solution of Equation 10.13 has an interesting geometric interpretation. As we saw in Figure 2.1, the solution of a two-dimensional ordinary least squares problem finds the smallest sum of squared distances between the data points and the regression line. When allowing for errors in the model matrix, the solution of Equation 10.13 instead minimizes the *orthogonal* distances between the data points and the regression line, illustrated in Figure 10.1 below.

Algorithms for solving the above problem were developed by Gleser [66], and in the numerical analysis literature by Golub and Van Loan [69] as the *Total Least Squares* method. But statisticians have long thought about accounting for errors in data—for example, see the papers by Adcock [2], Pearson [129] and Koopmans [76] from the late 19th and early 20th centuries. Much more recently, Sabine Van Huffel and her colleagues have thoroughly studied and expanded these ideas to cover many edge cases (like rank-deficiency), see

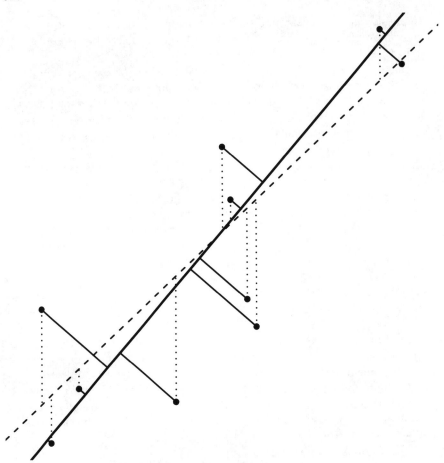

FIGURE 10.1: Comparison of ordinary least squares fit/residuals (dashed lines) and total least squares fit/residuals from Equation 10.13 (solid lines).

the survey by Markovsky and Van Huffel [162] for details. Like imputation, dealing with errors in variables can be somewhat domain-specific. Even so, we are somewhat surprised that this issue is not often considered.

10.7   Implementation and notes

As we have seen in several examples, sparse matrix operations are supported by the R package **Matrix** [19]. There is extensive support for multiple storage formats and a multitude of matrix operations. The package also supports storage of dense matrices, allowing the construction of code that can handle both dense and sparse matrices. A relatively small number of R packages support passing data via a sparse matrix, although this does include some of the most popular packages such as **glmnet** [61], **xgboost** [35], and **lme4** [17].

Several packages exist for performing chunk-wise computations over a large dataset. The **biglm** package supports linear and generalized linear models of data supplied by a generic iterator object using the same approach we developed in Section 10.4 [108]. The **biglasso** package uses the same approach to fit ℓ_1-penalized linear and logistic regression [186]. Additionally, packages provide the basic architecture for writing code to run over sections of a dataset. The **foreach** package allows for running a snippet of code across any number of various backends [174]. The **iotools** package provides functions to parse streaming data and iterate over the rows of large data files [11].

For feature hashing, the R package **FeatureHashing** can be used to convert a data frame into a set of hashed features [179]. As used in our examples, it is also possible to use **digest** to construct hashed functions manually [53]. The latter is useful because the package **FeatureHashing** is missing options that allow for a set of predefined hash values, as shown in our function `casl_fhash_cache`.

10.8 Application

10.8.1 Data and chunks

The application provided here draws from an important class of predictive modeling in text mining. Given a set of documents, we want to build a model that is able to predict the author of a text using information drawn solely from the text itself. One interesting aspect of authorship detection is that it may make use of both stylistic and topic features found in the writing style of a particular author. We chose this topic here because textual data inputs provide an excellent example of using sparse matrices, parallel processing, and feature hashing. The example is kept small in order to make the dataset easy to distribute and replicate. However, there are many freely available datasets of text that are very large and can be used with the same methods.

The corpus we will work with includes short snippets of text from two authors: Sir Arthur Conan Doyle and Jane Austen. The document here (converting new lines into spaces):

"Be Jove, if the Commandant let them come weak, he should be court-martialled." "Sure we're in God's hands, anyway," said his wife, in her soothing, Irish voice. "Kneel down with me, John, dear, if it's the last time, and pray that, earth or heaven, we may not be divided." "Don't do that! Don't!"

From Doyle's novel *The Tragedy of the Korosko*, one of the few that does not include his character Sherlock Holmes. And here, a similar snippet from Austen's *Northanger Abbey*:

> No matter which has it, so that there is enough. I hate the idea of one great fortune looking out for another. And to marry for money I think the wickedest thing in existence. Good day. We shall be very glad to see you at Fullerton, whenever it is convenient.

In order to successfully compare predictive models, we have split the data into training and testing sets such that all the passages from a given novel are either all in the training set or all in the testing set. Of course, this does not stop the model from using recurring character names such as "Watson," but it does avoid many such conflicts.

To test the chunk-wise and parallel applications developed in this chapter, we have stored the dataset into five separate files. The code can be easily adapted to datasets with data split into an arbitrarily large number of such files.

10.8.2   Logistic regression over row chunks

Our first step is to fit a logistic regression by processing the data in chunks. The features we use here can be extracted from the raw data using functions from the **stringi** package. We find the overall length of the passage and count the number of occurrences of several special characters and capital letters.

```
fproc <- function(input) {
  nd <- data.frame(author = input$author)
  nd$len <- stringi::stri_length(input$text)
  nd$spaces <- stringi::stri_count(input$text, fixed = " ")
  nd$comma <- stringi::stri_count(input$text, fixed = ",")
  nd$quest <- stringi::stri_count(input$text, fixed = "?")
  nd$excla <- stringi::stri_count(input$text, fixed = "!")
  nd$quote <- stringi::stri_count(input$text, regex = "[\'\"]")
  nd$caps <- stringi::stri_count(input$text, regex = "[A-Z]")
  return(nd)
}
```

The output of the function `fproc` is a new data frame containing just the response variable and the features that will be used in our model.

We will make use of the **biglm** package to fit a logistic regression to the authorship data. The function provided by the package requires an input with a very specific format. Specifically, we need to provide a function that successively returns chunks of the dataset. Calling the function with the option `reset` equal to true should start the iteration from the beginning and the function should return NULL when the input has been exhausted. In order to make this possible, we first need to define a *functional*, a function that returns another function. Within this function a global variable is defined, `iter`, that indicates which file is next in the iteration.

```
cnames <- c("author", "train", "text")
read_data_chunk <- function(fbase, max_size,
                                 fproc = identity) {
  iter <- 1
  max_size <- max_size
  function(reset=FALSE, train_id = "train"){
    if(reset){
      iter <<- 1
      return(0)
    } else {
      if (iter > max_size) {
        ret <- NULL
      } else {
        ret <- read.csv(sprintf(fbase, iter), header=FALSE,
                        col.names = cnames)
        ret <- ret[ret$train == train_id,]
        ret <- fproc(ret)
        iter <<- iter + 1
      }
      return(ret)
    }
  }
}
```

Internally, the function applies our function `fproc` before returning the result in order to have access to all of the features needed for the model. It also by default returns just training data, but the testing data can be selected if needed. Notice that we have used a new symbol here: `<<-`. This command changes the variable `iter` that exists in the parent functions space.[4]

To test our code `read_data_chunk`, we will create a new instance of the data reader and look at the head of the output.

```
gen <- read_data_chunk("data/author_c%d.csv", 5, fproc)
head(gen())
```

	author	len	spaces	comma	quest	excla	quote	caps
1	1	360	63	3	0	3	6	11
2	1	389	82	5	1	1	8	9
3	1	567	102	5	1	0	4	12
4	1	344	63	5	1	0	13	12
5	1	292	48	6	1	0	9	9

---

[4]We are entering into some advanced R behavior here. For complete details consult the manuals *R Internals* and *R Language Definition*. See https://cran.r-project.org.

| 6 | 1 208 | 37 | 2 | 1 | 0 | 10 | 5 |

The files are sorted by author, so all of these inputs are from author number 1. We see the seven derived variables present in the data that have been added by fproc.

Now it is possible to use the function bigglm from the **biglm** package. Its call structure is similar to the function stats::glm, but as mentioned it expects a function to be given to the data input rather than a single local data frame. The output also looks similar to the stats::glm function, but some model metrics are missing due to the difficulty in computing them without full access to the data.

```
library(biglm)
ff <- author ~ len + spaces + comma + quest + excla +
    quote + caps
model <- bigglm(ff, data = gen, family = binomial())
summary(model)
```

```
Large data regression model: bigglm(ff, data = gen,
                                    family = binomial())
Sample size =  5812
              Coef    (95%     CI)      SE       p
(Intercept)  1.1491  0.9443  1.3540 0.1024 0.0000
len         -0.0144 -0.0165 -0.0123 0.0010 0.0000
spaces       0.0746  0.0632  0.0861 0.0057 0.0000
comma       -0.0482 -0.0719 -0.0245 0.0118 0.0000
quest       -0.1262 -0.2133 -0.0392 0.0435 0.0037
excla       -0.3997 -0.4823 -0.3172 0.0413 0.0000
quote        0.0910  0.0692  0.1127 0.0109 0.0000
caps        -0.0534 -0.0688 -0.0380 0.0077 0.0000
```

Note that care needs to be taken when describing the formula object that is given to bigglm. Functions such as stats::poly (unless raw=TRUE) and scale that depend on the entire dataset to function correctly will run without error but the model matrices will not be consistently constructed. Also, factors can only be included if they contain the same values across all chunks. Anything explicit, such as log or sqrt, can be applied without issue.

We see from the logistic regression that all of the p-values are significant. How predictive is the model on the test set? Here, we grab the first chunk of testing data:

```
gen(reset = TRUE)
input <- gen(FALSE, "test")
pred <- predict(model, newdata = input)
mean((pred > 0) == input$author)
```

```
[1] 0.7096774
```

The original dataset has balanced classes, so a 70% classification rate is significantly better than random guessing. By adding more variables we can increase this rate significantly.

10.8.3 Dimensionality reduction with parallel processing

For the next task, we will apply dimensionality reduction using techniques from Chapter 9. Specifically, we will first produce a 25-dimensional representation using linear PCA. These factors will then be used in a 2-dimensional t-SNE model to produce a visualization of the terms used in the texts.

Our goal is to write code that can use parallel processing over the chunks of the data. To do this, we will use the excellent library **foreach**, which provides a general framework for working with multiple sequential and parallel backends. Each of these backends is contained in a separately maintained package. Here, to use multiple cores on a single machine, we will use the **doMC** (multicore) package. To start, load the package and *register* how many parallel processes should be used at a time.

```
library(doMC)
registerDoMC(cores = 2)
```

The number of cores should not exceed the number of cores on the machine running the code.

Our goal is to perform dimensionality reduction on the term frequency matrix for our documents. We will not make use of feature hashing until the next section, but we need to somehow keep the vocabulary of terms consistent between runs. To do this, we will grab just the first block of data and construct a vocabulary from all documents contained within it.

```
input <- read.csv(sprintf("data/author_c%d.csv", 1), FALSE,
                  col.names = cnames,
                  as.is = TRUE)
tok <- stringi::stri_trans_tolower(input$text)
tok <- stringi::stri_split_boundaries(tok, type = "word",
                                      skip_word_none = TRUE)
vocab <- unique(unlist(tok))
```

Now, we use the `foreach::foreach` function to iterate over the five the files of data. The function must be passed to the statement that is executed for each parallel chunk using the binary operator `%dopar%`. To simplify construction of the term frequency matrix, we use the package **cleanNLP** [10].

```
library(cleanNLP)
library(foreach)
cnlp_init_tokenizers()
```

```
res <- foreach::foreach( i = seq_len(5) ) %dopar%
{
  input <- read.csv(sprintf("data/author_c%d.csv", i), FALSE,
                 col.names = c("doc_id", "train",
                                      "text"), as.is = TRUE)
  anno <- cnlp_quick(input$text)
  anno$lemma <- stringi::stri_trans_tolower(anno$token)
  X <- cnlp_utils_tfidf(anno, vocab = vocab,
                   type = "tf", tf_weight = "raw")
  X <- Matrix(X, sparse = TRUE)
  return(crossprod(X))
}
```

The objects returned from the code are the chunk-wise values $X_k^t X_k$. They are returned as an R list and stored in a sparse matrix format. They are square matrices with over 10 thousand rows and columns and would be fairly large when stored in a dense format, at a total of about 4 gigabytes.

As we saw in Section 10.3, the matrix $X^t X$ can be produced by adding together the chunk equivalents. We can perform this operation without breaking the sparse structure of the matrix.

```
XtX <- Reduce('+', res)
```

Next, recall that principal components can be constructed using the singular value decomposition. Further, to get the first k dimensions we need only the largest k singular values and their corresponding singular vectors. The **irlba** package provides fast algorithms for computing these using sparse matrices without breaking the sparse structure.

```
library(irlba)
obj <- svdr(XtX, k = 25)
```

Now, with the first 25 columns of the matrix X, we can once again cycle through the data to compute the first 25 principal components. We will assume here that the data has been sufficiently reduced so that all of it can be stored together in memory.

```
res <- foreach::foreach( i = seq_len(5) ) %dopar%
{
  input <- read.csv(sprintf("data/author_c%d.csv", i), FALSE,
                 col.names = cnames,
                 as.is = TRUE)
  anno <- cnlp_quick(input$text)
  anno$lemma <- stringi::stri_trans_tolower(anno$token)
  X <- cnlp_utils_tfidf(anno, vocab = vocab,
                   type = "tf", tf_weight = "raw")
```

```
  X <- Matrix(X, sparse = TRUE)
  X <- X %*% obj$v
  return(list(y = input$author, X = X))
}


y <- do.call(c, lapply(res, getElement, "y"))
X <- do.call(rbind, lapply(res, getElement, "X"))
```

We now have a local dataset with a binary response y and the first 25 principal components stored as the matrix X.

With the data provided locally, the t-SNE algorithm is applied just as it would with any other dataset.

```
library(Rtsne)
X <- as.matrix(X)
Z <- Rtsne(X, perplexity = 20)[["Y"]]
```

The results are shown in Figure 10.2. While the dimensionality reduction does not perfectly separate the classes, we do see strong patterns where one or the other author dominates as well as regions where both authors are projected to similar regions.

10.8.4 Feature hashing

Our final task will be to apply feature hashing to the dataset, return all of the sparse matrices locally, and fit an elastic net model to the classification task. To access the data, we will again use `foreach::foreach`, pairing it with the functions `casl_fhash` and `casl_fhash_matrix` to create the feature hashed datasets.

```
res <- foreach::foreach( i = seq_len(5) ) %dopar%
{
  input <- read.csv(sprintf("data/author_c%d.csv", i), FALSE,
                    col.names = cnames,
                    as.is = TRUE)
  X <- casl_fhash_matrix(casl_fhash(input$text, k = 4L), k = 4L)
  return(list(y = input$author, X = X,
              train = input$train))
}
```

Here, we set the hash size to be 16^4, or 65536.

Next, the output of the `foreach::foreach` call needs to be combined row wise into a local dataset.

```
y <- do.call(c, lapply(res, getElement, "y"))
X <- do.call(rbind, lapply(res, getElement, "X"))
train <- do.call(c, lapply(res, getElement, "train"))
```

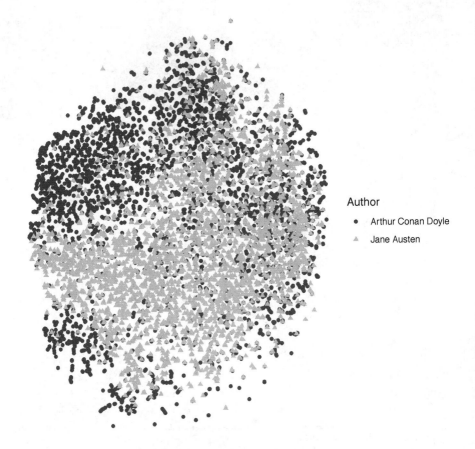

FIGURE 10.2: A two-dimensional t-SNE projection of the term frequency matrix from a set of novels broken into short snippets from Jane Austen and Arthur Conan Doyle. The data was first reduced to 25 dimensions using linear principal component analysis. Color and shape are used to distinguish between the two classes; the data was sorted by author (Austen snippets are represented on top of the Doyle snippets) to make it clear how the data are distributed.

```
train <- (train == "train")
```

Finally, we can apply the function `glmnet` on the training data and test the power of the model to classify chunks of data on the entire testing set.

```
out <- glmnet::cv.glmnet(X[train,], y[train],
                         family = "binomial")
pred <- predict(out, newx = X, type = "class")
```

	$k=1$	$k=2$	$k=3$	$k=4$	$k=5$
$\|\widehat{\beta}\|_0$	15	179	706	761	787
Train Rate	72.99%	91.12%	99.86%	99.93%	99.93%
Test Rate	73.32%	82.99%	88.72%	89.37%	89.76%

TABLE 10.1: The table shows the size of the estimated elastic net model and training and testing classification rates of a term frequency matrix with feature hashing. The columns indicate differing hash sizes, given as 16^k.

```
tapply(pred == y, train, mean)
```

```
    FALSE       TRUE
0.8975515 0.9993118
```

While the model is clearly overfit to the training set (only 2 mistakes are made over the entire training set), the predictive power is also nearly 90% on the testing set. This is a marked improvement compared to using only the aggregate features in the logistic regression from Section 10.8.2. Table 10.1 shows how the predictive power changes with the size of the hash. Note that even a hash size of 16 outperforms the model without word frequencies, and predictive power is nearly unchanged after the hash size includes 16^4 (4096) values.

10.9 Exercises

1. Write a function `smat_transpose` that takes a data frame representation of a sparse matrix as a coordinate list and computes its transpose.

2. Write a test suite of functions to verify that `casl_sparse_add` and `casl_sparse_multiply` work as expected.

3. Modify the sparsity of the simulated data matrix `Xd` on page 304 to make the matrix ten-times more sparse. How do the sizes of `Xs` and `qrs` change as a result? What if you increase the sparsity by a factor of 3?

4. Implement a function that performs PCA linear regression with a matrix X stored as a sparse matrix.

5. The function `Rprof` in the **utils** package can keep track of R's memory usage. Use this to verify that our function `casl_sparse_irwls` does not create a dense version of the data matrix.

6. Use `Rprof` to verify that the function `irlba` does not create a dense version of its input if given a sparse matrix as an input.

7. Write a function to compute the classification rate of a model produced by the function `irls_incremental` without storing y or anything else of size n. That is, compute the classification values on each chunk and intelligently combine the results.

8. Rewrite the function `word_to_index` to allow for any hash size h less than 16^7 (this should be smaller than the system's maximum integer size).

9. In the application, add the features computed by `fproc` to the hashed term frequencies with a hash size of 16^2. Fit an elastic net model on the training set and predict the testing classification rate. Does this improve the overall classification rate?

10. Recompute the elastic net model with a hash size of 16 (do not include the `fproc` features). Select a value of λ using the one standard deviation rule. Which features seem to have the strongest signal?

11. Read in the first set of data in Section 10.8.4 and determine which words are mapped to each of the hash values. What words are associated with the most prominent features?

12. Add manual indicators for the most prominent words in the feature hashing code by pre-constructing a partial hash table. Modify the function `casl_fhash_matrix` to account for this and re-run the elastic net using just the hash size of 16. How much do these custom defined terms effect the predictive power of the model on the testing set?

A

Linear algebra and matrices

A.1 Vector spaces

An algebraic *field* $\mathcal{F}$ is a set of objects called *scalars* on which the usual arithmetic operations (addition, subtraction, multiplication, division) are defined. This book almost always works with the field of real numbers $\mathbb{R}$. Other interesting fields include the complex numbers, the integers, and the rational numbers. Most of the definitions in this section explicitly use real-valued scalars but can usually be extended to other fields.

A *vector space* $\mathcal{V}$ over a field $\mathcal{F}$ is a set of objects called *vectors* that is closed under addition and left-multiplication by scalars from $\mathcal{F}$. In particular, vector addition in $\mathcal{V}$ is commutative and associative, $\mathcal{V}$ contains an additive identity element (simply written 0), and additive inverses exist for every element $x \in \mathcal{V}$ written $-x$ so that $x + (-x) = 0$. Scalar multiplication distributes over $\mathcal{V}$ as follows. For all $a, b \in \mathcal{F}$ and $x, y \in \mathcal{V}$, $a(x+y) = ax + ay$ and $(a+b)x = ax + bx$. The scalar multiplicative identity applies to vectors: $1x = x$.

The most common vector space we deal with in this book is the set of n-tuples of real values written as $\mathbb{R}^n$. Vector addition in this case is simply component-wise scalar addition. Other important vector spaces used by this book include the set of real-valued continuous functions on an interval in $\mathbb{R}$ and the set of polynomials with real coefficients up to a specific degree.

If $a, b \in \mathcal{F}$ and $x, y \in \mathcal{V}$, then $ax + by$ is a (particular) *linear combination* of x and y. The *span* of a set of vectors is the set of all linear combinations of those vectors, which is itself a vector space. A set of vectors $\mathcal{S}$ is said to span a vector space $\mathcal{V}$ if any vector in $\mathcal{V}$ can be written as a linear combination of vectors in $\mathcal{S}$.

Linear dependence and independence are of particular importance in this book (and in general). A set of vectors $x_1, x_2, \ldots, x_n$ is *linearly dependent* if there exists a linear combination $a_1 x_1 + a_2 x_2 + \cdots + a_n x_n = 0$ for scalars $a_1, a_2, \ldots, a_n$ not all equal to 0. For instance, the following set of vectors in $\mathbb{R}^3$ is linearly dependent:

$$
\begin{aligned}
x_1 &= (1, 1, 0) \\
x_2 &= (2, 0, 2) \\
x_3 &= (8, 2, 6),
\end{aligned}
$$

because $2x_1 + 3x_2 - x_3 = 0$. A set of vectors that is not linearly dependent is unsurprisingly called *linearly independent*. A linearly independent set of vectors that span a vector space $\mathcal{V}$ is called a *basis* for $\mathcal{V}$. Every vector space has a basis, but bases need not be unique. For instance, the set

$$\begin{aligned} x_1 &= (1,0,0) \\ x_2 &= (0,1,0) \\ x_3 &= (0,0,1) \end{aligned}$$

is a basis for $\mathbb{R}^3$ (often called the unit basis), but so is

$$\begin{aligned} y_1 &= (1, 1/2, 1/3) \\ y_2 &= (1/2, 1/3, 1/4) \qquad \text{and so is} \qquad \\ y_3 &= (1/3, 1/4, 1/5), \end{aligned} \qquad \begin{aligned} z_1 &= (1/\sqrt{2}, -1/\sqrt{2}, 0) \\ z_2 &= (1/\sqrt{2}, 1/\sqrt{2}, 0) \\ z_3 &= (0, 0, 1). \end{aligned}$$

Bases define coordinate systems of the vector space. A vector may be written using coordinates from any basis. For instance, the vector $(1,0,0) = 1x_1 + 0x_2 + 0x_3$ (in the unit basis) is the same as the vector $9y_1 - 36y_2 + 30y_3$, or coordinates $(9, -36, 30)$ using the y_i basis vectors above. A change of basis is simply a change of the coordinate system. Two vectors with identical coordinates in one basis will have identical coordinates in another basis. We avoid always indicating a basis and use the unit basis to write vector coordinates in this book, explicitly showing coordinate transformations when required.

The number of elements of a basis of $\mathcal{V}$ is called the *dimension* of $\mathcal{V}$. For instance, the dimension of $\mathbb{R}^3$ is 3. Some vector spaces may have infinite dimension, for instance the vector space of continuous real-valued functions over an interval.

Vector norms play an essential role in this book. Let $\mathcal{V}$ be a vector space over the field $\mathbb{R}$—for example $\mathcal{V} = \mathbb{R}^n$. A vector *norm* on $\mathcal{V}$, $\|\cdot\| : \mathcal{V} \to \mathbb{R}$ is a real-valued function of a vector in $\mathcal{V}$ that meets the following criteria for all $a \in \mathbb{R}$ and all $x, y \in \mathcal{V}$:

- $\|x\| \geq 0$.

- $\|x\| = 0$ if and only if $x = 0$.

- $\|ax\| = |a| \|x\|$.

- $\|x + y\| \leq \|x\| + \|y\|$.

A vector space equipped with a norm is called a *normed vector space*. Norms generalize the concept of length to vectors. An important example is the Euclidean norm on the vector space $\mathbb{R}^n$. Let $x \in \mathbb{R}^n, x = (x_1, x_2, \ldots, x_n)$. Then the Euclidean norm of x is

$$\|x\| = \sqrt{x_1^2 + x_2^2 + \cdots + x_n^2}.$$

Note that the subscripts in the above expression index the scalar entries within the vector (not multiple vectors). Subscripts should hopefully be understood from their context in this book; we try to clarify them as required. The Euclidean norm may be written $\|x\|_2$ to distinguish it from other norms when necessary.

An *inner product* of any two vectors x, y in a vector space $\mathcal{V}$ over the field $\mathbb{R}$ is a real-valued function $\langle x, y \rangle \to \mathbb{R}$ with the following properties:

- $\langle x, x \rangle \geq 0$.

- $\langle x, x \rangle = 0$ if and only if $x = 0$.

- $\langle x, y \rangle = \langle y, x \rangle$.

- For all $a \in \mathbb{R}$, $a \langle x, y \rangle = \langle ax, y \rangle$.

- For all $x, y, z \in \mathcal{V}$, $\langle x + y, z \rangle = \langle x, z \rangle + \langle y, z \rangle$.

A vector space $\mathcal{V}$ equipped with an inner product is called an *inner product space*. The *dot* (a. k. a. scalar) product on $\mathbb{R}^n$ is an important example. If $x, y \in \mathbb{R}^n, x = (x_1, x_2, \ldots, x_n), y = (y_1, y_2, \ldots, y_n)$, then the dot product of x and y is

$$\langle x, y \rangle = x_1 y_1 + x_2 y_2 + \cdots + x_n y_n.$$

This may remind R users of the `crossprod` function. Inner products generalize ideas from Euclidean geometry like angles and orthogonality to vector spaces. And inner products can be used to define norms. For instance, the Euclidean norm is simply the square root of the dot product of a vector with itself. The above definitions only apply to real-valued vector spaces, but they can be modified to work for complex-valued ones too.

Similarly to results in Euclidean geometry, right angles form the core of many fundamental techniques used in this book. Two vectors x, y in an inner-product space $\mathcal{V}$ are called *orthogonal* with respect to the inner product if $\langle x, y \rangle = 0$, generalizing the geometric notion of perpendicularity. A set of vectors $x_1, x_2, \ldots, x_n$ is called *orthonormal* if

$$\langle x_i, x_j \rangle = \begin{cases} 1 & \text{if } i = j, \\ 0 & \text{otherwise.} \end{cases}$$

Orthogonal (and orthonormal) vectors are linearly independent, and thus their span forms a basis. Nearly every computational method relies at least in part on manipulating sets of orthonormal vectors.

A.2 Matrices

Along with vectors, matrices form the core computational objects used by most algorithms in this book. A *matrix* is a special kind of linear function called a linear operator or linear transformation from one finite-dimensional vector space to another, relative to particular bases in each space. Explicitly noting the bases is usually not necessary—we write vector and matrix elements with respect to unit bases by convention.

For example, a linear operator $\mathbf{X}$ from $\mathbb{R}^2$ to $\mathbb{R}^3$ is a function $\mathbf{X} : \mathbb{R}^2 \to \mathbb{R}^3$ such that $\mathbf{X}(a_1 v_1 + a_2 v_2) = a_1 \mathbf{X}(v_1) + a_2 \mathbf{X}(v_2)$ for scalars a_1, a_2 and vectors $v_1, v_2 \in \mathbb{R}^2$. The matrix $X \in \mathbb{R}^{3\times 2}$, a rectangular array of three rows and two columns, corresponds to the function $\mathbf{X} : \mathbb{R}^2 \to \mathbb{R}^3$ when $u = \mathbf{X}(v)$ if and only if $u = Xv$. Note that function evaluation is written as a matrix-vector product.

It is sometimes convenient to think of an n row by p column matrix $X \in \mathbb{R}^{n\times p}$ as either a set of p column vectors, each of length n; or alternatively as a set of n row vectors, each of length p. Thinking in terms of column vectors, the matrix-vector product is a linear combination of the columns of the matrix using the scalar entries of the vector. Thinking of a matrix as a collection of row vectors, the matrix-vector product produces a vector whose entries are defined by inner products of the vector with each row.

For example, consider equivalent formulations for a 3×2 matrix X times a vector $v = (v_1, v_2) \in \mathbb{R}^2$:

$$
X = \begin{pmatrix} 1 & 2 \\ 0 & 1 \\ 1 & 0 \end{pmatrix}, \quad Xv = v_1 \begin{pmatrix} 1 \\ 0 \\ 1 \end{pmatrix} + v_2 \begin{pmatrix} 2 \\ 1 \\ 0 \end{pmatrix} = \begin{pmatrix} \langle (1,2), v \rangle \\ \langle (0,1), v \rangle \\ \langle (1,0), v \rangle \end{pmatrix}.
$$

Matrices can be multiplied from the left by a scalar, corresponding to multiplying each entry of the matrix by the scalar. Matrices of the same shape can be added. Matrix addition is defined by entry-wise scalar addition, inheriting commutativity and associativity from the real numbers, and corresponds to function addition of linear operators. The additive identity is a matrix of all zeros.

Multiplication of two matrices corresponds to function composition (remember, matrices are functions). Let X be an $n \times p$ matrix made up of n row vectors $x_1, x_2, \ldots, x_n \in \mathbb{R}^p$. Let V be a $p \times q$ matrix made up of q column vectors $v_1, v_2, \ldots, v_q \in \mathbb{R}^p$. Their matrix product results in a new $n \times q$ matrix composing X and V,

$$
XV = \begin{pmatrix} \langle x_1, v_1 \rangle, & \langle x_1, v_2 \rangle, & \cdots, & \langle x_1, v_q \rangle \\ \langle x_2, v_1 \rangle, & \langle x_2, v_2 \rangle, & \cdots, & \langle x_2, v_q \rangle \\ \vdots & \vdots & \vdots & \vdots \\ \langle x_n, v_1 \rangle, & \langle x_n, v_2 \rangle, & \cdots, & \langle x_n, v_q \rangle \end{pmatrix}.
$$

Multiplicative identity is provided by the square *identity matrix* I of appropriate size, with ones along its main diagonal and zeros elsewhere. When necessary, the size of the identity matrix used in a particular computation may be noted by a subscript. For instance if $X \in \mathbb{R}^{n\times p}$ I_n denotes an $n \times n$

identity matrix and I_p a $p \times p$ one, then $I_n X = X = X I_p$. Matrix multiplication is associative (with real-valued scalars) but generally not commutative. For example

$$\begin{pmatrix} 1 & 0 \\ 0 & 2 \end{pmatrix} \begin{pmatrix} 1 & 2 \\ 3 & 4 \end{pmatrix} \neq \begin{pmatrix} 1 & 2 \\ 3 & 4 \end{pmatrix} \begin{pmatrix} 1 & 0 \\ 0 & 2 \end{pmatrix}.$$

The *transpose* of a real-valued $n \times p$ matrix X, written X^t, is a $p \times n$ matrix whose rows are exchanged for columns and vice versa. For example

$$\begin{pmatrix} 1 & 2 \\ 3 & 4 \\ 5 & 6 \end{pmatrix}^t = \begin{pmatrix} 1 & 3 & 5 \\ 2 & 4 & 6 \end{pmatrix}$$

when $X = X^t$, X is called *symmetric*. Symmetric matrices are necessarily square.

We frequently think of vectors as single-column matrices or single-row matrices, defaulting to a column vector when not explicitly specified. This enables the useful concept of the transpose of a vector. For instance for the vector $x \in \mathbb{R}^3$,

$$x^t = \begin{pmatrix} x_1 \\ x_2 \\ x_3 \end{pmatrix}^t = (x_1, x_2, x_3).$$

Notice in particular that $x^t x = x_1^2 + x_2^2 + x_3^2 = \langle x, x \rangle$, the usual dot vector inner product. The product $x x^t$ results in a 3×3 matrix called an *outer product*, familiar to R users as the `outer` function.

The *range* of a matrix $X \in \mathbb{R}^{n \times p}$ is the set $\{y \in \mathbb{R}^n : y = Xv\}$ for all $v \in \mathbb{R}^p$. The *null space* of X is the set $\{v \in \mathbb{R}^p : Xv = 0\}$. The *matrix rank* of X is the largest number of linearly independent columns of X, written $\text{rank}(X)$. The $\text{rank}(X) \le \min(n, p)$. Remarkably, $\text{rank}(X) = \text{rank}(X^t)$.

Let X be a square $n \times n$ matrix. If the column vectors of X are orthonormal, then $X^t X = I$, the identity matrix. A square matrix with orthonormal columns is called an *orthonormal matrix*. The columns form a coordinate basis of $\mathbb{R}^n$ whose directions are separated by right angles—in other words, simply a rotation of the usual unit basis. For example, consider a matrix X with orthonormal columns:

$$X = \begin{pmatrix} 1/\sqrt{2} & -1\sqrt{2} \\ 1/\sqrt{2} & 1/\sqrt{2} \end{pmatrix}.$$

Figure A.1 illustrates the rotation by plotting the column vectors of X along with the usual unit basis vectors.

Multiplying a vector by X rotates the entries of the vector to the new coordinate system, and does nothing else. In particular, the Euclidean norm of the vector is not changed. This is a useful enough result to state formally.

If $X \in \mathbb{R}^{n \times n}$ is an orthonormal matrix,
then $\|Xv\| = \|v\|$ for all $v \in \mathbb{R}^n$. (A.1)

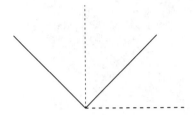

FIGURE A.1: Coordinates from matrix X with orthonormal columns (solid lines) compared to the standard unit basis vectors (dashed). X is a rotation matrix.

Although the singular value decomposition outlined in Chapter 2 is the main tool for analyzing algorithms, other matrix decompositions are useful and widely used in applications like R. We already mentioned the LU decomposition, a variation of Gaussian elimination for solving linear systems that breaks a square matrix X into a lower-triangular matrix L and an upper-triangular one U so that $X = LU$, sometimes including a diagonal scaling matrix D, $X = LDU$, and sometimes with a column permutation matrix to re-order the columns for improved numerical stability in floating point arithmetic.

When $X = X^t$ is symmetric, then $X = LU = (LU)^t = U^t L^t = X^t$ and we see that $U = L^t$ in the LU decomposition, so we only need to bother to compute one of them. The important *Cholesky decomposition* takes advantage of this to very efficiently decompose symmetric matrices, again along with possible scaling and column permutations, accomplished in R with the `chol` function.

The QR decomposition decomposes any matrix $X \in \mathbb{R}^{n \times p}$ into an $n \times p$ matrix Q with orthonormal columns, and a $p \times p$ upper-triangular matrix R. It is easy to solve linear systems and ordinary least squares problems with such a matrix thanks to the structure of R and the properties of Q. R uses QR decompositions under the hood to solve linear models in the `lm` and `glm` families of functions.

Compared to the SVD, the QR decomposition does not in general provide an easy, clear estimate of rank or numerical rank of a matrix. Thus the SVD is generally preferred for algorithm analysis. The QR decomposition is less computationally expensive than the SVD, however.

B

Floating Point Arithmetic and Numerical Computation

Most of the computational methods in this book work with floating point numbers and other tools from numerical analysis. This section presents a concise introduction to computation with floating point arithmetic. See [80] and [69] for many more details.

B.1 Floating point arithmetic

We typically compute with approximations of real numbers called the *floating point numbers*. Approximation is necessary because computers have limited memory and processing power. Unlike the real numbers, there are only finitely many floating point numbers. There is a largest floating point number and a smallest positive floating point number. More interestingly, the floating point numbers are not uniformly distributed among the reals. For example, there are twice as many floating point numbers between one and two than there are between two and four.

A set of normalized base-2 floating point numbers can be specified by four integers: a sign $s \in \{0, 1\}$, precision t, and lower and upper exponent range $[L, U]$ as follows

$$\{(-1)^s (1 + m) \times 2^e\} \cup \{0\},$$

where the exponent e is bounded by $L \le e \le U$ and the mantissa $0 \le m < 1$ (also called the *fraction*) can be represented with at most t binary digits. In other words, $2^t m$ is an integer in the interval $[0, 2^t)$. Additional special values like signed zeros, signed infinities, error flags, and subnormalized floating point numbers are also specified in floating point arithmetic standards and implementations. Typical values for double precision floating point numbers are $t = 52$, $L = -1022$, and $U = 1023$. Other precisions are common today, especially on modern GPU architectures used for machine learning problems that frequently employ single and even half-precision floating point arithmetic. And R language implementations may use extended-precision floating point for some operations.

The IEEE-754 standard specifies formats and operations for floating point arithmetic that most systems follow today. But not all computer architectures exactly follow the standard, and adherence to the standard may vary for a given application depending on how it was compiled and which numerical libraries it depends on. Even within the standard there is some leeway in carrying out computation. Thus, results from otherwise identical computations can vary from system to system. Algorithms must keep this in mind, and good algorithms are designed to minimize significant numeric variation across reasonable floating point implementations.

Simply representing a real number in a computer using double precision arithmetic may introduce error because the floating point numbers only approximate most real numbers. The following R example illustrates that the real number 1/10 is only approximated in floating point.

```
0.3 / 0.1 - 3
```

```
[1] -4.440892e-16
```

Golub and Van Loan [69] use a model of floating point arithmetic to help quantify numerical errors. Let $\mathcal{G}$ be the set of real numbers lying within the range of the floating point numbers, $\mathcal{G} = \{x \in \mathbb{R} : 0 \leq |x| \leq \texttt{xmax}\}$, where xmax is the largest double precision floating point number. This is called .Machine\$double.xmax in R and is just under 2^{1024}. Define the function $fl(x)$ to be the nearest double precision floating point number to $x \in \mathcal{G}$ with ties handled by rounding away from zero. This function satisfies

$$fl(x) = x(1 + \epsilon), \qquad |\epsilon| \leq \mathbf{u},$$

where $\mathbf{u} = 2^{-52}$ is the unit roundoff, also called *machine epsilon* or .Machine\$double.eps in R (in double precision). $\mathbf{u}$ is the distance between 1 and the next largest floating point number. Let "op" denote any of the four arithmetic operations. If a op $b \notin \mathcal{G}$, then a floating point exception occurs. *Overflow* occurs when $|a$ op $b| > $ xmax; a special signed infinity result is returned in this case. *Underflow* occurs when $|a$ op $b| < 2^{-1022}$ (called .Machine\$double.xmin in R), and a special signed zero result is returned in that case[1] Other floating point exceptions typically return the special value NaN, for example the quotient 0/0 in R.

Small numerical representation errors can accumulate over the course of a program. Unintuitively, errors may accumulate differently depending on the order of the operations, sometimes compounded by the non-uniform distribution of floating point numbers among the reals. In other words, floating point arithmetic is not always associative, illustrated by the following example.

```
(.Machine$double.eps/2  + 1)  - 1
```

---

[1]Many, but not all, implementations extend the number of double precision numbers near zero using subnormal floating point numbers, so the smallest representable floating point number may be less than this in practice.

```
[1] 0
```

```
.Machine$double.eps/2  + (1  - 1)
```

```
[1] 1.110223e-16
```

In extreme cases catastrophic loss of accuracy can occur. The function e below estimates the exponential function evaluated at a value x using the first n terms of its Taylor expansion. This simple function is accurate for many positive values of x. But it can catastrophically fail for negative values of x due to accumulated errors from cancellation.

```
e <- function(x, n) sum(x^seq(0,n) / factorial(seq(0,n)))
e(1, 25) - exp(1)
```

```
[1] 0
```

```
e(-25, 25) - exp(-25)
```

```
[1] -2834107793
```

The next examples illustrate accumulation of roundoff error, lack of associativity, and internal use of mixed-precision arithmetic in R by simply summing up a set of one million double precision numbers sampled from a normal distribution. The example uses the **Rmpfr** multi-precision arithmetic package for R to compute a very accurate summation (using 80 decimal digits) to use as a reference for comparison. First, notice a serious failure of associativity in double precision floating point sums.

```
library(Rmpfr)
set.seed(1)
x <- rnorm(1e6)
sum80 <- sum(mpfr(x, 80))      # (reference sum)
sum80 - Reduce('+', x)         # (double precision sum)
```

```
[1] -5.513121450112295693357645e-12
```

```
sum80 - Reduce('+', sort(x)) # (same sum, but re-ordered)
```

```
[1] -1.864680074608235350007435e-8
```

Unlike the standard double precision addition function + used above, R's sum function (and some other R functions) defaults to internal use of extended precision arithmetic depending on the compiler, compiler flags, and system architecture. It is much more accurate than +.

```
sum80 - sum(x)
```

```
[1] -6.415247971519252856431592e-15
```

Dekker[44] shows that it is possible to emulate extended precision arithmetic using only double precision numbers to yield more accurate solutions.

```
qsum <- function(...) {
  add2 <- function(x, y) {
    r <- x[1] + y
    if(abs(x[1]) > abs(y)) s <- x[1] - r + y + x[2]
    else s <- y - r + x[1] + x[2]
    z <- r + s
    c(z, r - z + s)
  }
  a <- Reduce(add2, unlist(list(...)), init=c(0,0))
  a[1] + a[2]
}
```

```
sum80 - qsum(x)
```

```
[1] 6.901793860817490022796505e-16
```

Somewhat surprisingly, this approach is slightly more accurate than R's native extended precision sum (but slower).

B.2   Computational effort

It is useful to establish conventions for reasonably estimating the amount of computational effort expended by algorithms in order to compare their performance in terms of speed and energy efficiency. This is especially important for large-scale computations that require considerable effort to compute. One simple approach simply counts the number of scalar floating point operations, called *flops*, used by the algorithm to compute its result.

Most core computational methods admit many different algorithmic implementations. The various implementations may exhibit different numerical stability, might perform better on certain processor architectures over others, or might be better suited to parallel computation, etc.

For example, the QR decomposition of an $n \times p$ matrix X can be computed using many very different algorithm implementations, including: the Gram–Schmidt orthogonalization process; the more numerically stable modified Gram–Schmidt (MGS); Givens rotations; or Householder reflections. The

MGS method requires $2np^2$ flops to compute the Q matrix. Householder QR, by comparison, needs about $4np^2 - 4p^3/3$ flops to compute its Q matrix. Thus, when n is sufficiently larger than p, MGS requires about only half the total flops of Householder to compute its result. However, it can be shown that MGS produces a matrix Q that satisfies $Q^T Q \approx I + \mathbf{u}\kappa_2(X)$, while Householder QR produces a Q that satisfies $Q^T Q \approx I + \mathbf{u}$. That is, MGS is less numerically stable than Householder for this purpose and should only be used when the condition number of X is known to be small.

Similarly, the Golub–Reinsch SVD algorithm for least squares problems takes about 9 times the number of flops as Householder QR. Despite the extra computation the SVD method is typically very slightly less accurate than Householder QR—some extra roundoff error accumulates during the computation. However, SVD-based least squares solvers have the important advantage of reliable detection of rank deficiency, unlike Householder QR. Thus for very badly conditioned problems, SVD-based least squares solutions can produce more accurate solutions than QR-based methods.

Perhaps surprisingly, even the most basic operations like general dense matrix multiplication admit many different algorithmic implementations, for instance by looping over rows and columns of the operands in one of six possible orders. Those approaches all consume the same number of flops, but access data in different patterns, which may be more or less favorable on particular computing architectures. However, Strassen [151] developed a *divide and conquer* method that reduces the overall flop count of matrix multiplication. Generic dense matrix multiplication of two square n by n matrices takes on the order of $O(n^3)$ flops normally. (The big-O notation used here simply means that there exists a positive constant number M such that the actual flop count is less than Mn^3.) Strassen showed that this flop count is not optimal, and related approaches continue to be improved down to, as of 2014, $O(n^{2.3728639})$. Remarkably, the lower bound of the asymptotic computational complexity of matrix multiplication is unknown and remains an open problem in mathematics! It is worth noting that these methods, although asymptotically more optimal, are not in general practical for most problems—their savings only kick in when n is so large as to be intractable.

Counting flops is a fair, but relatively crude way to account for computational effort because it ignores many important aspects of computing like data transfers. For instance, sparse matrix operations have much lower flop counts than their dense cousins but each flop can be much more costly to compute in terms of total CPU operations because their data may not be aligned sequentially in memory and caches. Also, most computer architectures today elide multiple floating point operations into a smaller number of processor operations through fused operators like fused multiply-add and vectorization within the CPU. Taking advantage of such specialized hardware instructions requires using tuned, high-performance numerical libraries.

Generally speaking, the most efficient floating point operations are the ones that are never computed. The best algorithms use whatever tricks they can

Algorithm	Flops
General matrix vector product plus vector $Ax + y$	$2np$
General dense matrix multiplication AB	$2npq$
Sparse matrix vector product Cx (worst case)	mp
LU decomposition $X = LU$	$np^2 - p^3/3$
Cholesky decomposition $S = LL^T$	$n^3/3$
Polynomial interpolation using V directly	$10n^2$
Polynomial interpolation using V via FFT	$10n \log_2(n)$
Least squares $\min_x \|Ax - y\|$ using LU	$2n^3/3$
Least squares $\min_x \|Ax - y\|$ using Householder QR	$4n^3/3$
Least squares $\min_x \|Ax - y\|$ using MGS QR	$2n^3$
Least squares $\min_x \|Ax - y\|$ using Golub–Reinsch SVD	$12n^3$

TABLE B.1: Typical flop counts of algorithms for general dense matrices $A \in \mathbb{R}^{n \cdot p}$ and $B \in \mathbb{R}^{p \cdot q}$, sparse matrix $C \in \mathbb{R}^{n \cdot p}$ with m non-zero elements, symmetric positive definite matrix $S \in \mathbb{R}^{n \cdot n}$, and Vandermonde matrix $V \in \mathbb{R}^{(n+1) \cdot (n+1)}$.

to compute their result with the fewest number of flops. This often involves exploiting (possibly latent) mathematical structure in the problem (symmetry, sparsity, block structure, low rank), and computing only what is required to get the solution. Consider a principal components decomposition of an $n \cdot p$ real-valued matrix X where only the first 10 components are required. Several algorithms exist to compute truncated decompositions like that in approximately $O(np)$ flops instead of the usual $O(n^3)$ flops required of a full SVD-based solution, a potentially enormous savings.

Bibliography

[1] ABADI, M., BARHAM, P., CHEN, J., CHEN, Z., DAVIS, A., DEAN, J., DEVIN, M., GHEMAWAT, S., IRVING, G., AND ISARD, M. Tensorflow: A system for large-scale machine learning.

[2] ADCOCK, R. Note on the method of least squares. *The Analyst 4* (1877), 183–184.

[3] AGGELOU, G., AND TAFAZOLLI, R. Rdmar: A bandwidth-efficient routing protocol for mobile ad hoc networks. In *Proceedings of the 2nd ACM International Workshop on Wireless Mobile Multimedia* (New York, NY, USA, 1999), ACM, pp. 26–33.

[4] AKYILDIZ, I., SU, W., SANKARASUBRAMANIAM, Y., AND CAYIRCI, E. Wireless sensor networks: A survey. *Computer Networks 38*, 4 (2002), 393–422.

[5] ALIZADEH, F. Interior point methods in semidefinite programming with applications to combinatorial optimization. *SIAM Journal on Optimization 5*, 1 (1995), 13–51.

[6] ALLAIRE, J. J., AND CHOLLET, F. *keras: R Interface to Keras*, 2017. R package version 2.0.8.

[7] ALLAIRE, J. J., AND TANG, Y. *tensorflow: R Interface to TensorFlow*, 2017. R package version 1.4.

[8] AMD, A. Core math library (acml). 25.

[9] ANDERSON, E., BAI, Z., BISCHOF, C., BLACKFORD, L. S., DEMMEL, J., DONGARRA, J., DU CROZ, J., GREENBAUM, A., HAMMARLING, S., AND MCKENNEY, A. *LAPACK Users' Guide*. SIAM, New York, NY, 1999.

[10] ARNOLD, T. A tidy data model for natural language processing using cleanNLP. *The R Journal 9*, 2 (2017), 1–20.

[11] ARNOLD, T., KANE, M., AND URBANEK, S. iotools: High-performance i/o tools for R. *The R Journal 9*, 1 (2017), 6–13.

[12] ARNOLD, T. B., AND TIBSHIRANI, R. J. *genlasso: Path Algorithm for Generalized Lasso Problems*, 2014. R package version 1.3.

343

[13] BAGLAMA, J., REICHEL, L., AND LEWIS, B. W. *irlba: Fast Truncated SVD, PCA and Symmetric Eigendecomposition for Large Dense and Sparse Matrices*, 2016. R package version 2.1.2.

[14] BAI, Z.-J., CHAN, R., AND LUK, F. Principal component analysis for distributed data sets with updating. In *Advanced Parallel Processing Technologies*, vol. 3756. Springer Berlin, Heidelberg, 2005, pp. 471–483.

[15] BARRACHINA, S., CASTILLO, M., IGUAL, F. D., MAYO, R., AND QUINTANA-ORTI, E. S. Evaluation and tuning of the level 3 cublas for graphics processors. In *Parallel and Distributed Processing* (2008), IEEE, pp. 1–8.

[16] BARRON, A. Approximation and estimation bounds for artificial neural networks. *Machine Learning 14*, 1 (1994), 115–133.

[17] BATES, D., MÄCHLER, M., BOLKER, B., AND WALKER, S. Fitting linear mixed-effects models using lme4. *Journal of Statistical Software 67*, 1 (2015), 1–48.

[18] BATES, D., AND MAECHLER, M. *MatrixModels: Modelling with Sparse and Dense Matrices*, 2015. R package version 0.4-1.

[19] BATES, D., AND MAECHLER, M. *Matrix: Sparse and Dense Matrix Classes and Methods*, 2017. R package version 1.2-8.

[20] BECK, A., AND TEBOULLE, M. Mirror descent and nonlinear projected subgradient methods for convex optimization. *Operations Research Letters 31*, 3 (2003), 167–175.

[21] BENDEL, R. B., AND AFIFI, A. A. Comparison of stopping rules in forward "stepwise" regression. *Journal of the American Statistical Association 72*, 357 (1977), 46–53.

[22] BEYGELZIMER, A., KAKADET, S., LANGFORD, J., ARYA, S., MOUNT, D., AND LI, S. *FNN: Fast Nearest Neighbor Search Algorithms and Applications*, 2013. R package version 1.1.

[23] BJÖRK, Å. *Numerical Methods for Least Squares Problems*. SIAM, New York, NY, 1996.

[24] BLACKFORD, L. S., CHOI, J., CLEARY, A., D'AZEVEDO, E., DEMMEL, J., DHILLON, I., DONGARRA, J., HAMMARLING, S., HENRY, G., AND PETITET, A. *ScaLAPACK Users' Guide*. SIAM, New York, NY, 1997.

[25] BONTEMPI, G., AND LE BORGNE., Y. An adaptive modular approach to the mining of sensor network data. In *Proceedings of the Workshop on Data Mining in Sensor Networks* (2005), SIAM Press, pp. 3–9.

[26] BORGNE, Y.-A. L., AND BONTEMPI, G. Unsupervised and supervised compression with principal component analysis in wireless sensor networks. *Accepted at the Workshop on Knowledge Discovery, colocated with the 13th International Conference on Knowledge Discovery and Data Mining* (2007).

[27] BORGNE, Y. L., SANTINI, S., AND BONTEMPI, G. Adaptive model selection for time series prediction in wireless sensor networks. *International Journal for Signal Processing (Elsevier pub.), Special Issue on Information Processing and Data Management in Wireless Sensor Networks* (September 2007). Accepted for Publication.

[28] BOTTOU, L. Large-scale machine learning with stochastic gradient descent. In *Proceedings of COMPSTAT*. Springer, 2010, pp. 177–186.

[29] BOURLARD, H., AND KAMP, Y. Auto-association by multilayer perceptrons and singular value decomposition. *Biological Cybernetics 59*, 4-5 (1988), 291–294.

[30] BOYD, S., AND VANDENBERGHE, L. *Convex Optimization*. Cambridge University Press, London, 2004.

[31] BÜHLMANN, P., AND VAN DE GEER, S. *Statistics for High-Dimensional Data: Methods, Theory and Applications*. Springer Science & Business Media, New York, NY, 2011.

[32] BURRI, N., AND WATTENHOFER, R. Dozer: Ultra-low power data gathering in sensor networks. In *Proceedings of the 6th International Conference on Information Processing in Sensor Networks* (2007), ACM Press, pp. 450–459.

[33] CHANG, J., AND TASSIULAS, L. Energy conserving routing in wireless ad-hoc networks. In *Proceedings of the 19th Annual Joint Conference of the IEEE Computer and Communications Societies* (2000), vol. 1, IEEE Press, pp. 22–31.

[34] CHEN, B., JAMIESON, K., BALAKRISHNAN, H., AND MORRIS, R. Span: An energy-efficient coordination algorithm for topology maintenance in ad hoc wireless networks. In *Proceedings of the 7th Annual International Conference on Mobile Computing and Networking* (New York, NY, USA, 2001), ACM, pp. 85–96.

[35] CHEN, T., HE, T., BENESTY, M., KHOTILOVICH, V., AND TANG, Y. *xgboost: Extreme Gradient Boosting*, 2018. R package version 0.6.4.1.

[36] CHEN, W., WILSON, J., TYREE, S., WEINBERGER, K., AND CHEN, Y. Compressing neural networks with the hashing trick. In *International Conference on Machine Learning* (2015), pp. 2285–2294.

[37] COMER, D. *Internetworking With TCP/IP*, fifth ed., vol. 1: Principles Protocols, and Architecture. Prentice Hall, Princeton, NJ, 2006.

[38] COSTA, J. A., PATWARI, N., AND HERO, A. O. Distributed weighted-multidimensional scaling for node localization in sensor networks. *ACM Transactions on Sensor Networks* (2006).

[39] COVER, T. M., AND THOMAS, J. A. *Elements of Information Theory*. John Wiley & Sons, New York, NY, 2012.

[40] DAUBECHIES, I. *Ten Lectures on Wavelets*. SIAM, New York, NY, 1992.

[41] DAUPHIN, Y. N., PASCANU, R., GULCEHRE, C., CHO, K., GANGULI, S., AND BENGIO, Y. Identifying and attacking the saddle point problem in high-dimensional non-convex optimization. In *Advances in Neural Information Processing Systems* (2014), pp. 2933–2941.

[42] DAVIS, T., HAGER, W., AND DUFF, I. *SuiteSparse*, 2014. C++ Library.

[43] DAVIS, T. A. *Direct Methods for Sparse Linear Systems*. SIAM, New York, NY, 2006.

[44] DEKKER, T. J. A floating-point technique for extending the available precision. *Numerische Mathematik 18*, 3 (1971), 224–242.

[45] DENG, J., DONG, W., SOCHER, R., LI, L.-J., LI, K., AND FEI-FEI, L. Imagenet: A large-scale hierarchical image database. In *Computer Vision and Pattern Recognition* (2009), IEEE, pp. 248–255.

[46] DESHPANDE, A., GUESTRIN, C., MADDEN, S., HELLERSTEIN, J., AND HONG, W. Model-based approximate querying in sensor networks. *The VLDB Journal The International Journal on Very Large Data Bases 14*, 4 (2005), 417–443.

[47] DESHPANDE, A., NATH, S., GIBBONS, P., AND SESHAN, S. Cache-and-query for wide area sensor databases. In *Proceedings of the 2003 ACM SIGMOD International Conference on Management of Data* (2003), ACM Press, pp. 503–514.

[48] DIAMANTARAS, K., AND KUNG, S. *Principal Component Neural Networks: Theory and Applications*. John Wiley & Sons, Inc., New York, NY, 1996.

[49] DONG, Q. Maximizing system lifetime in wireless sensor networks. In *Proceedings of the 4th International Symposium on Information Processing in Sensor Networks* (2005), IEEE Press, p. 3.

[50] DONOHO, D. Compressed sensing. *IEEE Transactions on Information Theory 52*, 4 (2006), 1289–1306.

[51] DUARTE, M., AND HEN HU, Y. Vehicle classification in distributed sensor networks. *Journal of Parallel and Distributed Computing 64*, 7 (2004), 826–838.

[52] DUARTE, M. F., SARVOTHAM, S., BARON, D., WAKIN, M. B., AND BARANIUK, R. G. Distributed compressed sensing of jointly sparse signals. In *Proceedings of the 39th Asilomar Conference on Signals, Systems and Computation* (2005), pp. 1537–1541.

[53] EDDELBUETTEL, D. *digest: Create Compact Hash Digests of R Objects*, 2018. R package version 0.6.15.

[54] EFRON, B., HASTIE, T., JOHNSTONE, I., AND TIBSHIRANI, R. Least angle regression. *The Annals of Statistics 32*, 2 (2004), 407–499.

[55] FARLEY, B., AND CLARK, W. Simulation of self-organizing systems by digital computer. *Transactions of the IRE Professional Group on Information Theory 4*, 4 (1954), 76–84.

[56] FISHER, R. A. The case of zero survivors. *Annals of Applied Biology 22* (1935), 164–165.

[57] FLETCHER, R., AND REEVES, C. M. Function minimization by conjugate gradients. *The Computer Journal 7*, 2 (1964), 149–154.

[58] FORSYTHE, G. E., MOLER, C. B., AND MALCOLM, M. A. Computer methods for mathematical computations.

[59] FREEDMAN, D., PISANI, R., AND PURVES, R. *Statistics*. WW Norton & Company, New York, NY, 2014.

[60] FRIEDMAN, J., HASTIE, T., AND TIBSHIRANI, R. *The Elements of Statistical Learning*. Springer, New York, NY, 2001.

[61] FRIEDMAN, J., HASTIE, T., AND TIBSHIRANI, R. Regularization paths for generalized linear models via coordinate descent. *Journal of Statistical Software 33*, 1 (2010), 1–22.

[62] GASTPAR, M., DRAGOTTI, P. L., AND VETTERLI, M. The Distributed Karhunen-Loève Transform. *IEEE Transactions on Information Theory 52*, 12 (2006), 5177–5196.

[63] GE, R., HUANG, F., JIN, C., AND YUAN, Y. Escaping from saddle pointsâĂŤonline stochastic gradient for tensor decomposition. In *Conference on Learning Theory* (2015), pp. 797–842.

[64] GELMAN, A., AND LOKEN, E. The garden of forking paths: Why multiple comparisons can be a problem, even when there is no "fishing expedition" or "p-hacking" and the research hypothesis was posited ahead of time. *Department of Statistics, Columbia University* (2013).

[65] GLAUDELL, R., GARCIA, R. T., AND GARCIA, J. B. Nelder-mead simplex method. *Computer Journal 7*, 4 (1965), 308–313.

[66] GLESER, L. J. Estimation in a multivariate errors in variables regression model: large sample results. *The Annals of Statistics* (1981), 24–44.

[67] GLOROT, X., AND BENGIO, Y. Understanding the difficulty of training deep feedforward neural networks. In *Proceedings of the Thirteenth International Conference on Artificial Intelligence and Statistics* (2010), pp. 249–256.

[68] GOEMAN, J. J. L1 penalized estimation in the cox proportional hazards model. *Biometrical Journal, 52* (2010), 1–14.

[69] GOLUB, G. H., AND VAN LOAN, C. F. *Matrix Computations*, vol. 3. JHU Press, New York, NY, 2012.

[70] HASTIE, T. *gam: Generalized Additive Models*, 2017. R package version 1.14-4.

[71] HASTIE, T., AND EFRON, B. *lars: Least Angle Regression, Lasso and Forward Stagewise*, 2013. R package version 1.2.

[72] HAYASHI, F. *Econometrics*. Princeton University Press, Princeton, NJ, 2000.

[73] HE, K., ZHANG, X., REN, S., AND SUN, J. Delving deep into rectifiers: Surpassing human-level performance on imagenet classification. In *Proceedings of the IEEE International Conference on Computer Vision* (2015), pp. 1026–1034.

[74] HE, K., ZHANG, X., REN, S., AND SUN, J. Identity mappings in deep residual networks. In *European Conference on Computer Vision* (2016), Springer, pp. 630–645.

[75] HEAD, M. L., HOLMAN, L., LANFEAR, R., KAHN, A. T., AND JENNIONS, M. D. The extent and consequences of p-hacking in science. *PLoS Biology 13*, 3 (2015), e1002106.

[76] HENDRY, D. F., AND MORGAN, M. S. *The Foundations of Econometric Analysis*. Cambridge University Press, London, 1997.

[77] HIGHAM, N. J. *Accuracy and Stability of Numerical Algorithms*, vol. 80. SIAM, New York, NY, 2002.

[78] HILL, M. D. What is scalability? *ACM SIGARCH Computer Architecture News 18*, 4 (1990), 18–21.

[79] HINTON, G. E., OSINDERO, S., AND TEH, Y.-W. A fast learning algorithm for deep belief nets. *Neural Computation 18*, 7 (2006), 1527–1554.

[80] HORN, R. A., AND JOHNSON, C. R. *Matrix Analysis.* Cambridge University Press, London, 1990.

[81] HUANG, L., NGUYEN, X., GAROFALAKIS, M., JORDAN, M., JOSEPH, A., AND TAFT, N. In-network pca and anomaly detection. In *Proceedings of the 19th conference on Advances in Neural Information Processing Systems* (2006), MIT Press.

[82] HUYNH, T., AND HONG, C. A novel hierarchical routing protocol for wireless sensor networks. In *Proceedings of the International Conference on Computational Science and Its Applications* (2005), pp. 339–347.

[83] HYVARINEN, A., KARHUNEN, J., AND OJA, E. *Independent Component Analysis.* J. Wiley New York, New York, NY, 2001.

[84] ILYAS, M., MAHGOUB, I., AND KELLY, L. *Handbook of Sensor Networks: Compact Wireless and Wired Sensing Systems.* CRC Press, Inc., New York, NY, 2004.

[85] IOFFE, S., AND SZEGEDY, C. Batch normalization: Accelerating deep network training by reducing internal covariate shift. In *International Conference on Machine Learning* (2015), pp. 448–456.

[86] JAIN, A., AND CHANG, E. Adaptive sampling for sensor networks. *ACM International Conference Proceeding Series* (2004), 10–16.

[87] JAMES, G., WITTEN, D., HASTIE, T., AND TIBSHIRANI, R. *An Introduction to Statistical Learning*, vol. 112. Springer, New York, NY, 2013.

[88] JELASITY, M., CANRIGHT, G., AND ENGØ-MONSEN, K. Asynchronous distributed power iteration with gossip-based normalization? In *Proceedings of Euro-Par* (2007), vol. 4641 of *Lecture Notes in Computer Science*, Springer, pp. 514–525.

[89] JOLLIFFE, I. *Principal Component Analysis.* Springer, New York, NY, 2002.

[90] JONES, C. E., SIVALINGAM, K. M., AGRAWAL, P., AND CHEN, J. C. A survey of energy efficient network protocols for wireless networks. *Wireless Networks 7*, 4 (2001), 343–358.

[91] KARATZOGLOU, A., SMOLA, A., HORNIK, K., AND ZEILEIS, A. kernlab – an S4 package for kernel methods in R. *Journal of Statistical Software 11*, 9 (2004), 1–20.

[92] KARGUPTA, H., HUANG, W., SIVAKUMAR, K., PARK, B., AND WANG, S. Collective principal component analysis from distributed, heterogeneous data. *Proceedings of the 4th European Conference on Principles of Data Mining and Knowledge Discovery* (2000), 452–457.

[93] KARUSH, W. Minima of Functions of Several Variables with Inequalities as Side Constraints. Master's thesis, Dept. of Mathematics, The University of Chicago, 1939.

[94] KING, G., HONAKER, J., JOSEPH, A., AND SCHEVE, K. Analyzing incomplete political science data: An alternative algorithm for multiple imputation. *American Political Science Review 95*, 1 (2001), 49–69.

[95] KRIJTHE, J. H. *Rtsne: T-Distributed Stochastic Neighbor Embedding using Barnes-Hut Implementation*, 2015. R package version 0.13.

[96] KRIZHEVSKY, A., SUTSKEVER, I., AND HINTON, G. E. Imagenet classification with deep convolutional neural networks. In *Advances in Neural Information Processing Systems* (2012), pp. 1097–1105.

[97] KUHN, H. W., AND TUCKER, A. W. Nonlinear programming. In *Proceedings of the 2nd Berkeley Symposium* (1951), Berkeley: University of California Press, pp. 481–492.

[98] KUMAR, P. R., AND GUPTA, P. The capacity of wireless networks. *IEEE Transactions on Information Theory 46*, 2 (March 2000), 388–404.

[99] LAKHINA, A., CROVELLA, M., AND DIOT, C. Diagnosing network-wide traffic anomalies. In *Proceedings of the 2004 conference on Applications, technologies, architectures, and protocols for computer communications* (2004), ACM Press, pp. 219–230.

[100] LAKHINA, A., PAPAGIANNAKI, K., CROVELLA, M., DIOT, C., KOLACZYK, E., AND TAFT, N. Structural analysis of network traffic flows. *Proceedings of the joint international conference on Measurement and modeling of computer systems* (2004), 61–72.

[101] LE BORGNE, Y., AND BONTEMPI, G. Unsupervised and supervised compression with principal component analysis in wireless sensor networks. In *Proceedings of the Workshop on Knowledge Discovery from Data, 13th ACM International Conference on Knowledge Discovery and Data Mining* (2007), ACM Press, pp. 94–103.

[102] LECUN, Y., BOTTOU, L., BENGIO, Y., AND HAFFNER, P. Gradient-based learning applied to document recognition. *Proceedings of the IEEE 86*, 11 (1998), 2278–2324.

[103] LECUN, Y. A., BOTTOU, L., ORR, G. B., AND MÜLLER, K.-R. Efficient backprop. In *Neural Networks: Tricks of the Trade* (2012), Springer, pp. 9–48.

[104] LEE, H., GROSSE, R., RANGANATH, R., AND NG, A. Y. Convolutional deep belief networks for scalable unsupervised learning of hierarchical

representations. In *Proceedings of the 26th Annual International Conference on Machine Learning* (2009), ACM, pp. 609–616.

[105] LEMARÉCHAL, C., NEMIROVSKII, A., AND NESTEROV, Y. New variants of bundle methods. *Mathematical Programming 69*, 1-3 (1995), 111–147.

[106] LI, J., AND ZHANG, Y. Interactive sensor network data retrieval and management using principal components analysis transform. *Smart Materials and Structures 15* (December 2006), 1747–1757.

[107] LOH, P.-L., AND WAINWRIGHT, M. J. High-dimensional regression with noisy and missing data: Provable guarantees with non-convexity. In *Advances in Neural Information Processing Systems* (2011), pp. 2726–2734.

[108] LUMLEY, T. *biglm: Bounded Memory Linear and Generalized Linear Models*, 2011. R package version 0.8.

[109] LUO, X., ZHENG, K., PAN, Y., AND WU, Z. A tcp/ip implementation for wireless sensor networks. In *IEEE International Conference on Systems, Man and Cybernetics* (2004), vol. 7, pp. 6081–6086.

[110] MAATEN, L. V. D., AND HINTON, G. Visualizing data using t-sne. *Journal of Machine Learning Research 9*, Nov (2008), 2579–2605.

[111] MADDEN, S., FRANKLIN, M., HELLERSTEIN, J., AND HONG, W. TAG: A tiny aggregation service for ad-hoc sensor networks. In *Proceedings of the 5th ACM Symposium on Operating System Design and Implementation* (2002), vol. 36, ACM Press, pp. 131 – 146.

[112] MADDEN, S., FRANKLIN, M., HELLERSTEIN, J., AND HONG, W. TinyDB: An acquisitional query processing system for sensor networks. *ACM Transactions on Database Systems (TODS) 30*, 1 (2005), 122–173.

[113] MAMMEN, E., AND VAN DE GEER, S. Locally adaptive regression splines. *The Annals of Statistics 25*, 1 (1997), 387–413.

[114] MARDIA, K., KENT, J., AND BIBBY, J. *Multivariate Analysis*. Academic Press New York, New York, NY, 1979.

[115] MARKOVSKY, I. *Low Rank Approximation: Algorithms, Implementation, Applications*. Springer Science & Business Media, New York, NY, 2011.

[116] MATLOFF, N. *The Art of R programming: A Tour of Statistical Software Design*. No Starch Press, New York, NY, 2011.

[117] MATTHIES, H., AND STRANG, G. The solution of nonlinear finite element equations. *International Journal for Numerical Methods in Engineering 14*, 11 (1979), 1613–1626.

[118] McCullagh, P., and Nelder, J. A. *Generalized Linear Models*, vol. 37. CRC Press, 1989.

[119] McCulloch, W. S., and Pitts, W. A logical calculus of the ideas immanent in nervous activity. *The Bulletin of Mathematical Biophysics 5*, 4 (1943), 115–133.

[120] Meinshausen, N., and Bühlmann, P. High-dimensional graphs and variable selection with the lasso. *The Annals of Statistics* (2006), 1436–1462.

[121] Minsky, M. L., and Papert, S. *Perceptrons: An Introduction to Computational Geometry*. MIT, Boston, MA, 1969.

[122] Miranda, A. A., Le Borgne, Y., and Bontempi, G. New routes from minimal approximation error to principal component. *Neural Processing Letters* (2007).

[123] Motulsky, H. J. Common misconceptions about data analysis and statistics. *British Journal of Pharmacology 172*, 8 (2015), 2126–2132.

[124] Nelder, J., and Wedderburn, R. Generalized linear models. *Statist. Soc A 1972* (1972).

[125] Nielsen, F., and Garcia, V. Statistical exponential families: A digest with flash cards. *arXiv Preprint arXiv:0911.4863* (2009).

[126] Olston, C., Loo, B., and Widom, J. Adaptive precision setting for cached approximate values. *ACM SIGMOD Record 30*, 2 (2001), 355–366.

[127] OâĂŹLeary, D. P. Robust regression computation using iteratively reweighted least squares. *SIAM Journal on Matrix Analysis and Applications 11*, 3 (1990), 466–480.

[128] Pattem, S., Krishnamachari, B., and Govindan, R. The impact of spatial correlation on routing with compression in wireless sensor networks. In *Proceedings of the Third International Symposium on Information Processing in Sensor Networks* (2004), ACM Press, pp. 28–35.

[129] Pearson, K. On lines and planes of closest fit to points in space. *The London, Edinburgh, and Dublin Philosophical Magazine and Journal of Science 2*, 11 (1901), 559–572.

[130] Petrovic, D., Shah, R., Ramchandran, K., and Rabaey, J. Data funneling: routing with aggregation and compression for wireless sensor networks. *Sensor Network Protocols and Applications, 2003. Proceedings of the First IEEE. 2003 IEEE International Workshop on* (2003), 156–162.

[131] POLASTRE, J., SZEWCZYK, R., AND CULLER, D. Telos: Enabling ultra-low power wireless research. In *Proceedings of the 4th International Symposium on Information Processing in Sensor Networks* (2005), pp. 364–369.

[132] PRADHAN, S., AND RAMCHANDRAN, K. Distributed source coding using syndromes (discus): design and construction. *IEEE Transactions on Information Theory 49*, 3 (2003), 626–643.

[133] RADFORD, A., METZ, L., AND CHINTALA, S. Unsupervised representation learning with deep convolutional generative adversarial networks. *arXiv Preprint arXiv:1511.06434* (2015).

[134] RAGHUNATHAN, V., AND SRIVASTAVA, C. Energy-aware wireless microsensor networks. *Signal Processing Magazine, IEEE 19*, 2 (2002), 40–50.

[135] RAO, C. R., AND TOUTENBURG, H. Linear models. In *Linear Models*. Springer, 1995, pp. 3–18.

[136] REINSCH, C. H. Smoothing by spline functions. *Numerische mathematik 10*, 3 (1967), 177–183.

[137] RIGBY, R. A., AND STASINOPOULOS, D. M. Generalized additive models for location, scale and shape. *Applied Statistics 54* (2005), 507–554.

[138] ROBBINS, H., AND MONRO, S. A stochastic approximation method. *The Annals of Mathematical Statistics* (1951), 400–407.

[139] ROSENBLATT, F. The perceptron: A probabilistic model for information storage and organization in the brain. *Psychological Review 65*, 6 (1958), 386.

[140] ROUSSEEUW, P., AND MOLENBERGHS, G. Transformation of non positive semidefinite correlation matrices. *Communications in Statistics-Theory and Methods 22*, 4 (1993), 965–984.

[141] SAMANIEGO, F. J. *Stochastic Modeling and Mathematical Statistics: A Text for Statisticians and Quantitative Scientists*. CRC Press, New York, NY, 2014.

[142] SANDERSON, C. Armadillo: An open source c++ linear algebra library for fast prototyping and computationally intensive experiments.

[143] SANKAR, A., AND LIU, Z. Maximum lifetime routing in wireless ad-hoc networks. In *Proceedings of the 23rd Annual Joint Conference of the IEEE Computer and Communications Societies* (2004), vol. 2, IEEE Press, pp. 1089–1097.

[144] SCAGLIONE, A., AND SERVETTO, S. On the interdependence of routing and data compression in multi-hop sensor networks. *Wireless Networks 11*, 1 (2005), 149–160.

[145] SERMANET, P., EIGEN, D., ZHANG, X., MATHIEU, M., FERGUS, R., AND LECUN, Y. Overfeat: Integrated recognition, localization and detection using convolutional networks. *arXiv Preprint arXiv:1312.6229* (2013).

[146] SHAH, R. C., AND RABAEY, J. M. Energy aware routing for low energy ad-hoc sensor networks. In *Proceedings of the IEEE Conference on Wireless Communications and Networking* (March 2002), vol. 1, pp. 350–355.

[147] SHALIZI, C. Advanced data analysis from an elementary point of view, 2013.

[148] SRIVASTAVA, N., HINTON, G. E., KRIZHEVSKY, A., SUTSKEVER, I., AND SALAKHUTDINOV, R. Dropout: a simple way to prevent neural networks from overfitting. *Journal of Machine Learning Research 15*, 1 (2014), 1929–1958.

[149] STOER, J., AND BULIRSCH, R. *Introduction to Numerical Analysis.* Springer, New York, NY, 2002.

[150] STOJMENOVIC, I., AND LIN, X. Power-aware localized routing in wireless networks. *IEEE Trans. Parallel Distrib. Syst. 12*, 11 (2001), 1122–1133.

[151] STRASSEN, V. Gaussian elimination is not optimal. *Numerische Mathematik 13*, 4 (1969), 354–356.

[152] SZEGEDY, C., ZAREMBA, W., SUTSKEVER, I., BRUNA, J., ERHAN, D., GOODFELLOW, I., AND FERGUS, R. Intriguing properties of neural networks. *arXiv Preprint arXiv:1312.6199* (2013).

[153] THOMPSON, B. *Stepwise Regression and Stepwise Discriminant Analysis Need Not Apply.* Sage Publications, 1995.

[154] TIBSHIRANI, R. Regression shrinkage and selection via the lasso. *Journal of the Royal Statistical Society. Series B (Methodological)* (1996), 267–288.

[155] TIBSHIRANI, R., BIEN, J., FRIEDMAN, J., HASTIE, T., SIMON, N., TAYLOR, J., AND TIBSHIRANI, R. J. Strong rules for discarding predictors in lasso-type problems. *Journal of the Royal Statistical Society: Series B (Statistical Methodology) 74*, 2 (2012), 245–266.

[156] TIBSHIRANI, R. J. Adaptive piecewise polynomial estimation via trend filtering. *The Annals of Statistics 42*, 1 (2014), 285–323.

[157] TREFETHEN, L. N., AND BAU III, D. *Numerical Linear Algebra*, vol. 50. SIAM, New York, NY, 1997.

[158] TSYBAKOV, A. B. Introduction to nonparametric estimation. revised and extended from the 2004 french original. translated by vladimir zaiats, 2009.

[159] TULONE, D., AND MADDEN, S. PAQ: Time series forecasting for approximate query answering in sensor networks. In *Proceedings of the 3rd European Workshop on Wireless Sensor Networks* (2006), Springer, pp. 21–37.

[160] TURK, M. A., AND PENTLAND, A. P. Eigenfaces for recognition. *Journal of Cognitive Neuroscience* (1991).

[161] VAN DER MAATEN, L. Accelerating t-SNE using tree-based algorithms. *Journal of Machine Learning Research 15*, 1 (2014), 3221–3245.

[162] VANHUFFEL, S., AND MARKOVSKY, I. Overview of total least-squares methods. *Signal Processing 87*, 10 (2007), 2283–2302.

[163] VAPNIK, V., AND CHERVONENKIS, A. On the uniform convergence of relative frequencies of events to their probabilities. *Theory of Probability and Its Applications 16*, 2 (1971), 264.

[164] VARSHNEY, M., AND BAGRODIA, R. Detailed models for sensor network simulations and their impact on network performance. In *Proceedings of the 7th ACM International Symposium on Modeling, Analysis and Simulation of Wireless and Mobile Systems* (2004), ACM Press, pp. 70–77.

[165] VENABLES, W. N. *An Introduction to R: A Programming Environment for Data Analysis*, 2006.

[166] VENABLES, W. N., AND RIPLEY, B. D. *Modern Applied Statistics with S*, fourth ed. Springer, New York, NY, 2002. ISBN 0-387-95457-0.

[167] VOSOUGHI, A., AND SCAGLIONE, A. Precoding and decoding paradigms for distributed data compression. *IEEE Transactions on Signal Processing* (2007).

[168] WAND, M. *KernSmooth: Functions for Kernel Smoothing*, 2015. R package version 2.23-15.

[169] WANG, E., ZHANG, Q., SHEN, B., ZHANG, G., LU, X., WU, Q., AND WANG, Y. Intel math kernel library. In *High-Performance Computing on the Intel Xeon Phi*. Springer, 2014, pp. 167–188.

[170] WANG, Y., WAN, C., MARTONOSI, M., AND PEH, L. Transport layer approaches for improving idle energy in challenged sensor networks. In *Proceedings of the 2006 SIGCOMM Workshop on Challenged Networks* (2006), ACM Press, pp. 253–260.

[171] WASSERMAN, L., AND ROEDER, K. High dimensional variable selection. *Annals of Statistics 37*, 5A (2009), 2178.

[172] WEBB, A. *Statistical Pattern Recognition*. Hodder Arnold Publication, New York, NY, 1999.

[173] WEINBERGER, K., DASGUPTA, A., LANGFORD, J., SMOLA, A., AND ATTENBERG, J. Feature hashing for large scale multitask learning. In *Proceedings of the 26th Annual International Conference on Machine Learning* (2009), ACM, pp. 1113–1120.

[174] WESTON, S. *foreach: Provides Foreach Looping Construct for R*, 2017. R package version 1.4.4.

[175] WHALEY, R. C., PETITET, A., AND DONGARRA, J. J. Automated empirical optimizations of software and the atlas project. *Parallel Computing 27*, 1 (2001), 3–35.

[176] WICKHAM, H. The split-apply-combine strategy for data analysis. *Journal of Statistical Software 40*, 1 (2011), 1–29.

[177] WOOD, S. N. Fast stable restricted maximum likelihood and marginal likelihood estimation of semiparametric generalized linear models. *Journal of the Royal Statistical Society (B) 73*, 1 (2011), 3–36.

[178] WOOLDRIDGE, J. M. *Introductory Econometrics: A Modern Approach*. Nelson Education, New York, NY, 2015.

[179] WU, W., AND BENESTY, M. *FeatureHashing: Creates a Model Matrix via Feature Hashing with a Formula Interface*, 2015. R package version 0.9.1.1.

[180] WU, X., KHAN, M. A. U., CHO, J., LEE, S., AND LEE, Y.-K. Energy-efficient clustering with fast data compression in sensor networks. In *Proceedings of the 2006 International Conference on Hybrid Information Technology* (Washington, DC, USA, 2006), IEEE Computer Society, pp. 403–408.

[181] XIAO, H., RASUL, K., AND VOLLGRAF, R. Fashion-MNIST: A novel image dataset for benchmarking machine learning algorithms, 2017.

[182] XIONG, Z., LIVERIS, A., AND CHENG, S. Distributed source coding for sensor networks. *Signal Processing Magazine, IEEE 21*, 5 (2004), 80–94.

[183] XU, Y., HEIDEMANN, J., AND ESTRIN, D. Geography-informed energy conservation for ad hoc routing. In *Proceedings of the 7th Annual International Conference on Mobile Computing and Networking* (New York, NY, USA, 2001), ACM, pp. 70–84.

[184] YAO, Y., AND GEHRKE, J. The cougar approach to in-network query processing in sensor networks. *ACM SIGMOD Record 31*, 3 (2002), 9–18.

[185] YOUN, H. Y., YU, C., AND LEE, B. Routing algorithms for balanced energy consumption in ad hoc networks. In *The Handbook of Ad-hoc Wireless Networks* (Boca Raton, FL, USA, 2003), CRC Press, pp. 415–428.

[186] ZENG, Y., AND BREHENY, P. The biglasso package: A memory- and computation-efficient solver for lasso model fitting with big data in r. *ArXiv e-prints* (2017).

[187] ZHAO, F., AND GUIBAS, L. *Wireless Sensor Networks: An Information Processing Approach.* Morgan Kaufmann, New York, NY, 2004.

[188] ZHOU, Y., WILKINSON, D., SCHREIBER, R., AND PAN, R. Large-scale parallel collaborative filtering for the netflix prize. In *International Conference on Algorithmic Applications in Management* (2008), Springer, pp. 337–348.

[189] ZOU, H., AND HASTIE, T. Regularization and variable selection via the elastic net. *Journal of the Royal Statistical Society: Series B (Statistical Methodology) 67*, 2 (2005), 301–320.

Index

ℓ_1-norm, 181
ℓ_2-norm, 48

absolute error, 25
activation function, 210
arm, 139
autoencoders, 282

B-spline, 100
backfitting, 159, 160
backpropagation, 213
bandwidth, 81, 82
basis expansion, 76, 155, 164
Bayesian regression, 53
bayesm, 139
best-split, 3
biglasso, 200, 321
biglm, 32, 321, 324
bottleneck layer, 283
brglm, 139
bundle methods, 186

canonical form, 128
casl, 7
    casl_am_backfit(), 160
    casl_am_lm_basis(), 166
    casl_blockwise_irwls(), 308
    casl_cnn_sgd(), 246
    casl_fhash(), 315
    casl_glenet(), 198
    casl_glm_irwls(), 132
    casl_glm_nr(), 130
    casl_lenet(), 189
    casl_lm_pca(), 62
    casl_lm_ridge(), 51
    casl_nlm1d_local(), 86
    casl_nlm1d_poly(), 78

    casl_nlm1d_trunc_power_x(), 91
    casl_nn_sgd(), 220
    casl_nn_sgd_mu(), 230
    casl_ols_chol(), 20
    casl_ols_svd(), 18
    casl_sparse_irwls(), 305
    casl_spectral_clust(), 276
    casl_tsne(), 280
chain rule, 215
Cholesky decomposition, 19, 305, 336
class, 104
classification, 124
cleanNLP, 325
compressed sparse row format, 299
concentration of measure, 158
condition number, 24, 29, 45, 48, 101
convex optimization, 186
convolutional neural network, 239
coordinate descent, 188, 198
coordinate list, 316
cross-validation, 52, 79, 83, 165, 193, 202
curse of dimensionality, 156

digest, 312
doMC, 325
dropout, 229

early stopping, 228
effective degrees of freedom, 86
elastic net, 186
Epanechnikov kernel, 82
epoch, 213
estimator bias, 43
exponential family, 124, 128, 159

feature hashing, 311

feature space, 263
FeatureHashing, 321
FNN, 104, 113
foreach, 321, 325
Fourier basis, 89

gam, 171
gamlss, 139, 148, 171
generalized linear models, 125, 198
genlasso, 200
glm2, 139
glmnet, 64, 139, 200, 303
graph Laplacian, 272

hard-threshold, 184
hash collision, 317
Hessian matrix, 16, 49
high-dimensional regression, 192
hyperparameter, 8, 50

inner product, 264
iotools, 321
irlba, 64, 269
iteratively reweighted least squares,
        131

Jenson's inequality, 157

k-nearest neighbors, 81, 115, 156, 274
Karush–Kuhn–Tucker conditions,
        194
keras, 249, 283
kernel, 81, 262
kernel function, 264
kernel principal component analysis,
        266
kernel regression, 81
kernel trick, 264
kernlab, 283, 286
KernSmooth, 104
Kullback–Leibler divergence, 278

L-BFGS, 186
LAPACK, 17
lars, 200
LASSO, 181

learning rate, 220
least angle regression, 187
linear model, 11
linear smoothers, 85
link function, 124
lme4, 320
loadings matrix, 270
local regression, 85, 86
local support, 101
LOESS, 88
log-likelihood, 125
logit, 124
loss function, 7
LU-decomposition, 305

MASS, 64, 139, 147
Matrix, 64, 302
MatrixModels, 32, 139
MCMCglmm, 139
mgcv, 139, 171
model initialization, 226
momentum, 229
multi-class regression, 123, 138, 232
MurmurHash, 312

Nadaraya–Watson, 82
natural cubic splines, 93
negative binomial, 147
negative examples, 259
Newton–Raphson, 125, 129
nnet, 138, 232
non-linear, 75, 170, 210, 262
non-linear least squares, 75
non-parametric regression, 75, 158
normal equations, 15, 125, 198
nycflights13, 172

one-hot encoding, 233
ordinary least squares, 13, 43, 85
orthogonal projection, 22
orthonormal basis, 77
overfitting, 152

penalized, 200
penalized regression, 180
perplexity, 278

polynomial regression, 77
polynomial regressions, 77
pooling layer, 248
positive definite matrix, 19
posterior distribution, 54
principal component analysis, 56, 262
pseudoinverse, 17

QR decomposition, 22, 23, 135, 304

radial basis function, 266
rectified linear unit, 210
reference implementation, 1
regression spline, 91
residual vector, 14, 49
ridge regression, 46
Rtsne, 283

safeBinaryRegression, 139
score function, 125
similarity matrix, 272, 273
singular value decomposition, 17, 44,
        50, 135, 266, 304
smoothing spline, 95, 164
soft-threshold, 184, 189
softmax, 233
sparse matrix, 298
sparse matrix fill-in, 302
spectral clustering, 273
splines, 104
split-apply-combine, 307
stats, 32, 104
stepwise regression, 179
stochastic gradient descent, 211, 308
stringi, 315
supervised learning, 2, 7, 76
symmetric k-nearest neighbors, 274

t-SNE, 277
tensorflow, 249
Tikhonov regularization, 46
total variation, 43, 44
triangular system, 20
truncated power basis, 91

unsupervised learning, 261

Vapnik-Chervonenkis dimension, 8
variable selection, 179
variance-covariance matrix, 44

xgboost, 320

Printed and bound by PG in the USA